MATTHEW BOULTON

John Flaxman, Bust of Matthew Boulton, white Carrara marble, from
the Boulton Monument, St Mary's Church, Handsworth, Birmingham
(Parochial Church Council of St Mary's Handsworth. Photograph by Jerome Turner).

Matthew Boulton
Enterprising Industrialist of the Enlightenment

Edited by

KENNETH QUICKENDEN
Birmingham Institute of Art and Design,
Birmingham City University, UK

SALLY BAGGOTT
University of Birmingham, UK

MALCOLM DICK
University of Birmingham, UK

Routledge
Taylor & Francis Group

LONDON AND NEW YORK

First published 2013 by Ashgate Publishing

2 Park Square, Milton Park, Abingdon, Oxon OX14 4RN
711 Third Avenue, New York, NY 10017, USA

Routledge is an imprint of the Taylor & Francis Group, an informa business

First issued in paperback 2016

British Library Cataloguing in Publication Data
Matthew Boulton : enterprising industrialist of the
 Enlightenment. -- (Science, technology and culture,
 1700-1945)
 1. Boulton, Matthew, 1728-1809. 2. Industrialists--Great
 Britain--Biography. 3. Mechanical engineers--Great
 Britain--Biography. 4. Enlightenment--Great Britain.
 I. Series II. Quickenden, Kenneth. III. Baggott, Sally.
 IV. Dick, Malcolm.
 338'.04'092-dc23

Library of Congress Cataloging-in-Publication Data
Quickenden, Kenneth.
 Matthew Boulton : enterprising industrialist of the Enlightenment / edited by Kenneth
Quickenden, Sally Baggott, and Malcolm Dick.
   pages cm. -- (Science, technology, and culture, 1700-1945)
 Includes bibliographical references and index.
 ISBN 978-1-4094-2218-1 (hardcover) 1. Boulton, Matthew, 1728-1809. 2. Inventors--Great
Britain--Biography. 3. Industrialists--Great Britain--Biography. 4. Enlightenment. I. Baggott,
Sally. II. Dick, Malcolm. III. Title.
 TJ140.B6Q53 2013
 338.092--dc23
 [B]

                                                                              2012022000

ISBN 978-1-4094-2218-1 (hbk)
ISBN 978-1-138-24785-7 (pbk)

Science, Technology and Culture, 1700–1945

Series Editors

David M. Knight
University of Durham

and

Trevor Levere
University of Toronto

Science, Technology and Culture, 1700–1945 focuses on the social, cultural, industrial and economic contexts of science and technology from the 'scientific revolution' up to the Second World War. It explores the agricultural and industrial revolutions of the eighteenth century, the coffee-house culture of the Enlightenment, the spread of museums, botanic gardens and expositions in the nineteenth century, to the Franco-Prussian war of 1870, seen as a victory for German science. It also addresses the dependence of society on science and technology in the twentieth century.

Science, Technology and Culture, 1700–1945 addresses issues of the interaction of science, technology and culture in the period from 1700 to 1945, at the same time as including new research within the field of the history of science.

Also in the series

Sir James Dewar, 1842–1923
A Ruthless Chemist
J.S. Rowlinson

Engineer of Revolutionary Russia
Iurii V. Lomonosov (1876–1952) and the Railways
Anthony Heywood

Selling Science in the Age of Newton
Advertising and the Commoditization of Knowledge
Jeffrey R. Wigelsworth

Contents

List of Figures and Tables

Figures

Tables

Notes on Contributors

Dr Jim Andrew, formerly Keeper, Birmingham Museum of Science and Industry and currently Collections Advisor to Thinktank and Birmingham Museums and Art Gallery.

Dr Sally Baggott, formerly Librarian and Curator, The Birmingham Assay Office, currently Research Development Manager, College of Arts and Law, University of Birmingham.

George Demidowicz, formerly Head of the Conservation and Archaeology Team, Coventry City Council.

Dr Malcolm Dick, Director of the Centre for West Midlands History, University of Birmingham.

Barbara Fogarty has recently completed an MPhil on Boulton and Eginton's Mechanical Paintings at the University of Birmingham.

Dr Irina Gouzévitch, Centre Maurice Halbwachs, École des Hautes Études en Sciences Sociales, Paris.

Professor Liliane Hilaire-Pérez, Professor in the Université Paris Diderot-Paris VII.

Professor Peter M. Jones, Professor of French History, School of History and Cultures, University of Birmingham.

Dr Val Loggie, formerly Curator of Soho House, has completed a PhD in 2010 on images of Boulton and the Soho Manufactory jointly supervised by the University of Birmingham and Birmingham Museums and Art Gallery.

Professor Joseph Melling, Professor of Medical and Industrial History and Director of the Wellcome-funded Centre of Medical History, University of Exeter.

Professor David Philip Miller, Professor of History and Philosophy of Science, School of History and Philosophy, University of New South Wales.

Dr Peter Northover and Nick Wilcox, Materials Science-Based Archaeology Group, Department of Materials, University of Oxford.

Professor Kenneth Quickenden, Research Professor, Birmingham Institute of Art and Design, Birmingham City University.

Professor Jennifer Tann, Professor Emerita of Innovation, University of Birmingham.

Sue Tungate, has recently completed a PhD on Boulton's Soho Mint, jointly supervised by the University of Birmingham and Birmingham Museums and Art Gallery.

Bernard Vaisbrot, Lecturer in the Yiddish Language, Université Paris VIII.

Foreword

Nicholas Goodison

I took several groups of visitors to the wonderful exhibition at the Birmingham Museum in 2009 marking the bi-centenary of the death of Matthew Boulton. Some already knew something of his work, particularly of his production of silver, Sheffield Plate and ormolu, of which there was a glittering display. Most, however, were astonished by the scope and extent of what they saw. They wondered how they could have known so little about a man who played such a key role in the eighteenth-century development of industry, commerce, consumer taste, technology, and the study of the natural sciences. They had all heard of James Watt, but knew little of the Birmingham engineer and entrepreneur who so successfully promoted Watt's improvements to the steam engine and was thus one of the central characters in the drama of the industrial revolution.

Ignorance of Boulton's key role in British history is surprising, bearing in mind the survival of the very extensive archives of his businesses and interests. The Matthew Boulton Papers and the papers of the Boulton and Watt businesses are a rich source of evidence not just about Boulton's economic activities but about many aspects of eighteenth and nineteenth-century life and intellectual activity during the period of the Enlightenment. They are a veritable gold mine waiting for historians to continue the work of excavating their treasure.

My own discovery of the papers in the late 1960s changed the course of my research. I studied them in order to learn more of the connection between Boulton and his friend the Derby clock maker John Whitehurst, on whose life and work I was planning to write. I was becoming interested particularly in the application of industrial techniques to the manufacture of objects of decorative art. I was astonished to find letters, which had never been published, referring in detail to the manufacture of the movements that Whitehurst made for Boulton's resplendent clock-cases (the 'geographical' and 'sidereal' clocks, both featured in the 2009 exhibition). And the more I looked into the archives, the more I realised that there was a wealth of information about the designing, making and marketing of ormolu ornaments that no-one had previously discovered. I had struck gold.

I am not the only writer who has experienced the excitement of discovery in these extraordinary archives, which are increasingly catching the attention of historians. The wide-ranging conference organised in 2009 by The Birmingham Assay Office, Birmingham City University and the University of Birmingham attracted scholars from far and wide. This volume of collected papers is one of the fruits of the conference. The papers exemplify the variety of subjects that studies of the Boulton archives provoke. I hope that their publication will help people to understand more of Matthew Boulton's role in the Industrial Revolution and the Enlightenment in a national and an international context, will stimulate more research into his life and work and broader aspects of late eighteenth-century history, lead to more publications, and generally maintain the excitement that I and others feel for the man and for what he achieved.

Sir Nicholas Goodison is the author of *Matthew Boulton: Ormolu* (2002), a revision of his *Ormolu: The Work of Matthew Boulton* first published in 1974. He chaired the Appeal for the restoration of Boulton's house at Soho. He was a member of the Advisory Board for the exhibition 'Matthew Boulton: Selling What All the World Desires' in 2009, and wrote sections of the catalogue. He was formerly Chairman of the London Stock Exchange, The Courtauld Institute of Art, The Burlington Magazine and the National Art Collections Fund, and is President of the Furniture History Society.

Acknowledgements

Kenneth Quickenden, Sally Baggott, Malcolm Dick

A number of organisations and individuals contributed to *Matthew Boulton: Enterprising Industrialist of the Enlightenment*. Our initial thanks as editors go to The Birmingham Assay Office, Birmingham Institute of Art and Design at Birmingham City University and the Centre for West Midlands History at the University of Birmingham for organising the conference: '"Where Genius and the Arts Preside": Matthew Boulton and the Soho Manufactory 1809–2009'. The contributions to this book are based on presentations which were made at this event. The success of the conference was due to the presence of scholars from inside and outside the UK and the interaction of academics, heritage professionals, independent scholars and postgraduate students. It was not possible to include papers from all those who spoke at the conference, but we would acknowledge everyone who contributed to the event and who helped to forge a fertile intellectual atmosphere in which the importance of studying Boulton and his life and times was recognised.

A particular personal acknowledgement goes to Professor Peter Jones of the University of Birmingham who conceived the idea of the conference and helped to steer the event through to its launch. Without the conference there would be no publication, and Peter has helped to shape this book in many ways. Dr Richard Clay of the University of Birmingham also played an important role as a member of the management team which planned the conference alongside Peter and the editors of this volume.

Our thanks are also due to others. Jerome Turner of Birmingham Institute of Art and Design photographed the Matthew Boulton Monument by John Flaxman in St. Mary's Church, Handsworth, Birmingham, which we have used for the front cover. Canon Brian A. Hall, Rector, and the Parochial Church Council of St. Mary's, Handsworth, Birmingham kindly gave permission for photography of the Monument. The Birmingham Assay Office Charitable Trust, Birmingham Central Library, Birmingham Museums and Art Gallery, the British Museum, the London Science Museum and Alec Cobbe have also kindly given permission to use and reproduce materials in their collections. Sian Roberts and her staff in Birmingham Archives and Heritage, Birmingham Central Library have provided a great deal of help with archival research and checking references. In particular, Fiona Tait's unrivalled knowledge of the Archives of Soho has been invaluable.

Our special appreciation goes to the contributors to this volume who have responded positively and efficiently to requests for information, changes to text and corrections. The book's scholarly worth is dependent entirely on the insights they have provided and the new material they have uncovered. We are also extremely grateful to Sir Nicholas Goodison for writing the Foreword to the book.

Finally, our publishers, Ashgate, their Commissioning Editor, Emily Yates, their Senior Editor, Aimée Feenan and anonymous peer reviewers, who have sensitively guided *Matthew Boulton: Enterprising Industrialist of the Enlightenment* from the initial idea through to final production.

Chronology: Matthew Boulton

Date	Event
1728	Born in Birmingham, 14 September. His parents were Matthew Boulton Senior and Christiana Peers.
c. 1742	Leaves Rev. John Hausted's school in Deritend, near Birmingham and joins his father's button, buckle and 'toy' making business at Snow Hill, Birmingham.
1749	Marries Mary Robinson of Lichfield.
1759	Death of his wife and father. Boulton takes over his father's business.
1760	Marries Ann Robinson, Mary's sister and secures access to £8,000, Ann's dowry, which is used for business investment.
1761	Boulton leases 13 acres at Handsworth, Staffordshire, near Birmingham, which includes Soho House. He rebuilds the watermill and starts to build the Soho Manufactory.
1762	Forms partnership with John Fothergill.
Early 1760s	Meetings of the Lunar Society begin. Dr William Small and Dr Erasmus Darwin are key figures with Boulton.
1764	Following the death of her brother, Ann's wealth increases to £28,000. Boulton starts to produce Sheffield Plate.
By 1765	Boulton produces japanned wares.
1766	Boulton and Ann move into Soho House. Production at Snow Hill entirely transferred to the Soho Manufactory.
1767	Production of tortoiseshell and gilt 'toys' begins.
1768	Birth of Boulton's only daughter Anne. Production of silver plate and ormolu starts. Boulton helps to organise the first of Birmingham's Triennial Music Festivals to benefit the new General Hospital.
1770	Birth of Boulton's son, Matthew Robinson.
1771	Boulton visits King George III and Queen Charlotte.
1773	The Birmingham Assay Office established.
1775	Start of partnership between James Watt and Boulton to build steam engines.
c1776	Boulton starts production of mechanical paintings.
1780	James Watt's copying machine patented. Dr James Keir temporarily helps to manage the Soho Manufactory.
Early 1780s	Production at Soho of japanned wares and mechanical paintings phased out.
1782	End of the partnership with Fothergill following the latter's death.
1783	Death of Boulton's wife, Ann.
1785	Boulton elected a fellow of the Royal Society. He agrees to produce Aimé Argand's new tubular wick oil lamp.
1785–86	The Albion Mill, the first steam-powered corn mill established in London.

Date	Event
1788	Building of the Soho Mint.
1793	Becomes chairman of Birmingham's Theatre Proprietors Committee.
1794	James Wyatt starts to remodel Soho House.
1794	Serves as High Sheriff of Staffordshire.
1794	Chairman of the Governors of the Birmingham Dispensary. Purchases the freehold of the Soho estate and enlarges it.
1795–96	Construction of the Soho Foundry, Smethwick, Staffordshire, the first purpose-built factory in England to produce steam engines. The management of the engine business is handed over to Matthew Robinson and James Watt junior.
1797	Contract from Royal Mint to strike copper coinage at the Soho Mint.
1798–99	Second Soho Mint established.
1805	Boulton strikes a medal to commemorate the Battle of Trafalgar.
1809	Boulton dies at Soho House on 17 August and is buried at St Mary's Church, Handsworth on 24 August.
1810	Steel 'toy' and button business released to another company, the first of many companies to rent or lease workshops at the Soho Manufactory from Boulton's son and grandson.
1812	London's Royal Mint powered with equipment from Soho.
1815	Matthew Robinson Boulton purchases an estate at Great Tew, Oxfordshire.
1819	Death of James Watt.
1829	Death of Boulton's daughter, Ann.
1834	Matthew Robinson sells the silver and Sheffield Plate business.
1841	Matthew Piers Watt Boulton takes over the running of the Soho Mint from his father, Matthew Robinson Boulton.
1842	Death of Boulton's son, Matthew Robinson.
1848	Following the death of James Watt junior, Boulton, Watt and Co. is sold, and this ends any Boulton family connection with the business.
1849	The first leases for building plots on the main Soho estate sold by Matthew Piers Watt Boulton.
1850	Matthew Piers Watt Boulton ceases to live at Soho House.
1853	Demolition of the Soho Manufactory begins.
1858	All manufacturing at the Soho Manufactory ceases.
1895	The Soho Foundry is taken over by W.T. Avery Ltd. who manufacture weighing machinery.
1909	Commemoration in Handsworth of the centenary of Boulton's death.
1963	The freehold of Soho House and the surrounding land sold to the City of Birmingham by the trustees of Boulton's great grandson, Matthew Ernest Boulton.
1995	Birmingham Museums and Art Gallery start to restore Soho House, which is opened to the public in the same year.
2009	Commemoration in Birmingham of the bicentenary of Boulton's death by an exhibition at Birmingham Museums and Art Gallery, several publications and an international conference supported by The Birmingham Assay Office, Birmingham City University and the University of Birmingham.

Chapter 1

Introduction:
Matthew Boulton – Enterprising
Industrialist of the Enlightenment

Kenneth Quickenden, Malcolm Dick, Sally Baggott

Matthew Boulton: Enterprising Industrialist of the Enlightenment represents the culmination of research into and thinking about Matthew Boulton. It does more than shed light on the life and times of a Birmingham-based manufacturer and entrepreneur who died over 200 years ago. The publication raises questions about the ways in which individual biography, industry, science, art, commerce and employment interconnect. It illuminates the history of the Industrial Revolution in Great Britain, which has shaped the economic and social history of the world in recent centuries, and the Enlightenment, which saw the application of rational thinking to the natural world, to politics, economics and social organisation. In particular, the book increases our knowledge and understanding of the 'Midlands Enlightenment' of the late eighteenth and early nineteenth centuries, when entrepreneurs such as Boulton brought reason to bear on the practical worlds of technology, production and the arts in what was, perhaps, the most scientifically and industrially advanced region in the world during the period.[1]

An outline of Boulton's life is provided in this book's Chronology, but in what ways was he important? He was born in 1728 in Birmingham, then a small and relatively insignificant metal-working and market town and died in 1809, by which time Birmingham had become a major international centre – for manufacturing. As an industrialist, Boulton contributed to this transformation. He was a businessman primarily, who inherited his father's 'toy' business and transformed it into a major enterprise.[2] He created the largest factory in Britain at

---

[1]   For a discussion of the West Midlands see: Peter Jones, *Industrial Enlightenment: Science, Technology and Culture in Birmingham and the West Midlands 1760–1820* (Manchester, 2008), pp. 22–69 and Philip K. Wilson, Elizabeth A. Dolan and Malcolm Dick, 'Introduction', in Wilson, Dolan and Dick (eds), *Anna Seward's Life of Erasmus Darwin* (Studley, 2010), pp. 1–49.

[2]   'Toys' were small, inexpensive metal items which were made for personal and domestic use, such as buckles, buttons, snuffboxes and tweezers. The term does not refer to playthings made for children as in the modern use of the word. See Rita McLean, 'Introduction', in Shena Mason (ed.), *Matthew Boulton: Selling what all the World desires*

Soho, near Birmingham, the first mechanically powered mint, also at Soho, and a purpose-built foundry at Smethwick to produce steam engines – a location which marked the take-off of the mechanical engineering industry.[3] He also applied scientific principles to manufacturing in partnership or collaboration with other men, the best-known being the inventor, James Watt. He was interested in mass production and the division of labour, and he pioneered industrial training at the Soho Manufactory. Boulton was also a promoter of the applied arts; for example, through the production of silverware, Sheffield Plate and ormolu.[4] He employed designers and absorbed decorative ideas from home and abroad, but he also knew that a successful manufacturer required more than well-designed products. He understood the need for marketing by displaying his wares in catalogues, employing agents and engaging with the tastes of consumers, particularly the aristocracy and rising middle classes. He was also fully aware of the importance of an effective transport infrastructure, particularly canals, to secure raw materials easily for his factories and move his finished goods quickly and cheaply to their markets. Boulton appreciated the importance of political activity and a legal framework to support his manufacturing activities. He used patent legislation to protect his inventions, campaigned successfully against opposition from London to establish an assay office in Birmingham for testing the standard of locally-produced silver, and, in the face of hostility from the Royal Mint, obtained a government contract to manufacture coins. He was also a key figure in the gestation of the 'Midlands Enlightenment', providing friendship and hospitality to scholars and thinkers in Britain, Europe and North America, and he was one of the most active members of the Lunar Society. This was not an organised society with a management structure or minutes, but a network or gentlemen's club which flourished in the West Midlands region from the mid 1760s to the early 1800s. As well as Boulton, it included scientists, writers and industrialists such as Erasmus Darwin, Thomas Day, Richard Lovell Edgeworth, James Keir, Joseph Priestley, James Watt, Josiah Wedgwood, John Whitehurst and William Withering.[5] In short, Boulton's

---

(New Haven and London, 2009), pp. 1–6, 2, 3–4; Kenneth Quickenden, 'Matthew Boulton's Silver and Sheffield Plate', in ibid., pp. 41–46, 41; Sylvia Crawley, 'The Birmingham Toy Trade', in ibid., pp. 121–122.

[3]   See Chapter 5 in this volume for images of the Soho Manufactory.

[4]   Sheffield Plate refers to a layered combination of silver and copper, which was used to manufacture items for domestic use such as candlesticks or trays. It was so-called because it was developed in Sheffield, England in the eighteenth century. Sheffield Plate was cheaper to manufacture than items made out of silver alone. On Sheffield Plate see Kenneth Quickenden, 'Silver, Plated and Silvered Products from the Soho Manufactory, 1780', *The Silver Society Journal* (Autumn, 1998), pp. 77–95. Ormolu is gilded decorative metalware where the decoration is frequently on a stone base. See Nicholas Goodison, *Matthew Boulton: Ormolu* (London, 2002), p. 151.

[5]   There are biographical entries for each of these individuals in the Oxford Dictionary of National Biography (ODNB). For the online edition see http://www.oxforddnb.com/

Figure 1.1   Soho House, Handsworth, Birmingham
Source: Photograph by Mohsen Keiany.

involvement in eighteenth and nineteenth-century economic and cultural life was indeed massive, as this volume reveals. His home, Soho House in Handsworth survives as a physical record of his significance (see Figure 1.1).

   This introduction sets out the framework for this volume. It details events connected with the 200th anniversary of his death in 2009, the secondary literature which has illuminated his life and work, and the range of primary sources which exist. It ends with a summary of the chapters contained in this publication.

Boulton 2009

In 2009 Boulton's importance in Britain was recognised during the commemorative events which celebrated the 200th anniversary of his death.[6] Nationally, the Royal Mail produced a series of eight postage stamps, 'Pioneers of the Industrial Revolution', which included a first-class stamp showing Matthew Boulton and the Soho Manufactory, and Sir Mervyn King, Governor of the Bank of England, announced plans to produce a new £50 banknote featuring both Boulton and Watt.[7] There were several exhibitions. 'Matthew Boulton: "Selling what all the world desires"' at the Gas Hall, Birmingham Museums and Art Gallery (BM&AG), featured objects, paintings and prints from collections in Birmingham and elsewhere to demonstrate Boulton's place as not only a local manufacturer of utilitarian and artistic objects, but also an individual of national and international significance.[8] The exhibition attracted 35,000 national and international visitors. A second exhibition, 'Matthew Boulton and the Art of Making Money' at the Barber Institute of Fine Arts on the campus of the University of Birmingham, focused on Boulton's production of coins, tokens and medals.[9]

In commemoration of Boulton's role in its establishment, The Birmingham Assay Office commissioned a silver candelabrum by the contemporary silversmith, Shona Marsh, which reinterpreted Boulton's silver in a distinctive way for the twenty-first century. During 2009, this was displayed at BM&AG and Birmingham's Museum of the Jewellery Quarter.[10] In the USA, the Speed Art Museum, Louisville, Kentucky, hosted an exhibition based on one of its principal collections: 'English Silver in the Age of Matthew Boulton: The James C. Codell, Jr. Collection'.[11]

Birmingham City University did much to enhance our understanding of Boulton's technology in 2009. It produced two films on the silver manufacturing techniques used at the Soho Manufactory in 1776–77 to make candlesticks and a tureen. The Jewellery Innovation Centre in the School of Jewellery exhibited replicates of Boulton candlesticks that were produced by using digital photographs from which a computer-aided design file was generated. The file was used to make a 3D resin print which became the pattern for mouldings from which white metal replicates were produced and subsequently silver plated. A 'Matthew Boulton Discovery Day' of talks, workshops and displays was held at Aston University

---

[6]   www.matthewboulton2009.org

[7]   http://www.royalmail.com/portal/stamps/jump1?catId=32300674&mediaId=93000750; http://www.bankofengland.co.uk/publications/news/2009/043.htm

[8]   Mason, *Matthew Boulton.*

[9]   Richard Clay and Sue Tungate (eds), *Matthew Boulton and the Art of Making Money* (Studley, 2009).

[10]   http://www.theassayoffice.co.uk/boulton_commission.html; http://www.visitheartofengland.com/news/article/?id=42

[11]   http://www.speedmuseum.org/past/past_exhibitions_2009.html

to enable members of the public to make comparisons between the technology of Boulton's own time and today. St. Philip's Cathedral staged a concert 'Hark I hear Musick', which included some of Matthew Boulton's favourite music and readings from the Matthew Boulton Papers. On the bi-centenary of Boulton's death, 17 August 2009, a wreath was laid at the statue of Boulton, Watt and William Murdoch in Broad Street, close to Birmingham's city centre.

Several academic dimensions of the Boulton commemorations have influenced this book. Before 2009, the University of Birmingham, in collaboration with BM&AG successfully applied to the Arts and Humanities Research Council (AHRC), the leading state-funded agency which supports academic research in the arts and humanities, to fund two PhD studentships to further research on Boulton.[12] Additionally, in partnership with Birmingham Archives and Heritage, The Birmingham Assay Office, Birmingham City University and BM&AG, the University of Birmingham successfully applied to the AHRC for funding to run six workshops at Birmingham venues during 2007 and 2008 in order to explore Boulton's significance. These events attracted academics, students, heritage professionals and independent scholars and featured contributors from Australia and the USA as well as the UK. On completion, the AHRC gave the project its highest grade. The workshops put in place networks and provided intellectual opportunities, which crystallised in the form of a bicentennial conference. This conference, '"Where Genius and the Arts Preside": Matthew Boulton and the Soho Manufactory 1809–2009' was organised by The Birmingham Assay Office, Birmingham City University and the University of Birmingham and took place in July 2009. It attracted over 100 participants from the UK, Europe, USA, Australia and Canada. The contributions to this volume can be considered the intellectual outcomes of this event; they convey new research, interpretations and insights from archaeology, economic and social history, the history of art and the history of science and build upon existing Boulton scholarship.

Boulton Literature

The secondary literature devoted to Boulton is considerable but there is still no substantial biography, a surprising fact in view of the vast quantity of archival material which is now available to scholars. Boulton attracted comment while he was still alive, but it was only after his death, and more particularly in the twentieth century that he received detailed evaluation. In his lifetime, Boulton was a celebrity who drew other celebrities to visit him at Soho, such as Vice-Admiral Horatio Nelson, the Astronomer Royal, Sir William Herschel, and the Russian Ambassador, Count Alexander Woronzow.[13] Stebbing Shaw's *The History*

---

[12] These studentships were held by Val Loggie and Sue Tungate.

[13] Shena Mason, '"The Hôtel d'amitié sur Handsworth Heath": Soho House and the Boultons', in Mason, pp. 14–21, 21.

and Antiquities of Staffordshire (1798) provided an account of Boulton's Soho Manufactory, his use of steam power and the picturesque landscape he created around his estate.[14] In 1801, one publication lauded him as a man who exhibited the 'purest patriotism'.[15]

In the nineteenth century, as David Miller has argued, Boulton's reputation as a manufacturer dipped whilst that of Watt grew.[16] This was due in part to the Watt family strongly promoting his reputation through publications such as J.P. Muirhead's *The Life of James Watt with Selections from his Correspondence* (1858). The cult of the self-made inventor and the importance of the steam engine to Victorian economic and social life also tended to shift the focus of interest towards Watt. Samuel Smiles' *Lives of Boulton and Watt* (1865) was a double biography, but its main emphasis was on the steam engine and Watt. Though Boulton was praised as an entrepreneur and a chapter was devoted to the Soho Mint, little attention was paid to his other manufacturing activities. Only a few pages covered ormolu, silver and Sheffield Plate. Later studies of decorative metalwork in the nineteenth and early twentieth centuries make little or no reference to Boulton. In the nineteenth century, art scholarship was hierarchical; the study of painting and sculpture was privileged above that of mere decorative objects. When writers referred to the latter, they discussed items which were produced before industrialisation by craft techniques or objects made in London. Little attention was paid to Matthew Boulton's work both because he mainly produced decorative items rather than fine art and because he was a provincial rather than metropolitan figure.[17] Only in the twentieth century was there some redress when Frederick Bradbury in 1912 drew attention to the quality of Soho's Sheffield Plate.[18]

Boulton's importance as an entrepreneur was reassessed in the early twentieth century. This reflected the influence of J.A. Schumpeter's *The Theory of Economic Development*, first published in Germany in 1911.[19] As a result, studies paid less attention to inventive genius, which had preoccupied Smiles, and concentrated on the business skills which enabled inventions to be translated into reality. J. Lord writing in the 1920s, found Boulton to be 'the first great commercial intelligence'.[20] H.W. Dickinson's *Matthew Boulton*, published in 1936, argued that without Boulton's 'enthusiasm and enterprise Watt could hardly have succeeded

[14]   Stebbing Shaw, *The Antiquities and History of Staffordshire*, 2 vols (London, 1798–1801), 2, pp. 117–121.

[15]   *Public Characters of 1800–1* (4th edition, London, 1801), pp. 1–9, 1.

[16]   David Miller, 'Scales of Justice: Assaying Matthew Boulton's Reputation and the Partnership of Boulton and Watt', *Midland History*, 34, 1 (Spring 2009), pp. 58–76.

[17]   Charles James Jackson, *An Illustrated History of English Plate*, 2 vols (London, 1911), 2, p. 962.

[18]   Frederick Bradbury, *History of Old Sheffield Plate* (London, 1912), p. 47.

[19]   Miller, 'Scales of Justice', pp. 70–72.

[20]   J. Lord, *Capital and Steam Power 1750–1800* (London, 1923), p. 97, quoted in Miller, 'Scales of Justice', p. 71.

in bringing his improved steam engine before the world ...'.[21] Dickinson drew attention to Boulton's part in financing and marketing the steam engine, and to the fact that it was Boulton who pressed Watt to develop rotary motion.[22] It was Watt's 'sun and planet gears' that enabled the 'up-and-down' pumping engine to become a device which powered machinery.[23] Dickinson also noted Boulton's role in promoting Watt's copying press, producing ormolu and Sheffield Plate and establishing The Birmingham Assay Office.[24] Also in 1936, Arthur Westwood produced the first substantial history of The Birmingham Assay Office, which emphasised Boulton's political skills in overcoming resistance from the London silversmiths.[25] These studies helped to bring Boulton out from Watt's shadow, but they largely presented Boulton uncritically. During the 1930s, Boulton's importance as a manager also began to attract the attention of historians. Eric Roll in 1930 argued that Boulton's Soho Foundry exhibited advanced principles of accounting and costing by, for example, allowing for the depreciation of plant and the loss of interest on capital when drawing up accounts.[26] He also highlighted the processes of standardisation, sequencing of business operations and the use of piece-rates to pay workers rather than fixed contracts, though he attributed these changes to the sons of Boulton and Watt, Matthew Robinson Boulton and James Watt junior, when they took over their fathers' business.[27] Boulton was not always a successful financial manager; J.E. Cule in 1935 charted the varied financial fortunes of Boulton's businesses.[28] Sidney Pollard's general history of management in 1965 examined Boulton's use of the division of labour in his businesses to speed up production and manage his labour force.[29] Pollard also argued that it was simplistic to debate whether Boulton should be regarded as an entrepreneur or engineer since like other captains of industry in his generation, his role was multi-faceted.[30]

Additions to Boulton scholarship since the mid twentieth century have focused on specific aspects of his work rather than take a broad-brush approach to his entrepreneurial activities. Articles by Eric Robinson looked at Boulton's

[21] Dickinson, p. ix.
[22] Ibid., pp. 75–87.
[23] Ibid., pp. 113–114.
[24] Ibid., pp. 108, 64–74.
[25] Arthur Westwood, *The Assay Office at Birmingham Part 1: Its Foundation* (Birmingham, 1936).
[26] E. Roll, *An Early Experiment in Industrial Organization* (London, 1930), pp. 123, 244–252.
[27] Ibid., pp. 267, 178, 200–201.
[28] Cule, J.E., 'The Financial History of Matthew Boulton, 1759–1800' (University of Birmingham, unpublished M. Comm. thesis, 1935).
[29] Sidney Pollard, *The Genesis of Modern Management: A Study of the Industrial Revolution in Great Britain* (London, 1965), pp. 251, 265.
[30] Ibid., p. 254.

significance before his involvement with the steam engine by considering the production of metal hardware such as buttons and 'toys' and his skill in marketing these products.[31] Eric Delieb and Michael Roberts', *The Great Silver Manufactory* (1971), was the first exploration of Boulton's silver business at Soho.[32] No longer could national surveys ignore Boulton's contribution to British silver production.[33] Kenneth Quickenden has written several articles which have increased our knowledge of Boulton's manufacture of silver and various silver imitations including Sheffield Plate.[34] In 1974, Nicholas Goodison published a substantial study of Boulton's ormolu and explored his methods of production.[35] The book was the first to show that Boulton produced considerable quantities of this decorative ware.[36] Goodison revisited this subject in 2002 with a re-titled, more substantial and lavishly illustrated second edition.[37] These different publications consider Boulton's importance as a supplier of products to an aristocratic and aspirational middle-class market. Economic and social historians have attended to consumption during the period and placed greater stress on decorative items than traditional historians of art. For example, Neil McKendrick and Maxine Berg cite Boulton as an important influence on taste in their general studies of eighteenth-century consumer society.[38] It is clear, however, that there is still more room for exploring Boulton's activities as a manufacturer and purveyor of buttons, buckles and other items of metalware.[39]

Historians have also focused on minting and steam technology, and, in particular, they have added quantitative precision to our knowledge of the output of Boulton's businesses. During the 1960s, detailed studies of the Soho Mint began to appear.[40] Richard Doty provided the first major publication, in

---

[31]   Eric Robinson, 'Boulton and Fothergill, 1762–1782, and the Birmingham Export of Hardware', *University of Birmingham Historical Journal*, V11, 1 (1959), pp. 60–79; Eric Robinson, 'Eighteenth Century Commerce and Fashion, Matthew Boulton's Marketing Techniques', *Economic History Review*, 2nd series, XV1, 1 (1963), pp. 39–60.

[32]   Eric Delieb and Michael Roberts, *The Great Silver Manufactory: Matthew Boulton & the Silversmiths of Birmingham, 1760–90* (London, 1971).

[33]   See for example, Philippa Glanville, *Silver in England* (London, 1987), pp. 106–110.

[34]   Quickenden, *Boulton, Silver and Sheffield Plate* (London, 2009) provides a collection of seven essays which were originally published in *The Silver Society Journal*.

[35]   Nicholas Goodison, *Ormolu: The Work of Matthew Boulton* (London, 1974).

[36]   Ibid., p. 23.

[37]   Goodison, *Matthew Boulton: Ormolu* (London, 2002).

[38]   Neil McKendrick, John Brewer and J.H. Plumb, *The Birth of a Consumer Society: The Commercialization of Eighteenth-century England* (London, 1983), pp. 69–77; Maxine Berg, *Luxury and Pleasure in Eighteenth-Century Britain* (Oxford, 2005), pp. 162–169.

[39]   Diana Epstein and Millicent Safro, *Buttons* (London, 1991), pp. 28–29, is one general study which mentions Boulton briefly.

[40]   J.G. Pollard, 'Matthew Boulton and J.P. Droz', *The Numismatic Chronicle*, 8 (1968), pp. 241–265; Richard Margolis, 'Matthew Boulton's French Ventures of 1791

Figure 1.2   Attributed to Boulton and Fothergill, one of a pair of *Lyon-faced* candelabra, Sheffield Plate, probably either 1778 or 1780; Thomas and Lady Betty Cobbe, Newbridge House, Co. Dublin
Source: Cobbe Collection, courtesy of Alec Cobbe.

Figure 1.3   Boulton and Fothergill, candle vases, blue john and ormolu, ebonised wooden base, c. 1770
Source: Courtesy of Birmingham Museums and Art Gallery.

and 1792: Tokens for the Monnerons Frères of Paris', *The British Numismatic Journal*, 58 (1988), pp. 102–109.

which he considered the supply of technology to the Royal Mint and overseas, the improvement in dies to render counterfeiting more difficult and the massive quantity of coins produced.[41] In 1806–07, 166 million copper coins were produced for the British government and between 1786 and 1850, a total of two-thirds of a billion coins were minted by Boulton and his son and grandson.[42] George Selgin placed the Soho Mint in a wider context of other Birmingham industries and also discussed counterfeiting and the striking of coins in other localities.[43] A number of writers have increased our knowledge of Boulton and Watt steam engines. Jennifer Tann, in 1978, analysed the difficulties of exporting and protecting Watt's patents overseas,[44] and, with M.J. Breckin, she looked at how steam engine technology spread abroad.[45] Demand for engines from the sugar plantation economies in the Caribbean was especially strong.[46] J. Kanefsky and J. Robey in 1980 calculated the total number of engines produced by Boulton and Watt by 1800 at about 450, to which could be added 83 pirate engines and about 20 Newcomen engines adapted with a separate condenser without licence.[47] Richard Hills' three volume study of James Watt between 2002 and 2006, gave due attention to Boulton's entrepreneurial and technical contributions to Watt's steam engine.[48]

The need to understand Boulton's economic and artistic activities has developed alongside an interest in uncovering Boulton, the man. He had many enthusiasms, for balloons and astronomy, for example.[49] He understood how a factory building, the Soho Manufactory, might be employed as a symbol of taste and refinement, and

---

[41]   Richard Doty, *The Soho Mint and the Industrialisation of Money* (London, 1998).

[42]   Ibid., p. 297.

[43]   George Selgin, *Good Money, Birmingham Button Makers, the Royal Mint and the Beginnings of Modern Coinage, 1775–1821* (Michigan, 2008), p. 287.

[44]   Jennifer Tann, 'Marketing Methods in the International Steam Engine market: The Case of Boulton and Watt', *Journal of Economic History*, XXXVIII, 2 (June 1978), pp. 363–391.

[45]   Jennifer Tann and M. J. Breckin, 'The International Diffusion of the Watt Engine, 1775–1825', *Economic History Review*, 31 (1978), pp. 541–564.

[46]   Jennifer Tann, 'Steam and Sugar: The Diffusion of the Stationary Steam Engine to the Caribbean Sugar Industry', *History of Technology*, 19 (1997), pp. 63–84.

[47]   J. Kanefsky and J. Robey, 'Steam Engines in Eighteenth Century Britain: A Quantitative Assessment', *Technology and Culture* (London, 1980), pp. 161–186, 169. Newcomen was responsible for developing an atmospheric engine in about 1712, which utilised steam power and was widely used in the eighteenth century to pump water out of mines. Watt's separate condenser was an important technical development which turned the atmospheric engine into a steam-powered machine. For Newcomen see ODNB: http://www.oxforddnb.com/

[48]   Richard Hills, *James Watt*, 3 vols (Ashbourne, 2002–06); see especially 2, p. 47 and 3, p. 79.

[49]   J.J. Wolfe, *Brandy, Balloons and Lamps: Ami Argand, 1750–1803* (Carbondale, 1999); Andrew P. B. Lound, *Lunatick Astronomy: The Astronomical Activities of the Lunar Society Odyssey* (Birmingham, 2008).

Figure 1.4   Boulton, British halfpenny, reverse of Britannia, copper, 1806
Source: Courtesy of The Assay Office, Birmingham.

he used its setting in a polite landscape to project a series of positive messages to visitors about industrial production, the division of labour and steam power.[50] The best insight into his domestic and family life is provided by Shena Mason's 2005 biography of his daughter Anne. Mason confirmed the views of earlier writers that Boulton was kind to his family, valued close friendships and acted as a munificent host.[51] He also contributed to the social infrastructure of industrial Birmingham by helping to establish a theatre, which later became the Theatre Royal, and supporting the General Hospital.[52] Boulton provided his workers with an insurance society to offer financial support if they were unable to work because of injury,[53] but this may have had as much to do with his need to retain skilled workers as a charitable approach to labour-management relationships. His willingness to share the Enlightenment's commitment to spreading scientific knowledge, however, was limited: he freely divulged details about fossils, for example, but technical knowledge which might be useful to a rival manufacturer was more closely

[50]   Malcolm Dick, 'Discourses for the new industrial World: Industrialisation and the Education of the Public in late eighteenth-century Britain', *History of Education*, 37, 4 (July 2008), pp. 567–584.

[51]   Shena Mason, *The Hardware Man's Daughter: Matthew Boulton and his 'Dear Girl'* (Chichester, 2005), pp. 1–138.

[52]   Peter Jones, *Industrial Enlightenment*, pp. 65–68.

[53]   Dickinson, pp. 179–182.

guarded.[54] Boulton, unlike several of his colleagues such as Darwin, Priestley and Wedgwood did not support radical political or religious views.[55] His approach to one of the great causes of the time, anti-slavery was also somewhat opaque. He subscribed to the autobiography of the former slave and anti-slavery campaigner, Olaudah Equiano, but he also negotiated with merchants and landowners who benefited from the slave trade and the West Indies plantation system.[56] Though most of Boulton's Lunar friends were on record as opponents of slavery, at least in principle, he appears not to have expressed an opinion on this matter.[57] Boulton made much of his patriotism, as when he ordered the striking of medals at his own expense for distribution to the survivors of the Battle of Trafalgar.[58] However, he did not allow patriotism to stand in the way of profit. Boulton did not want it known that he had investments abroad, and he refused on one occasion to assist the government to prosecute a worker bound for Vienna in defiance of the Tools Act of 1719.[59] It is difficult to agree with earlier writers who described Boulton as highly principled.[60] Indeed, as Jennifer Tann remarked, Boulton was 'a business man first and an Englishman second'.[61]

Another approach which researchers have pursued is to locate Boulton within the context of the intellectual environment of the eighteenth century. Boulton's correspondence with individuals in the Lunar Society and other intellectuals further afield covers science, technology and manufacturing at a time when the process of industrialisation was accelerating. R.E. Schofield's substantial study of the Lunar Society in 1963 focuses on the practical significance of these men in applying scientific ideas to industrial technology.[62] Jenny Uglow's collective autobiography of the Lunar Men in 2002 looks more broadly at their contribution to education and involvement in politics as well.[63] Much of the recent debate about the Lunar Society has been boosted by studies of the Enlightenment. Roy Porter's studies of

---

[54]   Jones, *Industrial Enlightenment*, p. 158.

[55]   Schofield, pp. 360–362.

[56]   Jenny Uglow, *The Lunar Men: The Friends who made the Future* (London, 2002), p. 414.

[57]   Malcolm Dick, 'Joseph Priestley, the Lunar Society and Anti-slavery', in Dick (ed.), *Joseph Priestley and Birmingham* (Studley, 2005), pp. 65–79.

[58]   Nicholas Goodison, *Matthew Boulton's Trafalgar Medal* (Birmingham, 2007), p. 5.

[59]   J.R. Harris, *Industrial Espionage and Technology Transfer, Britain and France in the Eighteenth Century* (Aldershot, 1998), pp. 493–503.

[60]   Smiles, p. 178 wrote: 'Though he had a keen eye for business, Boulton regarded character more than profit. He would have no connexion with any transaction of a discreditable kind'. Dickinson, p. 197, regarded Boulton as 'high principled'.

[61]   Tann, 'Marketing Methods', p. 383.

[62]   Robert E. Schofield, *The Lunar Society of Birmingham: A Social History of Provincial Science and Industry in Eighteenth-Century England* (Oxford, 1963).

[63]   Jenny Uglow, *The Lunar Men: The Friends who made the Future 1730–1810* (London, 2002).

the British Enlightenment in 2001 and 2002 drew attention to the part played by English intellectuals, including members of the Lunar Society, in furthering the possibility of human progress.[64] The work of Joel Mokyr has been particularly important in this regard. In *The Gifts of Athena* (2002), he used the term 'Industrial Enlightenment' to refer to the forms of knowledge and ways of thinking, including interest in science, which, he believed, contributed to industrialisation in the late eighteenth and early nineteenth centuries.[65] He argued that the intellectual roots of industrialisation have not received adequate attention from historians.[66] He advocated a focus on the interface between thinkers and doers: 'the building of bridges between the sphere of knowledge and that of production, between *savants* and *fabricants*'.[67] In *The Enlightened Economy* (2009), Mokyr applied these notions to the study of the British Industrial Revolution and paid particular attention to Boulton as someone who encouraged the connection between rational and scientific ideas and their application to manufacturing.[68] In her magisterial exploration of the relationship between industry and art in the Enlightenment (2009), Celina Fox considered Boulton as a man who 'straddled the polite and mechanical arts', as a result of his intellectual associations and interest in science, technology and design.[69] The varied social networks of the 'Midlands Enlightenment', which connected Boulton and the knowledge economy of the West Midlands to the European and North American world, have been uncovered by Peter Jones.[70] Jones' *Industrial Enlightenment* (2008) remains the most significant study of Boulton's intellectual environment and a major contribution to the history of the period.[71]

---

[64]   Roy Porter, *Enlightenment: Britain and the Creation of the Modern World* (London, 2001), especially ch. 19, 'Progress'; Porter, 'Matrix of Modernity? The Colin Matthew Memorial Lecture', *Transactions of the Royal Historical Society*, 12 (2002), pp. 245–259.

[65]   Joel Mokyr, *The Gifts of Athena: Historical Origins of the Knowledge Economy* (Princeton, 2002).

[66]   Joel Mokyr, 'The Intellectual Origins of Modern Economic Growth', *Journal of Economic History*, 65, 2 (June 2005), pp. 285–351.

[67]   Joel Mokyr, 'The Great Synergy: The European Enlightenment as a factor in Modern Economic Growth', pp. 8–9: http://www.crei.cat/activities/sc_conferences/23/papers/mokyr.pdf (accessed 4 June 2011).

[68]   Joel Mokyr, *The Enlightened Economy: An Economic History of Britain 1700–1850* (New Haven and London, 2009).

[69]   Celina Jones, *The Arts of Industry in the Age of Enlightenment* (New Haven and London, 2009), p. 226.

[70]   Peter Jones, 'Industrial Enlightenment in Practice. Visitors to the Soho Manufactory, 1765–1820', *Midland History*, 33, 1 (Spring 2008), pp. 68–96; Jones, *Industrial Enlightenment*, pp. 1–237.

[71]   Jones was awarded the Business Archives Council Wadsworth Prize in 2009 for his outstanding contribution to the study of British business history: http://www.businessarchivescouncil.org.uk/activitiesobjectives/wadsworthprize/

The commemorative events of 2009 added significantly to the secondary literature. The lavishly illustrated catalogue of BM&AG's exhibition contained descriptions of paintings, prints and a number of short studies by Jenny Uglow, Shena Mason, Val Loggie, Nicholas Goodison, Kenneth Quickenden, Sally Baggott, Jim Andrew, Peter Jones, Sue Tungate, David Symons, George Demidowicz and Fiona Tait, to illuminate Boulton's life and activities.[72] The catalogue of the Barber Institute's exhibition was also well-illustrated. Sue Tungate's catalogue entries were accompanied by essays by David Symons on Boulton and the forgers, Peter Jones on the commercial tokens and medal coinage of the French Monneron brothers and Richard Clay, who argued that under Boulton's influence as a maker of medals and coins, Birmingham became 'the art capital of the world'.[73] A book on the landscape of Boulton's Soho estate contained contributions by Phillada Ballard, Val Loggie and Shena Mason.[74] Finally, *Matthew Boulton: A Revolutionary Player*, included detailed essays by Malcolm Dick, Peter Jones, Shena Mason, David Brown, Val Loggie, Fiona Tait, Olga Baird, Jim Andrew, George Demidowicz, Nicholas Goodison, Kenneth Quickenden, David Symons, Sue Tungate and Sally Baggott.[75] The contributors to these four books provide different perspectives on Boulton, without duplication: an indication of the depth and breadth of primary source material which is now becoming available to researchers.

Boulton Sources

Students of Boulton are well served by archives. Birmingham Archives and Heritage (BA&H) in Birmingham Central Library contains a huge amount of manuscript and printed sources relating to the personal, business and intellectual activities of Matthew Boulton, his family, associates and employees. The importance of this material extends beyond Boulton himself and is probably the most significant collection of British archival sources relating to the Industrial Revolution and Enlightenment networks in the late eighteenth and early nineteenth centuries. Between 1998 and 2003 the 'Archives of Soho Project' re-catalogued these archives to make them more accessible to researchers and students. These archives are comprised of three separate collections; the Matthew Boulton Papers (MS 3782) contain material relating to the Boulton family, the Soho Manufactory and the Soho Mint. These papers were loaned by the Boulton family to The Birmingham Assay Office in 1920, converted to a gift in 1926 and transferred to Birmingham Central Library in 1974. The other two collections include the papers

---

[72]   Mason (ed.), *Matthew Boulton.*

[73]   Richard Clay, 'How Matthew Boulton helped make Birmingham "the art capital of the world"', in Clay and Tungate (eds), pp. 39–55.

[74]   Phillada Ballard, Val Loggie and Shena Mason, *A Lost Landscape: Matthew Boulton's Gardens at Soho* (Chichester, 2009).

[75]   Malcolm Dick (ed.), *Matthew Boulton: A Revolutionary Player* (Studley, 2009).

of the Boulton and Watt steam engine business (MS 3147), which were given to the City in 1912 and the James Watt Family Papers. Part of the Watt papers (MS 3219) was held by the Muirhead branch of the family from the 1860s until they were donated to the Library in the 1930s. The remaining part was retained by the Watts and purchased for Birmingham Central Library in 1994.[76] There are archival materials relating to Boulton in other record offices at home and abroad as several contributions to this volume indicate. The materials in BA&H have been filtered through different processes of classification and being family and business papers, they reflect the approaches of those who have compiled, owned and deposited them. Despite their extent and value, these are managed archives.

Physical objects relating to Boulton are more widely dispersed in private and public collections. BM&AG hold particularly significant objects, including paintings of Boulton and examples of the complete range of his manufactures.[77] The interior of Soho House Museum in Handsworth, Birmingham, is based on visual representations and written accounts of the house during Boulton's life.[78] Thinktank, Birmingham's Museum of Science and Technology, also has a significant collection, including the Smethwick engine of 1779, the world's oldest working steam engine, which was manufactured by Boulton and Watt.[79] Further substantial collections can also be found at The Birmingham Assay Office,[80] and in London at the British Museum, the Science Museum and the Victoria and Albert Museum.[81] The Powerhouse in Sydney, Australia and the Speed Art Museum, Kentucky also contain major collections.[82]

Several websites provide access to digitised examples of source materials, where the originals are held by archives and museums. The 'Collections' section of BM&AG's website has examples of a range of objects held in Birmingham's local authority museums.[83] Many images relating to Boulton, Soho House and the products of the Soho Manufactory can be found on the *Digital Handsworth* site.[84] *Revolutionary Players*, by focusing on the West Midlands during the Industrial Revolution, provides a broader geographical focus and contains examples of

[76] Information supplied by Fiona Tait, Birmingham Archives and Heritage. For information about the Archives of Soho see also: http://www.birmingham.gov.uk/cs/Satell ite?c=Page&childpagename=Lib-Central-Archives-and-Heritage%2FPageLayout&cid=12 23092751138&pagename=BCC%2FCommon%2FWrapper%2FWrapper

[77] http://www.bmag.org.uk/

[78] http://www.bmag.org.uk/soho-house

[79] http://www.thinktank.ac/

[80] For examples of The Birmingham Assay Office's collection of silverware, coin and medals, see: www.revolutionaryplayers.org.uk – search for 'Assay Office'.

[81] http://www.britishmuseum.org/; http://www.sciencemuseum.org.uk/; http://www. vam.ac.uk/

[82] http://www.powerhousemuseum.com/; http://www.speedmuseum.org/

[83] http://www.bmagic.org.uk/

[84] http://www.digitalhandsworth.org.uk/

relevant Boulton material from The Birmingham Assay Office, BA&H and BM&AG, including Soho House, as well as a number of contextual essays in the 'Theme' section.[85]

The Chapters

Matthew Boulton: Enterprising Industrialist of the Enlightenment, makes available revised and expanded versions of a number of the papers which were delivered by academics, heritage professionals and research students at the 2009 conference, '"Where Genius and the Arts Preside": Matthew Boulton and the Soho Manufactory 1809–2009'. Some of these chapters reinterpret Boulton's significance, whereas others focus specifically on aspects of his activities or his changing reputation since his death.

Three chapters deal generally with how historians might approach Boulton. In 'Matthew Boulton, Birmingham and the Enlightenment', Peter Jones analyses the cultural relationship between Boulton and his place of birth. Boulton engaged with Birmingham and all it represented but he later became disillusioned and retreated to Soho House, where the 'hôtel de l'amitié sur Handsworth Heath' provided a locus for polite conversation and contact with the *savants* and *fabricants* of the Enlightenment. Jennifer Tann's 'Matthew Boulton: Innovator' argues that the stereotypical image of the Boulton and Watt partnership in which Watt is the engineering genius and Boulton, the entrepreneur, underplays the latter's hands-on approach. Boulton is portrayed as an innovator in applying science to industrial processes and systems, precision engineering and marketing. David Miller's 'Was Matthew Boulton a Scientist?' considers whether the historiographical exclusion of Boulton from the ranks of the scientists is appropriate. Traditionally, the term has been confined to the disinterested theoretician, rather than the practical individual. Bringing together a range of perspectives from the historiography of science, Miller argues that Boulton operated in the space between pure theory and entrepreneurship in a manner that fits well with contemporary definitions of scientific activity.

Several chapters deal with Soho and Boulton's interest in science and technology. George Demidowicz's archaeological analysis, 'The Origins of the Soho Manufactory and its Layout' investigates the planning and construction of the buildings and their integration with water and steam power. The chapter also considers those developments in relation to the later evolution of the Manufactory and its products. Jim Andrew's 'Boulton, Watt and Wilkinson: the Birth of the Improved Steam Engine' analyses Wilkinson's boring of the cylinders used in Boulton and Watt steam engines and advances our understanding of the relative costs of the Newcomen and Watt engines. He provides a new analysis of the reasons for the greater power and efficiency of Watt's engines through the

---

[85]   http://www.revolutionaryplayers.org.uk; http://www.sohomint.info/

application of thermodynamic theory rather than the traditional explanation of the introduction of the separate condenser. The chapter also furthers our understanding of Boulton's contributions to testing and improvements in engine design. Another chapter with a scientific focus is 'Matthew Boulton's Copper' by Peter Northover and Nick Wilcox, which provides a chemical analysis of the copper coins which Boulton minted. They offer an insight into the methods used to smelt copper and a technical comparison between the copper used in Boulton's coins and that used in other countries. One chapter which bridges the gap between technical analysis, the history of art and marketing is Barbara Fogarty's 'The Mechanical Paintings of Matthew Boulton and Francis Eginton', which examines the debates and technical challenges involved in the production of this little understood aspect of Boulton's work with Eginton. She argues that the process relied on aquatint prints to transfer tonal impressions from the copper plate to canvas which were then hand painted in oils.

Other chapters throw light on the political, social and cultural history of Boulton's industrial activities. In 'Samuel Garbett and Early Boulton and Fothergill Silver', Kenneth Quickenden questions the validity of evidence given by Garbett to the Parliamentary Committee in 1773, which has been largely accepted by previous historians, to support Birmingham's petition to obtain an assay office. Quickenden argues that the progress of Boulton's silversmithing by 1773 and its potential were not as great as Garbett maintained. The cultural significance of Boulton's endeavours to establish a local assay office is explored by Sally Baggott in 'Hegemony and Hallmarking: Matthew Boulton and the Battle for the Birmingham Assay Office'. Using theoretical perspectives from Cultural Studies, she considers the political and ideological forces at work during the campaign that were displayed in the tension between London's dominance in the silver trade of the eighteenth century and Birmingham's emergence as an industrial centre of silver production. Two chapters focus on Boulton's workforce, a subject which has been somewhat neglected hitherto. Jo Melling's 'Dark Satanic Millwrights: Forging Foremanship in the Industrial Revolution; Matthew Boulton and the Leading Hands of Boulton and Watt' analyses foremen in the steam engine business at the Manufactory and the Soho Foundry. Melling argues that much was expected of foremen although their roles were loosely defined and they were kept in check by Boulton's subtle management skills which contrasted with the more confrontational approach of Watt. Sue Tungate's 'The Workers at Matthew Boulton's Soho Mint' looks at the roles, recruitment, training, pay and management of die-engravers, shop-floor workers, engineers, mechanics and labourers at this establishment.

Two chapters present an international perspective. Liliane Perez and Bernard Vaisbrot's 'Matthew Boulton's Jewish Partners between France and England: Innovative Networks and Merchant Enlightenment' looks at the import of Boulton's 'toys' into France including the ways in which language, the flow of information between Soho and Paris and Boulton's involvement with Jewish partners shaped this trade. Secondly, Irina Gouzevitch's 'Enlightened Entrepreneur versus

Philosophical Pirate 1788–1809: Two Faces of the Enlightenment', considers the case of the Spanish engineer and spy, Augustin Betancourt and his visits to Soho. Gouzevitch suggests that Watt was keen to spy on Betancourt even though previously scholars suggested that espionage had only taken place in the opposite direction.

The final chapters deal with the promotion of Boulton. Val Loggie's 'Creating an Image: Portrait Prints of Matthew Boulton' concentrates on the commissioning and production of a mezzotint by S.W. Reynolds after a painted portrait by C.F. von Breda of 1792 and a line engraving by William Sharp after a painted portrait by Sir William Beechey of 1798. She argues that the distribution of prints by Boulton and his descendants helped to promote Boulton's reputation after his death. Lastly, Malcolm Dick's 'The Death of Matthew Boulton 1809: Ceremony, Controversy and Commemoration', analyses Matthew Robinson Boulton's attempts to protect and enhance his father's standing after his death: via funeral practices, obituaries, memoirs, medals and a memorial in Handsworth Church. The process was problematic and in comparison to the efforts of the Watt family, less successful. One hundred years after his death, there were only limited celebrations of Boulton, in contrast to the commemorations in 2009.

Boulton scholarship will continue because of the wealth of archival material and the many opportunities for increasing our knowledge of his personal and domestic life and the economic, political, social and cultural significance it provides. Archaeology, the study of material culture, scientific analysis and the development of historiography, all present new ways of understanding his life, work and context. *Matthew Boulton: Enterprising Industrialist of the Enlightenment* captures the current state of research and thinking but more work remains to be done.

Chapter 2

Matthew Boulton, Birmingham and the Enlightenment

Peter M. Jones

Most assessments of Matthew Boulton have little to say about his relationship to the town of his birth and even less about the cultural context in which his career as one of England's leading manufacturers of metal goods was played out. Instead, the focus tends to be placed on his role as a businessman and entrepreneur. His strengths and – increasingly – his weaknesses as a businessman are recounted; his far-sightedness in enabling James Watt to make commercially viable his improved steam engine is acknowledged, as is his contribution to the development of the factory system based on the division of labour. More recently, the campaigns he mounted to overcome the restrictive practices of assayers and moneyers which resulted in the establishment of an assay office in Birmingham and a mint at Soho have been a target of research as well. Only his capabilities as an engineer and technologist await the scrutiny of historians.

Yet it can be questioned how far these well-documented achievements help us to uncover the threads that gave shape and purpose to Boulton's eminently public career. Late in life, and no doubt with an eye to posterity, he intimated that it was not money, or influence but concern for reputation that had driven him forward: the ambition to be recognised as the foremost industrialist of his generation.[1] There is no reason to doubt this claim. Matthew Robinson Boulton came to the same, or very similar, conclusions about his father, as did his long-term business partner, James Watt.[2] It seems likely that the ultimate apotheosis of a baronetcy or,

---

[1] See Birmingham Archives and Heritage (BA&H), MS 3782/13/36, MB to M.R. Boulton, 18 December 1788; MS 3782/12/44, G. Gilbert to MB, Hackney, 30 June 1799; MS 3782/12/45, J.H. Redell to MB, n.d. (1 January 1800); MS 3782/12/48, Bracebridge to MB, Atherstone, 22 May 1803; MS 3782/12/52, W.T. Fitzgerald to MB, London, 4 January 1807; MS 3782/13/104, J. Bedingfield to MB, 11 February 1808; MS 3782/12/53, Lord Muncaster to MB, Ravenglass, 21 September 1808. Also Warren R. Dawson (ed.), *The Banks Letters* (London, 1958), Sir J. Banks to Sir E. Thomason, Soho Square, 6 July 1817 and Kenneth Quickenden, 'Boulton and Fothergill Silver' (University of London, Westfield College, unpublished PhD dissertation, 1989), p. 29.

[2] H.W. Dickinson, Matthew Boulton (Cambridge, 1936), p. 205. Also Malcolm Dick, 'Matthew Boulton: A Revolutionary Player', in Dick (ed.), *Matthew Boulton: A Revolutionary Player* (Studley, 2009), pp. 1–12, 2 and note 4.

at the very least, a knighthood *à la Richard Arkwright* only eluded him because the king cancelled his 1805 visit to the Midlands and headed to Weymouth instead.[3]

Nonetheless, people who encountered Boulton in his prime found him to be a rather complex individual: he could be steadfast and devious, generous and self-righteous by turns.[4] He could also be self-deprecating to the point of false modesty: 'if I glory in anything', he gratuitously informed one correspondent, 'it is in being one of those exporting importing Tradesmen whose connections are not circumscribed by the bounds of Europe & that can travil [sic] to every Capital city in it without any other letters of cred.' than those of his own writing'.[5] The attention paid to reputation seems to have been the preoccupation of an elderly man whose ill health had greatly curtailed his participation in the affairs of the world. It is unlikely to have been the over-riding rationale of his life.

Unfortunately, direct testimony to Boulton's ambitions, whether in life or in death, is hard to find. His was not a generation much given to introspection or self-examination. There are no 'ego-texts' allowing us to access his patterns of thought, and few of the men and women he moved among betrayed traces of pre-romantic sensibility either. The argument that follows, therefore, relies extensively on plausible reconstruction. It places Boulton in two contexts which undoubtedly mattered to him, even if we can find only scattered and oblique references to the fact in his personal papers and business correspondence. Juxtaposed, they offer a frame of reference for understanding his career that researchers have tended to overlook until now. By birth and by upbringing Boulton was a citizen of Birmingham, yet he spent much of his life trying, by process of assimilation, to make himself a citizen of the Republic of Letters.[6] In retracing his efforts to combine these roles – efforts that would unravel when civil disorders eclipsed Birmingham's social experiment with enlightenment – it is possible to explore his character from a less familiar angle.

I

Although not very much is known about Boulton's early years, there can be no doubting his Birmingham roots. Born in 1728, he entered the world as the second

---

[3]   BA&H, MS 3782/12/50, F. Deluc to MB, Windsor, 5 July 1805. Also *Memoir of Matthew Boulton by James Keir F.R.S., December 3 1809* (Birmingham, 1947), p. 8.

[4]   Franx-Xaver Swediaur with whom MB had convoluted business relations that turned sour would claim that his dominant trait was vanity, see BA&H, MS 3219/4/111, F.-X. Swediaur to J. Watt Snr, Paris, 4 October 1792. Also George Selgin, *Good Money: Birmingham Button Makers, the Royal Mint, and the Beginnings of Modern Coinage* (Ann Arbor, 2008), pp. 70–71; 102; 104–105.

[5]   BA&H, MS 3783/12/55, undated letter draft, c. 1782–85.

[6]   The contemporary characterisation of the intellectual community of eighteenth-century Europe.

son of a prospering button and buckle merchant who came to own a complex of workshops and a steel cementation furnace in the Snow Hill district of the town. The first-born male child having died in infancy, Matthew became the acknowledged heir to this business. His formal education probably ended at the age of 14 when he joined his father in the workshops – around 1742, therefore.[7] Seven years later he married into a well-to-do family of Lichfield mercers named Robinson whose trading profits had been invested in landed property. Having attained his majority, if not yet any appreciable degree of financial independence, he began to think about fashioning for himself a role that would extend beyond the usual social and intellectual horizons of a Birmingham 'toy' maker.[8] We know, from entries in his diaries, that from 1754 he began to repair omissions to his basic education and to set the trajectory that would mark out the rest of his life. Works by Herman Boerhaave and Voltaire were purchased, to be followed by those of Shakespeare, Locke, Swift and Pope's *An Essay on Man* (1733–34). For good measure, he also took out a subscription for Baskerville's lavish edition of Virgil's poetry and acquired a copy of Chambers's *Cyclopaedia*, a popular science compilation, together with several foreign language dictionaries.[9]

The real turning-point in his life came in 1759, however, for in that year both his wife and his father died in swift succession. Boulton found himself a 31-year-old widower in possession of one of the larger metal-working enterprises of the town, but in much less secure possession of his late wife's dowry. This latter problem was resolved by marrying his deceased wife's sister – a practice frowned upon by the Church but which seems to have been commonly resorted to notwithstanding. It is at this point that the well-documented public – and public-spirited – figure replaces the sparsely documented and otherwise unremarkable Birmingham button maker.

As if to give notice of his expanding ambitions, not to mention the very considerable financial resources now at his command, Boulton started to look for an out-of-town site where he could erect a large edifice to bring in-house the many operations associated with button and buckle manufacture that were currently dispersed along the back streets of Snow Hill. The Soho Manufactory located in the Staffordshire parish of Handsworth, just across the heath from Birmingham town, was the result. Having been about four years in construction, it opened to customers and visitors in 1766. The principal building, as Figure 1.1 makes plain, was intended from the outset to make a statement. Neo-Palladian in inspiration, it was designed as if to disguise its true identity like so many contemporary industrial

---

[7]   Shena Mason, *The Hardware Man's Daughter: Matthew Boulton and his 'Dear Girl'* (Chichester, 2005), p. 204.

[8]   'Toys' were the generic name given to Birmingham fashion goods such as buttons, buckles, watch chains, chatelaines, ornamented boxes etc. They were often made of gilt metal or cut steel.

[9]   Mason, p. 208.

premises on the continent of Europe.[10] If we examine the many promotional
engravings commissioned by Boulton, the factory vocation of the complex is
only hinted at in a blurred perspective of steep roofs, courtyards and chimney
stacks. There can be no doubt that Soho impressed by its uniqueness in the
industrialising landscape of the West Midlands, then. Yet for continental visitors,
the building would not have been altogether unfamiliar for it was reminiscent of
the architectural styles of the Industrial Enlightenment.[11] Only the Georgian sash
windows of the front façade might have appeared a little strange, plus the fact
that the edifice was constructed entirely of dark red Birmingham brick rather than
masonry blocks.

From 1766 the Soho Manufactory together with Soho House, the elegant family
residence situated only a short distance away, became the vehicles for Boulton's
transformation from Birmingham button maker to man of the Enlightenment.
Much of the evidence for this reincarnation can be found in the testimony of
the hundreds and thousands of individuals who came to visit his new industrial
premises over a span of close on four decades. Boulton's celebrity, we learn,
stemmed from more than simply his status as a thrusting English industrialist and
the proprietor of the country's largest single-site factory.[12] It flowed from the fact
that he was universally acknowledged to be a consummate purveyor of 'civility'.

Historians tend to regard the practice of civility as a cultural trait characteristic
of the eighteenth century, but it can be traced to gentry advice manuals of the
later sixteenth century. With this back projection it becomes easier to see how it
functioned – as a social marker or means of discriminating between those who
were gentlemen and those who were not. It is true that by the eighteenth century
the rules of civil behaviour tended to be tacit rather than encoded. They had also
been further refined with the addition of the value of 'politeness' – an attribute
which Tobias Smollett referred to as 'the art of making oneself agreeable'.[13] Thus
Boulton's Scots partner, James Watt could remark, in 1783, to a Swiss acquaintance
who was anxious to send his daughter for language training, that the learning of
common English civility would be feasible in Birmingham, but 'politeness must

[10]   The crystal glass works and the foundry at Le Creusot are examples. See Jean-
Robert Pitte, *Histoire du paysage français* (2 vols, Paris, 1983), vol. 1, p. 72.

[11]   For Industrial Enlightenment, see Joel Mokyr, *The Gifts of Athena: Historical
Origins of the Knowledge Economy* (Princeton, NJ, 2002), pp. 28–77 and Peter M. Jones,
*Industrial Enlightenment: Science, Technology and Culture in Birmingham and the West
Midlands, 1760–1820* (Manchester, 2008).

[12]   An observation frequently made by travellers and, with a Soho-based workforce
of up to a thousand, probably exact until the erection of the Manchester cotton mills in the
1800s. See Norman Scarfe, *Innocent Espionage: The La Rochefoucauld Brothers' Tour of
England in 1786* (Woodbridge, 1995), p. 111.

[13]   Tobias Smollett, *Travels through France and Italy. In two Volumes* (London, 1778),
vol. 1, p. 100.

be learned in better company than this town affords'.[14] Civility denoted a social frontier, then; a frontier that applicants to visit the Manufactory were at pains to acknowledge and respect.

The affable, urbane, hospitable and conversational Boulton, with an unflagging appetite for corralling visitors around his dinner table at Soho House well understood the rules of the game of civility. But he liked to interpret them in the utilitarian terms of the commercial society to which he belonged.[15] In other words, he endowed the practice of civility with transactional underpinnings. Whilst entertaining Europe's intelligentsia, he expected to receive something in return. Careful scrutiny of his behaviour also indicates that he viewed the practice as an enabling process. In his mind 'civility' was linked closely to the aspiration to 'civilise'. It represented less an insurmountable social barrier than an avenue which he hoped to equip his fellow citizens to travel along so that they in turn could become 'civil'. The citizens in question were of course the inhabitants of Birmingham: those men and women whose rudeness of manners and lack of refinement could be softened by inciting them to adopt civilised modes of behaviour.

In this endeavour Boulton drew on the pan-European Enlightenment and, more especially, on one of its main intellectual pillars: the empirical philosophy of Locke. When filtered through the preoccupations of the mercantile classes, though, the thought and practice of the continental Enlightenment tended to shed its militancy in matters religious and political. Expressed in provincial English terms, it became instead a secular creed of human betterment. The only French *philosophe* whose writings we can be reasonably confident that Matthew Boulton actually consulted was the Baron de Montesquieu. Even so, he seems to have shown more interest in Montesquieu's sociology than in his political science.[16] The casual free-thinking of French intellectuals that bordered on outright atheism repelled him. After all, it shocked even Dr Joseph Priestley and Dr Erasmus Darwin who passed for the closest native-grown approximations to the *philosophes* of France.

If we are to understand fully the Enlightenment embraced by Boulton as a meliorative project we need to look elsewhere, then. We find it in the palpable realisation that society in general and provincial urban culture in particular had embarked upon an ascending curve of improvement and material betterment, the likes of which the western world had not before experienced. The symptoms of this progress, as apprehended by men of Boulton's stamp, were several: the signal increase in the quantum of knowledge in circulation throughout Europe; the rapid

---

[14] BA&H, MS 3219/4/123, J. Watt to J.-A. Deluc, 26 June 1783.

[15] See Felicity Heal, *Hospitality in Early Modern England* (Oxford, 1990), pp. 392–400.

[16] 'Early in life Fortune give me the option of assumeing the character of an idle man, commonly called a Gen.tm but I rather chose to be of the class wh.ch Le baron Montisque discribes as the constant contributors to the purze of the common wealth rather than of another class which he says are always takeing out of it without contributing any thing towards it', BA&H, MS 3782/12/55, undated letter draft, c. 1782–85.

development of the associational habit in the shape of clubs, coffee houses and masonic lodges; the sharp increase in technological optimism rooted in reason and experiment; the expansion of a culture of consumerism in which all social classes participated to some degree; and, last but not least, the extraordinarily enabling social climate of England during the second half of the eighteenth century which acted as the lubricant of all these phenomena.

This latter point is often taken for granted, yet one has only to read a number of foreign visitor accounts and the tourist guides published for the use of visitors from overseas to grasp that eighteenth-century England offered a glimpse of a very different society from that of most of the rest of Europe. After a seven-week tour of England undertaken in 1782, the Prussian clergyman, Carl Philipp Moritz concluded, 'there is through all the ranks here not near so great a distinction between high and low as there is in Germany'.[17] On crossing the Channel from France in 1802, the Saxon law professor, Christian August Gottlieb Goede remarked that in Paris he had experienced little difficulty in judging social status from dress, whereas in England, 'it is scarcely possible to know a lord from a tradesman, a man of letters from a mechanic; and this seems to arise from the sovereignty of fashion in the metropolis'.[18] Later on that summer, Goede took the stagecoach to Birmingham and in due course he joined those seeking admission to Boulton's industrial emporium.

One thing that nearly all visitors to the town from mid century onwards, whether foreign or domestic, felt the urge to comment on was that it was indeed a candidate for the firm hand of enlightenment. Only the recently constructed play-house on New Street, with its pretty neo-classical façade elicited a favourable reaction from Goede.[19] Birmingham's resident elite, both mercantile and professional, would have agreed. For all its prowess in fashion-based metal working, Birmingham could scarcely be described as a wholesome place. It was unwholesome in the modern environmental sense, of course, but it was grossly deficient also in the pre-requisites of 'civility'. Coal smoke belched from a hundred chimneys, house windows were permanently covered in soot, the streets were either unpaved or at best poorly paved, elegant public buildings were notable by their absence, and there was scarcely anywhere for the beau monde to stroll and display themselves.

Migrants to the town such as William Hutton and James Bisset obliquely endorsed this verdict.[20] Yet they also drew attention to the extraordinary bustle and energy to be found in the souk-like alley ways of Digbeth and the older parts

---

[17]   *Travels of Carl Philipp Moritz in England. A Reprint of the English Translation of 1795* (London, 1924), p. 152.

[18]   C.A.G. Goede, *The Stranger in England or Travels in Great Britain* (3 vols, London, 1807), vol. 2, p. 84.

[19]   Ibid., 3, p. 93.

[20]   William Hutton, *An History of Birmingham* (3rd edition, Birmingham, 1795), pp. 31–2, 71–8, 99, 156–61, 304–310, 410–412; BA&H, MS 263924, J. Bisset commonplace book.

of the lower town. In fact, Bisset who arrived from Calvinist Scotland in 1776 was more inclined to evoke the raw cheerfulness of the town's rapidly expanding population. In jottings probably intended to serve as his memoirs, he recalled how much more fun the Sabbath was in Birmingham compared to the land of his birth; how he had been able to enjoy music in places of worship for the first time, and how he used to spend Sunday afternoons rowing on Hockley Pool together with other workmen.[21] Would Boulton have approved of hire boats, picnics and boisterous sports around the pool below his manufactory? It seems unlikely, for such working-class recreational activity jarred with the contemplative harmonies of his park – its walks, grottoes, cascades, temples, neo-classical statuary and ingenious water features. However, he controlled only a portion of the lake at the bottom of his gardens.[22]

II

It is here that we encounter, in microcosm, the paradox of Georgian civic culture. Plainly there existed an 'imagined' Birmingham and a 'real' Birmingham, and the task which men like Boulton set themselves was to reconcile the two phenomena. The imagined Birmingham was a construct of the Enlightenment and, for perhaps three decades, it looked as though it might be possible to give it some social substance. After bitter party-political and sectarian feuding at the time of the Hanoverian succession which had occasioned serious episodes of rioting (1714–15), stability returned and the inhabitants of Birmingham settled down to concentrate on the business of making money instead. They did so under the leadership of an elite that could no longer find a *raison d'être* in the prejudices and enthusiasms of Queen Anne's reign. From the late 1750s and 1760s, this elite energetically and self-consciously began the work of 'civilising' Birmingham. Urban improvement initiatives were launched; a cross-party consensus was nurtured around a scheme to endow the town with a general hospital; a music festival was introduced to raise the necessary funds; local booksellers began to publish the bestsellers of the continental Enlightenment; libraries and advertisements for subscription lectures appeared; Bull Street was paved; and as visitors to the workshops became more frequent, the local hostelries started to rename themselves 'hotels'.[23] Some small and temporary progress may even have been achieved in discouraging local blood sports: bull and bear baiting, cock fighting and bare-knuckle boxing. The key to these developments, it must be emphasised, was a process of cultural convergence among the town's well-to-do citizens. The elite were steadily regrouping in the name of civility

[21]   BA&H, MS 263924, ibid.

[22]   BA&H, MS 263924, ibid. Also Phillada Ballard, Val Loggie and Shena Mason, *A Lost Landscape: Matthew Boulton's Gardens at Soho* (Chichester, 2009), pp. 1–13, 25.

[23]   Hutton, p. 204.

and social utility. Cordial relations developed between the parish churches and
the town's meeting houses and, for a time, the new ecumenism even extended
to mixed dinner parties between Churchmen [Anglicans] and Dissenters.[24] The
rapprochement lasted from around 1760 until the middle of the 1780s.

The 'real' Birmingham was of course the other: the Birmingham which
civic leaders, whether enlightened Churchmen or Rational Dissenters, hoped to
attenuate, even to eliminate. Although they did not put it in so many words, they
had in mind the boom town whose population tripled to 75,000 in the space of
three decades (1765–92); the Birmingham choked with smoke and metallic odours;
the Birmingham from which they retreated to their country villas at sunset; the
Birmingham whose inadequate store of social capital was constantly under threat
of dilution from in-migration to supply the labour requirements of the workshops;
the Birmingham lacking effective institutions of local government and police.[25] In
short, the Birmingham of ale houses, gin shops, pawn brokers, drunks, prostitutes,
metal thieves, counterfeiters and shoddy goods merchants.

Boulton was no stranger to the 'real' Birmingham, and there can be little doubt
that one of his abiding ambitions was the desire to do something about it through
the medium of a social agenda rooted in civility and utility. However, at crucial
moments in his public career he gave up on his fellow citizens – usually when he
felt ill-used by them. On such occasions he would retreat, physically as well as
metaphorically, into an alternative world centred on Soho. In this world, the guests
invited to sup and converse in the elegant dining room of Soho House 'replaced'
the decidedly inelegant and, in his view, ungracious and ungrateful citizenry of
Birmingham. However, this reflex of withdrawal was chiefly a characteristic of
his later years – when events which he was powerless to influence began to gnaw
away at provincial England's Enlightenment.

This is to look ahead, however. On coming of age, the first task Boulton
set himself was to acquire the accoutrements of an enlightened gentleman.
Familiarity with Newtonian science, albeit in a dilute version, was fast becoming
a denominator of the status he coveted, and we know that his first excursions into
the field of natural philosophy focused on the subtle and mysterious phenomenon
of electricity.[26] In 1757, he began to purchase electrical apparatus, and in 1760
he showed off one of his experiments to Benjamin Franklin when he paid a visit
to Birmingham.[27] Acquaintance with Erasmus Darwin, William Small and Josiah
Wedgwood appears to have dated from the mid 1760s, and it caused his interest in

---

[24]   BA&H, MS 263924, J. Bisset, commonplace book; Hutton, pp. 170–172; Emily
Bushrod, 'The History of Unitarianism in Birmingham from the Middle of the Eighteenth
Century to 1893' (University of Birmingham, unpublished MA dissertation, 1954), pp.
127–150.

[25]   Hutton, pp. 142–153, 373–376; Jones, pp. 34–39, 60–62.

[26]   Jones, pp. 77–78.

[27]   Ibid., p. 77; Leonard W. Labaree et al. eds, *The Papers of Benjamin Franklin* (39
vols, New Haven, 1959–2008), vol. 10, p. 39; 12, p. 140.

the sciences to broaden and to move in more mechanical and utilitarian directions. After all, he was now the proprietor of a purpose-built manufactory whose energy requirements were heavily dependent on an immense waterwheel supplied from the Hockley brook mill pool. No doubt this lay at the back of his mind when pursuing experiments with Small on power from steam engines and conducting a correspondence with Watt. Although Boulton would have been familiar with Joseph Priestley's works on electricity and optics, direct contact with the doctor seems not to have been established until the decade following. However, it is clear from the style of address employed by Priestley in which he referred to Boulton as 'a friend of *science* as well as a great promoter of the *arts*' that the Birmingham manufacturer had by this time secured recognition as an experimental philosopher.[28]

Perhaps it was confidence born of this recognition that prompted Boulton to take the lead in organising the body which came to be known as the Lunar Society. This private society almost certainly started off life as an informal conversation group linking Boulton to Darwin and Small, and perhaps also to James Keir and Wedgwood, although the latter lived too far distant from Birmingham to take much part. The appearance in the Midlands in 1774 of Boulton's new partner, Watt, was probably significant as well. At all events it was Boulton who made the first move to put their philosophical encounters on a more permanent footing. Having suggested a meeting at Soho early in 1776, he told Watt, 'I then propose to submit many motions to the members respecting new laws and regulations, such as will tend to prevent the decline of a Society which I hope will be lasting'.[29] In fact, the members seem to have preferred to remain a loose grouping of convivial, drawing-room philosophers who visited each other's houses at regular intervals. The 'lunaticks' as Samuel Galton's butler once described them, must therefore have turned a deaf ear to Boulton's proposals.[30] The Birmingham philosophers' club, unlike Manchester's Literary and Philosophical Society which was constituted as such in 1781, stayed informal from beginning to end. But this did not affect their capacity to function, or to act as a vehicle for discussion of the latest ideas in science and technology. In 1776, they seem to have spent many an evening discussing the subject of heat, and there is evidence to show that the group was still meeting fairly regularly some 25 or 30 years later – though several of the members had quit the scenes by this time and Boulton was largely bed-bound.[31]

A consideration of Boulton's spiritual outlook will serve to underline further his credentials as an enlightened gentleman. Like many educated Englishmen of his generation, he regarded revealed religion, or rather the enthusiasms it was apt to occasion, as a fertile source of discord. A deist, he distanced himself

---

[28]   BA&H, MS 3782/12/24, J. Priestley to MB, Calne, 22 October 1775.

[29]   T. Whitmore Peck and K. Douglas Wilkinson, *William Withering of Birmingham M. D., F. R. S., F. L. S.* (Baltimore, 1950), p. 131.

[30]   Andrew Moilliet (ed.), *Elizabeth Anne Galton, 1808–1906: A Well-Connected Gentlewoman* (Hartford, 2003), p. 6.

[31]   Jones, pp. 82–94, figure 3 and table 3.

from conventional religious sensibility, whether of the neo-classical or the pre-romantic variety, once telling the daughter of an old friend recently deceased not to over-indulge in mourning ('a sin instead of a Virtue').[32] On the other hand, the commemorative possibilities of religion appealed to him – as they did to many of the *philosophes*.[33] He would have been familiar with William Shenstone's tomb-infested gardens at the Leasowes and when, in 1775, Small, his family doctor and closest confidant, passed away he had a monument to his memory erected in his garden. Did he also believe in religion as social cement? No doubt he did, but not with any discernible conviction. There is not the least trace of the evangelical fervour of the nineteenth-century mill owner in Boulton. In 1775 or 1776, he informed his partner, Watt that he had launched a chapel subscription at Soho, but we hear no more of the subject after that year and there is no sign that a work-place chapel was ever built.[34]

Yet, he could be highly respectful of the religious convictions of others – whether they were Anglican, Dissenting, or indeed Catholic. The regular, monthly meetings of the Lunar Society could not have taken place otherwise. A nominal Anglican, he attended services at Saint Paul's on the northerly outskirts of Birmingham, but not on a regular basis as far as can be judged. As his attitude to Canon Law on the subject of remarriage reveals, his religion was utilitarian in tincture. In matters of doctrine he was impatient of Original Sin and probably less than orthodox on the subject of the Trinity, but he would certainly have disapproved of Priestley's systematic enumeration of the Incarnation, the Atonement, Christ's divinity and so forth as being so many irrational 'corruptions of Christianity'.[35] On the morrow of Birmingham's civil disturbances he informed a London friend, perhaps rather smugly, 'I live peaceably & securely amidst the Flames, Rapin, plunder, anarchy & confusion of these Unitarians, Trinitarians, predestinarians & tarians of all sorts'.[36]

III

By 1791 – the year marred by several days of rioting following an ill-advised invitation dinner to celebrate the anniversary of the taking of the Bastille – the Enlightenment in Birmingham was no longer at the top of the agenda. This much is certain. The shadows had started to gather in the mid 1780s with disagreements over library provision and schooling causing rifts in the town's elite. These

---

[32]   BA&H, MS 3782/12/50, M. Boulton to Miss M. Linwood, June 1805.

[33]   See Joseph Clarke, *Commemorating the Dead in Revolutionary France: Revolution and Remembrance, 1789–1799* (Cambridge, 2007).

[34]   BA&H, MS 3219/4/66, MB to J. Watt Snr, n.d. Also Jones, p. 185.

[35]   See Joseph Priestley, *An History of the Corruptions of Christianity* (Birmingham, 1782).

[36]   BA&H, MS 3782/12/36, copy letter of MB to C. Dumergue, Soho, 10 August 1791.

tensions threatened to wreck the systems for social control that had been put in place since the 1760s. Boulton, we know, grew increasingly dissatisfied with his fellow citizens' behaviour, although the riots were as much a symptom as a cause of his dissatisfaction.[37] The mission to smooth away the town's robust plebeian culture was failing.

Back in the 1760s, when the outlook had seemed so much more promising, he had been in the forefront of the collective effort to raise funds for the new hospital. In the early 1770s another civic initiative – the bid to secure a licence for the New Street theatre – had been launched and, again, Boulton together with Small and the Presbyterian 'toy' manufacturer, John Taylor had gladly lent their names to the campaign. Boulton made the purpose of his involvement explicit in a letter to the Earl of Dartmouth, then Lord Privy Seal in Lord North's ministry, declaring that the theatre project would bring more visitor spending power to the town and promote the civilising of the general population. In fact, he subsidised his own workmen so that they could attend the theatre.[38] The recreation park opened at Ashted just outside Birmingham (subsequently renamed Vauxhall Gardens) can be seen as a further attempt to advance the civilising process.

But in the 1780s these efforts slowly turned sour. Sectarian squabbling which had been noticeable by its absence during the preceding 30 years reasserted itself; economic hardships linked to the trade cycle triggered an increase in social unrest and insubordination; traditional pastimes such as bull and bear baiting resumed; louts now began to trek out to Vauxhall Gardens in order to throw stones rather than admire and emulate their social superiors. In short, 'police', or rather the lack of it, became the main priority of the elite agenda. No doubt it was symptomatic of the times that Boulton's last big public commitment to his compatriots was a watch committee which organised and paid for nightly street patrols through the winter and spring months of 1789–90.[39]

Dismayed that even his fellow entrepreneurs were now defaulting on the obligations of civility and refusing respectable strangers access to their manufactories, he began to live *his* version of the Enlightenment at Soho instead. Here, in the controlled environment of the 'hôtel de l'amitié sur Handsworth Heath' (as he once dubbed Soho House), it really was possible to combine the responsibilities attaching to civility with the characteristic practices of an enlightened gentleman.[40] Indeed, Soho would emerge in the later 1780s as a bright star in the Europe-wide constellation of the Republic of Letters. The Lunar meetings took place at Soho House more often than not, and a veritable procession of eminent foreigners beat a path to his doorstep and dinner table. They included

[37] BA&H, MS 3147/3/38, M.R. Boulton to J. Watt Jr, London, 25 July 1800; MS 3782/12/46, copy letter of MB to E. Carless, 2 January 1801 and note 44 below.

[38] Mason, p. 30.

[39] BA&H, MS 386813, minute book of the Birmingham police committee, 1789–90.

[40] BA&H, MS 3219/4/1, MB to J. Watt, Soho, 7 February 1769.

enough internationally known *savants* to flatter a monarch let alone a Birmingham 'toy' manufacturer turned natural philosopher.

Boulton's cultural ambitions – for himself and for the town that witnessed his birth – were thus transposed during the last two decades of his life. The Enlightenment of his imagination took on a real presence not in the benighted streets and alleyways of Birmingham, but in escorted factory tours around his industrial premises followed by conversation and good cheer in the drawing room of Soho House. By the century's end Birmingham's mercantile and professional elite were starting to re-learn the habits of cooperation and collaboration, it is true. In 1802, the barrister John Morfitt reported an easing of the 'coldness generated by mutual suspicion' when the various fractions of the town's bourgeoisie came together to promote a theatrical entertainment for the benefit of both the Anglican and Nonconformist charity schools.[41] But Boulton never again took the lead in public affairs. He felt shabbily treated by his fellow manufacturers and the myriad craftsmen and women employed in the town's workshops. In 1800 they organised an attempt to block his plans to export coining mills to Saint Petersburg, fearful that diffusion of his new steam-driven technology would have an undermining effect on the local economy.[42] When a fire broke out in the Soho Mint building the following year, James Watt junior added a revealing aside in a letter to Boulton's son, 'the Birm.ᵐ people (*horesco referens*) wish it had been the whole manufactory & your father in the midst of it'.[43]

Mindful no doubt of setbacks such as these, Boulton senior penned a reproachful judgement on his compatriots which highlighted their failure to respond to the agenda he had laid out a quarter of a century earlier, 'when I reflect upon the Character, Dress & Manners of the lowest Order of the working people of Birmingham & compare them with those of a similar Class that I remember nearly 70 years ago I am sorry to say that the present Race have not improved in their Religion & Morals whatever they may have done in the Arts'.[44] A paid-up member of the Enlightenment generation, he was not alone in this inability to comprehend that a working population which manufactured articles of refinement with extraordinary dexterity and skill could nonetheless remain immune to the cultural messages enshrined in such articles. That said, however, it is hard to

[41]   Samuel Jackson Pratt, *Harvest-Home: consisting of supplementary Gleanings, original Drama and Poems, contributions of literary Friends and select republications* (3 vols, London, 1805), vol. 1, p. 282. Heneage Legge, the Earl of Dartmouth's son, also commented on 'the happy presage of returning harmony', see *Aris's Birmingham Gazette*, 31 May 1802.

[42]   BA&H, MS 3782/12/45, copy letter of MB to J. Smirnove, 23 April 1800. See also *Memorial of the Merchants and Manufacturers of Birmingham to His Majesty's Secretaries of State, 21 June 1800.*

[43]   BA&H, MS 3147/3/49, J. Watt to M.R. Boulton, Soho, 22 July 1800.

[44]   BA&H, MS 3782/12/48, copy letter of MB to Rev. Spencer Madan, Soho, 29 January 1803.

disentangle his general sense of dissatisfaction and disillusionment from the more immediate obstacles that some of the inhabitants of Birmingham were now keen to place in the path of his business. Again, in the spring of 1804, the news leaked out that Soho would shortly be despatching an entire steam-powered mint across the North Sea to Copenhagen.[45] This time it was Birmingham's coiners and metal workers who were up in arms, and they riposted with a classic piece of street theatre designed to remind Boulton that he should not be exporting the town's manufacturing livelihood. Workmen set two barrels of copper coins on a waggon and enacted a scene showing money pouring down a drain.

Still, there remained the visitors and the 'hôtel de l'amitié sur Handsworth Heath'. Here, just three miles outside the town, the promise – if no longer the certainty – of human progress lived on. Fanny Deluc, the same young Swiss woman whose father back in 1783 had sought Watt's advice regarding a civil English upbringing, recalled an encounter with the 72-year-old patriarch of Soho by two of her friends. Her re-telling of this meeting will serve as his epitaph. 'Of all the interesting Things to be seen at Soho, Mr Boulton himself is the most interesting. We passed an evening with him, & we wished he had never ceased to speak; we could have heard him constantly without having another wish; he is an inexaustible [sic] mine of precious knowledge, & of everything that is amiable; he treated us with the utmost kindness & and made us travel through a Country of *Wonders!*'[46]

---

[45] BA&H, MS 3782/12/49, Professor O.Warberg to MB, London, 4 June 1804; J. Knott to MB, High Street (Birmingham), 24 June 1804; E. Thomason to MB, Colmore Row (Birmingham), 30 June 1804.

[46] BA&H, MS 3782/12/45, F. Deluc to MB, Windsor, 7 September 1800.

Chapter 3
Matthew Boulton – Innovator

Jennifer Tann

Introduction

The stereotype of the Boulton and Watt (B&W) partnership as one in which the engineering genius (James Watt) was complemented by the entrepreneur with capital (Matthew Boulton) was neatly put by Samuel Timmins who stated that Watt's inventions would have been wasted 'but for the indomitable energy, the untiring hopefulness and the commercial genius of Matthew Boulton'. It is time that this polarised image of the partners is laid to rest.[1] All business partnerships are unique and while there are some parallels between B&W and, for example, Wedgwood & Bentley (ceramics) and McConnel & Kennedy (cotton spinning) amongst others, on closer examination Boulton by no means exemplifies the merchant/financier with a hands-off approach to manufacturing. Matthew Boulton was a highly creative problem solver who innovated in all of his business ventures. His innovations ranged from the application of science to industrial processes and systems, to precision engineering, new product development, business processes and marketing. Membership of the Lunar Society provided a social context for creativity and innovation in which puzzles and solutions could be explored without threat of ridicule or (usually) theft.[2] Over his active business life, Boulton invested capital and creative energy to innovate in steam engine manufacture, minting, buttons, buckles, 'toys', silver plate, Sheffield Plate, ormolu, rolled metals and mechanical paintings.

Financially cushioned by capital inherited from his father's mercantile business and by two advantageous marriages, Boulton was enabled to give full rein to his creative energies even when the business concerned proved not to be sustainable. To this extent he might be considered to have been, on occasion, a self-indulgent

---

[1] Samuel Timmins (ed.), *Birmingham and the Midland Hardware District* (1866, repr. Cass, 1967) pp. 218–219; Jennifer Tann, 'Boulton and Watt's organization of Steam Engine Production before the opening of Soho Foundry', *Transactions of the Newcomen Society*, 49, 1979, p. 51; Jim Andrew, 'Was Matthew Boulton a Steam Engineer?', in Malcolm Dick (ed.), *Matthew Boulton a Revolutionary Player* (Studley, 2009), pp. 107–115, p. 107.

[2] Robert E. Schofield, *The Lunar Society of Birmingham* (Oxford, 1963), pp. 84–85; Jenny Uglow, *The Lunar Men; The Friends who made the Future* (London, 2002). One dispute between Drs Darwin and Withering on the subject of digitalis was unresolved. Boulton threatened Wedgwood with becoming a potter but did not pursue it, Schofield, p. 84.

innovator, humouring his good aesthetic in his silver plate, Sheffield Plate and ormolu businesses, besides his choice of architectural design for Soho House and Soho Manufactory.

Boulton's innovations can be considered under three headings:

- those invented or initiated by Boulton and his business associates;
- those adopted from other sources;
- Boulton's crossover innovations which originated in one product context or technical application and were transferred to another.

Matthew Boulton was capable of holistic thinking with an ability to hold a range of different projects simultaneously, whilst acknowledging the significance of location, economic and social context. The disadvantage was a low boredom threshold manifest by a tendency to lose interest in a project and a desire to move onto something new as, for example, when he considered manufacturing clocks and pottery.[3] He could, however – and did – focus on detail; he was an excellent draughtsman and kept meticulous notes of experiments.[4]

Boulton's Creativity

In the processes of creativity and problem solving the distinction has been made between level and style.[5] All people are creative but differently.[6] Level is related both to potential and to motivation in a particular context; a person will use the level which seems appropriate to a given situation up to the maximum of their potential or capacity – on a scale from low to high. Thus a person with high motivation and a moderate level will exercise creativity effectively in many situations but, particularly in a demanding context, not so effectively as an individual with a high capacity. Style can be measured on a continuum from highly adaptive (a greater preference for structure, identifying a limited number of solutions, incremental progression and within a given paradigm) to innovative (less preference for structure, the proliferation of possible solutions, more tolerance of ambiguity, and comfortable with thinking outside the paradigm). E.M. Rogers identifies innovators as having 'almost an obsession' with venturesomeness which leads them out of local peer networks 'into more cosmopolite social relationships', sometimes maintained over considerable geographical distances.

---

[3]   H.W. Dickinson, *Matthew Boulton* (Cambridge, 1936), pp. 48–59.

[4]   Fiona Tait, 'Coinage, Commerce and Inspired Ideas: The Notebooks of Matthew Boulton', in Dick (ed.), *Boulton*, pp. 77–91.

[5]   M.J. Kirton, *Adaption-Innovation in the Context of Diversity and Change* (London, 2003), p. 169.

[6]   Ibid., p. 167.

The innovator plays a 'gatekeeper' role in the flow of new ideas into a group.[7] The combining of the two concepts of level and style suggests a way of contrasting, for example, an individual with a high creative capacity and an adaptive style who may have been an effective translator of an engineering sketch to a physical structure, or an implementer of a complex manufacturing system, with an individual possessing both a high level of creativity and an innovative cognitive style who could comprehend the total problem and proliferate solutions, often outside the existing paradigm. M.J. Kirton[8] employs the term 'innovator' and 'innovative' as cognitive descriptors rather than as usually applied in the literature,[9] and as used in this Chapter, namely as an outcome and a social process. Process is the operation progressing from presumed start to presumed conclusion and is the means by which an individual or group reaches a conclusion. Boulton's creative capacity was manifest in a variety of activities when, for example, planning the detailed layout of powered machinery in the Soho Manufactory, the production design of a mint, or when presented with a specific technical problem such as steam engine boiler design, engaging the design capabilities of architects in his plate enterprise, or his socio-political networking to achieve the passage of the Assay Office Bill.

Boulton's style is evidenced by the proliferation of creativity in his different businesses (Figure 3.1), besides his preference for holistic conceptualising, for taking a broad overview, whether imagining and bringing to fruition the Soho Manufactory, unique at the time in the hardware sector, or in envisaging supplying the entire world with engines, or recognising the significance of context, for example, the Cornish market for engines, or the potential for rotative power in the cotton textile industry. Boulton was a risk taker who could see a way to export an engine to France, when it was stopped by Customs at the height of the French wars, by a 'proper dose of the golden powder to the eyesight'; while on another occasion parts were shipped from different ports to avoid suspicion.[10]

He was a playful experimenter, in for example, making an 'immense' paper balloon which was filled with hydrogen gas and sent up in a field in Cornwall 'to the great delight of all concerned'.[11] He was also an experimenter of intense seriousness, focusing on detail when necessary, taking scrupulous care in documenting experiments, as evidenced in his letters to Watt and his personal

---

[7]   Ibid., p. 49; E. M. Rogers, *The Diffusion of Innovations* (New York, 1995), pp. 264–265.

[8]   Kirton, pp. 149–165.

[9]   For example Rogers, p. 16; Louis G. Tornatzky and Mitchell Fleischer, *The Process of Technological Innovation* (Lexington, 1990); Jennifer Tann, 'Space, Time and Innovation Characteristics; the Contribution of Diffusion Process Theory to the History of Technology', in Graham Hollister-Short and Frank A.J.L. James (eds), *History of Technology*, 17, 1995, pp. 143–164.

[10]   Jennifer Tann, 'Marketing Methods in the International Steam Engine Market: the case of Boulton and Watt', *The Journal of Economic History*, XXXVIII, 1978, pp. 363–391.

[11]   Schofield, pp. 250–254.

'Toys', Buttons, Buckles

Ormolu

Sheffield Plate

Silver Plate

Rolled Metals

Mechanical Paintings

Letter Copying

Mints

Coinage

Steam Engines

Figure 3.1   Boulton's business enterprises
Source: Diagram by Jennifer Tann.

notebooks. Boulton devised the idea of the engine counter which was affixed to the beam to record the number of strokes, developing the mechanism from a pedometer made by a firm in Liverpool.[12] It was Boulton who perceived the need for engine standardisation and, in 1782 urged Watt to comply, although Watt resisted and the beginnings of standardisation were not in evidence until two years later.[13]

In process terms, Boulton applied his creative problem solving in social contexts such as Lunar Society meetings, in which puzzles and solutions, together with opportunities for building upon the ideas of others, were presented. He welcomed the technical expertise of employees and was known to offer a job at Soho when an individual demonstrated creativity; for example, Murdock's alleged arrival at Soho wearing a turned wooden hat which excited Boulton's curiosity.[14] In the early years of the engine business B&W benefited from the creativity and skills of a small

---

[12]   H.W. Dickinson, *James Watt, Craftsman and Engineer* (Cambridge, 1937), p. 108.

[13]   Schofield, pp. 334–335; Dickinson, *Boulton,* pp. 115–116.

[14]   R.B. Prosser, *Birmingham Inventors and Inventions* (Birmingham, 1881, repr. Trowbridge, 1970), pp. 23–24.

group of excellent engineers including John Southern, William Murdoch, James Lawson and the Creighton brothers,[15] all of whom made important contributions to engine and other manufacturing technologies. He also engaged his creativity in the relative solitude of his home laboratory, writing detailed jottings in his notebooks.

The disadvantages of this hypothesised cognitive capacity and style preference for creative problem solving were a low boredom threshold with a tendency to lose interest in a project, seeking to move onto something else without compelling evidence that it would pay its way – or to continue with a business in the face of evidence that it was not breaking even because it met his needs for creative fulfilment.[16] The propensity to take risks had both positive and negative attributes. There was also the potential for Boulton's attention to be spread too thinly across projects at any given time with the possibility of a major business or technical error being made. The cognitively innovative person can be *perceived* as being less efficient, evidenced by a perceived lack of consistency, less predictability, and inattention to detail; yet this is an effective style for producing something radically different and, in Boulton's case, many different things, besides implementing improvements on what already existed in his businesses.

Innovation Framework

Innovation is a much-used concept with a range of meanings. The definition more typically employed by economists and some historians of technology[17] focuses on intrinsically new products, processes and technologies, termed in this chapter Producer Innovation (Figure 3.2). Button and buckle making were competitive industries in which Boulton's Producer Innovations raised product quality through consistency, strength and a good aesthetic, besides manufacturing process innovations. In one of his notebooks Boulton wrote 'Improvments made by MB in buckl(es) Better Steel, Scaled & Rolld fine without Grinding, Bobing in the Mill. Systimised the Sizes & forms of Buckles, Better patterns & better Dies, Better Hooks to Springs'.[18] He calculated the springs of silver buckles, noted the operations for making latchets (a kind of shoe buckle), besides writing-up

---

[15]   Dickinson, *Boulton*, p. 171; Schofield, p. 336; Jennifer Tann, *The Selected Papers of Boulton and Watt, Vol. 1, The Engine Partnership, 1775–1825* (London, 1981), p. 44; Jennifer Tann, 'Two Knights at Pandemonium; A Worm's Eye View of Boulton Watt & Co 1800–1820', in Graham Hollister-Short (ed.), *History of Technology*, 20, 1998, pp. 47–72.

[16]   Birmingham Archives and Heritage (BA&H), MS 3783/13/37, James Keir, 'Memoir of Matthew Boulton', 1809, p. 5.

[17]   For example, Rod Coombs, Paolo Saviotti, Vivien Walsh, *Economics and Technological Change* (Basingstoke, 1987), pp. 93–120; Gerhard Mensch, *Stalemate in Technology* (Cambridge, MA, 1975), p. 47; Nathan Rosenberg, *Inside the Black Box, Technology and Economics* (Cambridge, 1982), pp. 193–241.

[18]   Tait, in Dick (ed.), p. 86.

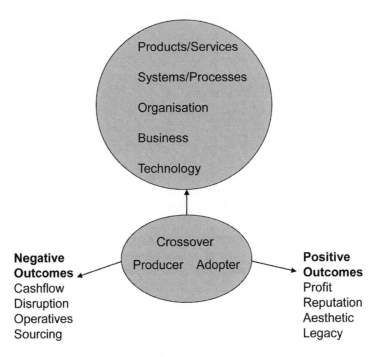

Figure 3.2   Innovation framework
Source: Diagram by Jennifer Tann.

chemical experiments on ways of precipitating silver and experiments with platina (an alloy of zinc and copper) buttons.[19]

The social science interpretation, Adopter Innovation, incorporates product development besides the adoption of technologies, systems and processes which, while not intrinsically new, are new to the user.[20] The basis for this interpretation being that, if the adopted item is new to the user, the intended outcomes (for example, profit, production problem solved, reputation, legacy) are likely to be similar to those associated with an intrinsically new object. Moreover the possible side effects (for example, disruption, financial difficulties, resistance, skills shortage) are also likely to be similar to those associated with intrinsically new items.

As early as 1757 Josiah Tucker noted 'almost every Master manufacturer hath a new invention of his own and is daily improving on those of others'.[21]

---

[19]   Ibid., p. 88.

[20]   Rogers, pp. 1–37.

[21]   Josiah Tucker, *Instructions for Travellers* (1757) quoted in Rosemary Ransome-Wallis, *Matthew Boulton and the Toymakers, Catalogue of Exhibition in Goldsmith's Hall* (London, 1982), p. 45.

This is a classic exposition of Crossover Innovation or 'borrowing brilliance'[22] in which the adopter modifies within the same product range, transfers the idea to a wholly different product, extends a range of 'borrowed' products, or explores new markets. Boulton, for example, while not the originator of Sheffield Plate, improved the product by sealing the edge of wares with sterling silver to conceal the rolled copper in the centre.[23]

As Boulton's business portfolio developed he was able to engage in Crossover Innovation between his businesses (Figure 3.2). Building on the foundations of the mercantile business, he saw marketing opportunities for the 'toy' business of buttons, buckles, watch chains and other articles, manufactured in his partnership with John Fothergill, accessing his already established European mercantile networks: 'I should be glad to work for all Europe in all things that they may have occasion for – gold, silver, copper plated, gilt, pinchbeck, steel, platina, tortoise steel or anything else that may become an article of general demand'.[24]

In terms of design for silver and plate goods, Boulton sought to adopt classical patterns: 'ye present age distinguishes itself by adopting the most elegant ornaments of the most refined Grecian artists; I am satisfyd in conforming thereto, and humbly copying their style, and making new combinations of old ornaments without presuming to invent new ones'.[25] Copying a design does not necessarily imply good execution but this is where Boulton excelled. His interests in experimental chemistry and metallurgy provided a sound basis from which to move into creative materials substitution in his various 'toy', silver and various silver-plating businesses. Boulton's scientific curiosity, coupled with social networking, brought a wide range of techniques to Soho, for example, French plating, close-plating, Sheffield Plate, and non-precious metals resembling silver such as 'white metal' and platina.[26]

Knowledge and skills were brought to Soho in various ways, sometimes through sending a piece acquired overseas to be analysed, sometimes by analysis at Soho, while on other occasions a technique was imported through hiring an artisan skilled in the craft.[27] Through these businesses Boulton was introduced to painters, sculptors, architects, and engravers who provided significant connections, knowledge and skills which Boulton exploited in his silver plate, Sheffield Plate

---

[22] David Kord Murray, *Borrowing Brilliance* (London, 2010).

[23] Dickinson, *Boulton*, pp. 53–53.

[24] Matthew Boulton (MB) to Wendler, July 1767, quoted in Dickinson, *Boulton*, p. 55.

[25] MB to Mrs Montague, 1772, quoted in Dickinson, *Boulton*, p. 54.

[26] French-plating was done on copper, brass, or iron, squares of silver foil being applied to the heated metal. Close-plating involved cleaning the article, dipping it into ammonium chloride and then into molten tin. Sheffield Plate (called 'plated wares' by Boulton) involved fusing a thin layer of silver onto copper on one or (from the 1760s) both sides. 'White metal' meant an alloy of tin, copper and antimony; Kenneth Quickenden, 'Silver and its Substitutes', in Dick (ed.), pp. 154–155.

[27] Quickenden, in Dick (ed.), pp. 152–156.

and ormolu and, later, minting, for example, bringing together a portrait painter (West) and Droz, the metal engraver, to produce the magnificent medal of George III in 1788. When Boulton undertook his first Government coinage contract in 1797, he issued explicit instructions for the standards of coins in both weight and diameter, perhaps informed by his earlier insistence on standardisation in steam engine manufacture.[28]

The establishment of the Birmingham Assay Office deployed Boulton's abilities in making connections.[29] While he claimed a high rate of damage to goods sent to Chester for hallmarking (probably overplaying the case), the parliamentary lobbying skills honed in seeking to introduce a Bill for the establishment of assay offices in Birmingham and Sheffield in 1772 were effectively deployed in attempting an extension to Watt's patent three years later, besides in seeking to defeat engine pirates in the 1790s. Boulton deployed his lobbying expertise in seeking to raise Government awareness and interest in the subject of counterfeit coinage at around the same time as he was working towards the establishment of the Assay Office. He was corresponding with Lord Dartmouth on the subject in 1772–1773[30] and, two years later, was discussing the application of steam power to coining.

His marketing expertise enabled Boulton to claim, in the negotiations with Watt prior to the commencement of the engine partnership, that he would wish 'to sell engines to all the world'.[31] And, for the first 10 years of the Boulton and Watt partnership, much of the active market research and sales promotion overseas was carried out by representatives of Boulton's other business interests. Clerks, and travellers for the hardware business were directed to carry out specific engine marketing tasks at no cost to the engine partnership. A hardware agent visiting Russia in 1777, for example, circulated papers about the engine at court, while an agent due to leave for Germany and the Low Countries in 1787 was given similar instructions.[32]

Steam Engines

Matthew Boulton's contributions to steam engine technology have been under-recognised. He was an experimenter with both mechanisms and larger engineering matters, as well as in chemistry. He had a laboratory at Soho House and his first

[28] Stebbing Shaw, *The History and Antiquities of Staffordshire*, II (London, 1801), p. 118.

[29] Jennifer Tann, *Birmingham Assay Office, 1773–1993* (Birmingham, 1993).

[30] Earl of Dartmouth to MB, 5 January 1773, quoted in A. Westwood, *The Assay Office at Birmingham. Part 1: Its Foundation* (Birmingham, 1936), p. 11.

[31] BA&H, MS 3219/4/66/1, MB to James Watt (JW) 7 February 1769.

[32] BA&H, MS 3782/12/24/132, P. Capper Jr to MB, 24 November 1777; Tann, 'Marketing Methods', p. 372.

notebook (1751–1759) records 'An experiment made by MB and Dr S(mall) upon steam'.[33] In 1766, before the engine partnership was even a serious possibility, he was corresponding with Benjamin Franklin[34] on the subject of steam power and 'playing' with a model engine. He discussed steam valves and where to introduce cold water to condense the steam and 'made many fruitless experiments on the subject'.[35]

Watt's first working engine was installed at the Soho Manufactory. Each early engine was a mobile laboratory in which B&W and their employees proposed developments. In 1775 Boulton was testing engine components for excessive friction and directing mechanics to make improvements and, between 1775 and 1778, he continued to make suggestions on ways of improving engine performance. He was not averse to offering practical help as, for example, in removing his jacket to assist with packing grease into the piston at Chasewater in 1788.[36] Boulton recorded the fuel consumption on test runs of the Soho engine and, in 1778, proposed a formula for calculating it.[37] His concern with boiler design was prompted by the desire to obtain optimum efficiency from Watt's engine (thereby enhancing the premium due to the partnership from 1/3 of the savings in fuel of the B&W engine compared with a Newcomen one). One of his notebooks is titled 'Boylers-1780' and contains dated experiments with diagrams of boilers mostly for Cornish engines. The monthly performance of Cornish engines was carefully noted with information on piston strokes, depth in fathoms, diameter of pumps, engine load, consumption of cylinder and total savings, as well as B&W's third share.[38] Boulton suggested the use of 'copper spheres' to increase the heating surface of boilers and, in 1780, proposed that iron or copper fire tubes be inserted in boilers, thus anticipating the Lancashire boiler which was introduced some years later.[39] He was interested in the expansive working principle, in which the steam supply was cut off before the end of the stroke, and conducted trials on the Soho engine[40] but this development, covered in Watt's 1782 patent, generated lower savings than anticipated and was more effectively implemented in the nineteenth century in Cornish and high pressure engines.[41]

---

[33]   Tait, in Dick (ed.), p. 87.

[34]   MB to Benjamin Franklin, 22 February 1766, quoted in Dickinson, *Boulton*, pp. 75–76.

[35]   Ibid.

[36]   Andrew, p. 110; R.L. Hills, 2, p. 106.

[37]   Tait, in Dick (ed.), pp. 86–87.

[38]   Ibid.

[39]   Dickinson, *Boulton*, p. 117.

[40]   Dickinson, *Watt*, p. 132.

[41]   Dickinson, *Boulton*, p. 118; John Farey, *A Treatise on the Steam Engine, Volume 2* (1827 repr. Newton Abbot, 1971).

Boulton recognised the business potential of rotative power generation, urging a reluctant Watt to apply himself to the problem.[42] He raised the subject again in 1779 and, two years later, made a model sun and planet engine, enabling him to confirm not only the business case but a practical solution to it. One Sunday early in 1781 Boulton reported to Watt 'I have reflected so much upon your last rotative motion that I could not refrain from making one this day instead of going to church … there is something so agreeable in a good Movement to one's mechanical feelings that I could not forbear playing with it all this afternoon'.[43] Boulton urged Watt to patent his methods for producing rotative motion 'the most likely line for the consumption of our engines (being) the application of them to mills, which is certainly an extensive field … the people in London, Manchester and Birmingham are steam mill mad'.[44] Watt's rotative patent was granted that year. In 1782 Boulton suggested a more straightforward method for attaching the sun and planet gears and it was this arrangement which became standard on sun and planet engines.[45] An important crossover innovation in rotative power generation was derived from Matthew Boulton's knowledge of clock manufacture and 'clockwork' which, when scaled-up, was transferable to industrial power transmission as 'millwork'.

The Soho Manufactory, the first factory to be powered by a Watt engine,[46] was a focus for engine applications testing in the early 1780s, in particular for tilt hammers and corn grinding, besides the industrial processes required in Boulton's own directly supervised Soho-based businesses. With the introduction of the rotative engine Boulton played a key role in targeting key innovating entrepreneurs in major industries, seeking to make them sector role models in the adoption of steam power. John Wilkinson had earlier pioneered the use of reciprocating engines at his Bradley and Broseley foundries and these provided practical evidence of the application of steam in the iron industry. In cotton spinning Richard Arkwright experimented with one or more alternative engine suppliers (with little success, to Boulton's ill-disguised delight), only purchasing a B&W engine in 1790 after the Robinsons had installed the first B&W engine for cotton spinning at their Papplewick Mill in Derbyshire.[47] Samuel Whitbread proved more amenable in brewing; Boulton's fellow Lunar Society member Josiah Wedgwood was the pottery industry role model while, following the

[42]   H.W. Dickinson and R. Jenkins, *James Watt and the Steam Engine* (Oxford 1927, repr. Ashbourne, 1981), p. 159.

[43]   MB to JW, 13 January 1781, quoted in Dickinson, *Boulton*, p. 113.

[44]   MB to JW, 21 June 1781, quoted in Dickinson, *Boulton*, p. 113.

[45]   R.L. Hills, *James Watt Vol. 3: Triumph Through Adversity, 1785–1819* (Ashbourne, 2006), p. 24.

[46]   This was the Kinneil engine brought to Birmingham by Watt which went through several re-buildings at the Soho Manufactory, Dickinson, *Boulton*, p. 84.

[47]   Arkwright enquired after a B&W engine in 1777, again in 1780, finally ordering one in 1790, Jennifer Tann, 'Richard Arkwright and Technology', *History*, 58, 1973, pp. 29–44.

establishment of their experimental corn mill at Soho, Boulton and Watt went into partnership with Samuel Wyatt and others to build the Albion steam corn mill in London.[48] The Soho Mint demonstrated a fully integrated steam-powered plant which undoubtedly influenced production design at the Royal Mint, besides overseas government mint purchasers.[49] The Soho-based developments, together with observations and reports from customers' factories, enabled B&W to estimate appropriate power to machine ratios in various industries with some authority.[50]

It was Boulton whose business sense, coupled with his integrative mind-set, recognised the need for the standardisation of engines and engine parts. All the early engines were custom-built, enabling the continuation of R&D, whilst ensuring an income stream. Standardisation did not imply that engine development was concluded but that the stage had been reached in which engine drawings prepared for one customer could be used in fulfilling a later order for another customer.[51] Boulton initiated keeping a stock of spare parts, for example sun and planet wheels which had a tendency to break. And in 1782 he urged the making of pattern cards for engine parts.[52] By 1795 engine production had been regularised to 14 engine sizes from 4–50 hp and the opening of the Soho Foundry provided the opportunity to plan for the manufacture of all engine parts.

Crossover innovations could benefit other businesses besides Boulton's. The initial organisation of steam engine production, in which Boulton and Watt manufactured a few precision parts at the Soho Manufactory and sub-contracted the rest of the engine parts, promoted crossover innovations from Soho to its suppliers. When Boulton perceived the potential of John Wilkinson's canon boring machine for boring steam engine cylinders and ordered an 18-inch cylinder for the Soho experimental engine, followed by encouragement to innovate a larger boring mill to bore larger engine cylinders, he not only solved a technical problem in B&W's engine production but enabled business diversification at Wilkinson's Bersham works: 'am preparing a machine ... to finish (cylinders) with greater truth ... in short nothing shall be wanting that is my way which can facilitate your engine'.[53] Such was Wilkinson's expertise at cylinder boring that, even in the case of a Scottish iron foundry customer, B&W wrote plainly: 'Wilkinson hath bored us several cylinders almost without error ... so that you must improve in boring or we must furnish the cylinder'.[54] And

---

[48] Jennifer Tann, *The Development of the Factory* (Cornmarket, 1970), pp. 71–94.

[49] B&AH, MS 3147/3/506/3, J. van Liender to JW, 12 May 1797.

[50] Tann, *Factory*, pp. 71–94.

[51] Schofield, pp. 334–335. A number of engine drawings in the portfolios have more than one customer's name on them.

[52] Ibid., p. 335.

[53] BA&H, MS 3147/3/533/3, John Wilkinson (JWi) to JW, 17 August 1775.

[54] BA&H, MS 3147/3/80/33, MB to Samuel Garbett (SG), 10 February 1776.

a week later: 'bore your cylinders as true as Wilkinson and then say there is no truth in me if we are not good customers to Carron'.[55]

Boulton's own metallurgical and engineering skills, besides his abilities as a networker, enabled him to identify the most appropriate engine parts suppliers, B&W specifying in some detail which supplier should make what. Special parts demanding precision founding and forging were obtained from the best supplier, wherever he was located. Piston rods were ordered from James Spedding of Whitehaven, for example, and pumps were sometimes made at Carron. Where possible, however, B&W used West Midlands suppliers: The Eagle Foundry, Birmingham, for heating cases, Izons of West Bromwich for tubes, William Whitmore of Birmingham for heating cases while, by the 1790s, Thomas Horton of West Bromwich had secured a virtual monopoly of boiler supply.[56] Only the heaviest parts, such as the flywheel, or with remotely located customers, were B&W prepared to allow a customer to specify a local ironfounder; Marshall & Benyon of Leeds, for example, obtained a flywheel, boiler and other parts locally. Quality was a potential problem with customer-sourced parts as Whitbread found with a flywheel purchased from a local founder in London.[57]

But, while outsourcing parts had the advantages of requiring less capital investment in manufacturing plant and materials, as well as providing opportunities to learn from suppliers, synchronising the arrival of parts for assembly at the customer's site proved difficult. And, in the early years of the partnership, the custom design of each engine could lead to orders for parts from a single supplier failing to arrive together. A frustrated Wilkinson remarked 'had all the order been given at once the whole would have been (delivered) together with the cylinder'.[58] Where parts failed, some customers, at least, had a clear idea where the responsibility lay, notwithstanding B&W's attempts to persuade the customer to deal directly with the supplier: 'Much time has elapsed and I am referred to J. Wilkinson. ... I expect ... that you will do your own business yourselves and not (be) leaving us to the factiousness and correspondence of persons with whom we have not, or wish to have, any concern'.[59] By the early 1790s, B&W could produce over 50 per cent of engine parts by value[60] at the Soho Manufactory and, with the opening of the Soho Foundry in 1796, the intention was to manufacture complete engines, an objective that was not fully achieved for some years.

---

[55]   BA&H, MS 3147B/3/80/34b, MB to SG, 18 February 1776.

[56]   Jennifer Tann, 'Suppliers of Parts; the relationship between Boulton and Watt and the Suppliers of Engine Components 1775–1795', *Birmingham and Warwickshire Archaeological Society Transactions*, 86, 1974, pp. 167–177.

[57]   BA&H, MS 3147/3/4/435/5, Marshall & Benyon to B&W, 30 July 1794; MS 3147/3/404/3, Yallowley (for Whitbread) to B&W, 28 May 1785.

[58]   BA&H, MS 3147/3/533/17, JWi to MB, 9 August 1776.

[59]   BA&H, MS 3147/3/416/76, Robert Mylne to B&W, 7 April 1795.

[60]   Tann, 'Organisation of Engine Production', p. 311.

Boulton's lobbying abilities, honed in negotiations for the extension of Watt's British patent in 1775, besides his delight in networking with members of the aristocracy to promote the sale of silver plate and luxury goods, provided a platform from which he could take the lead in pursuing overseas patent coverage for the Watt engine. Since patent protection had been such a central issue in the formulation of their domestic engine business policy, it is not surprising that Boulton sought patent or exclusive privilege rights in overseas countries from which the earliest enquiries were received. The first enquiries came from France, and Boulton approached, amongst others, members of the French government, a member of the aristocracy and a Paris banker to obtain an *arret* – an exclusive privilege. The fact that this failed to open up the French market did not deter the partners from seeking exclusive privileges elsewhere. A 15-year patent was granted for Holland in 1786; one was secured in Spain in 1790; attempts were also made to secure patents in the Austrian Empire, Prussia, America and possibly Italy.[61] But enforcement was well-nigh impossible and, when it suited them, Boulton and Watt, themselves, by-passed Soho-appointed sole selling agencies in the countries in which they had patent privileges, thus creating a disincentive to agents seeking further orders. However the partners' abilities in networking with overseas scientific communities and governments proved effective in securing some public works contracts for waterworks engines and national mints where cost was less of a consideration for the customer than the reputation of the supplier. Moreover, since no direct expense was incurred in gaining, or seeking to gain, monopolies in overseas markets, B&W's patent activities served to publicise the Watt engine and, viewed in these terms, were an effective form of advertising.

The Soho Manufactory

Physical expansion of Boulton's Snow Hill-based mercantile and 'toy' manufacturing business was constrained by its town-centre site and, in 1761, he acquired the leasehold of a watermill site on Handsworth Heath. The mill was demolished, being replaced by a multi-storey water-powered building erected with the assistance of John Wyatt.[62] Four years later Boulton began planning the Soho Manufactory, employing the architects Benjamin and William Wyatt, and the factory was completed in 1767.[63] This was the first vertically integrated factory for the manufacture of 'toys' and, within a few years for a variety of other products too. It was not the first multi-storey factory to be built, but earlier ones, such as the Derby silk mill, had housed a single process, while later cotton factories housed

---

[61]   Tann, 'Marketing', pp. 367–371.

[62]   George Demidowicz, 'Power at the Soho Manufactory and Mint', in Dick (ed.), pp. 116–131, p. 118.

[63]   Ibid., p. 118.

carding and spinning processes in a horizontally organised system.[64] In the case of Boulton's enterprises, while some of the processes were mechanised, many were not. The motive for vertical integration was to ensure both supervision and scale of production and, thereby, greater consistency and higher quality of product. Moreover labour was more likely to be assured by the offer of housing in the upper stories.[65]

In architecture Boulton was drawn to classical design; not for him the clumsy castle gothic favoured by Arkwright at Willersley Hall, Cromford, intended perhaps to hint at an imaginary landed ancestry. Boulton favoured English Palladianism with its simplicity, reasonableness and universal intelligibility. Lord Burlington, the arbiter of classical taste in England, had aimed to re-establish the rules of architecture in accordance with the principles of Palladio as interpreted by Inigo Jones. Burlington sought to create a national taste and style built on the spirit of national freedom 'a freedom resulting from the British constitutional government'.[66] These 'rules' opened up opportunities to eighteenth-century industrialists and their builders with the possibility of producing a passing fair imitation of Inigo Jones. What had emerged by the mid to late eighteenth century was a two, rather than three, dimensional concept of architecture; whereas 'Italian architecture must always be judged for its plastic values; an English eighteenth-century building should be seen from a distance like a picture'.[67] The setting of the Soho Manufactory, as contemporary images show, was ideal for such a perspective.[68]

The Soho Manufactory was one of the finest examples of English Palladianism in factory design. Before the building had been completed Fothergill, Boulton's 'toy' trade partner, reported: 'the buildings now begin to look so very sumptious [sic] as to engage the attention of all ranks of people'.[69] It is probable that Boulton was sufficiently swayed by the importance of having a regular front elevation of appropriate Palladian proportions that he was prepared to compromise the amount of light entering the building, on the advice of the architect William Wyatt, to the possible detriment of the activities which went on inside. Wyatt advised 'if, therefore, you can reconcile this matter of light my opinion is the form of ye outside will be more pleasing'.[70]

---

[64]   Tann, *Factory*, pp. 3–26.

[65]   George Demidowicz 'A walking tour of the Three Sohos', in Shena Mason (ed.), *Matthew Boulton: Selling what all the World Desires* (New Haven and London, 2009), pp. 99–107.

[66]   Rudolph Wittkower, *Palladio and English Palladianism* (London, 1974), p. 178.

[67]   Ibid., p. 174.

[68]   Val Loggie, 'Picturing Soho: Images of Matthew Boulton's Manufactory', in Mason (ed.), *Boulton*, pp. 22–30.

[69]   BA&H, MS 3782/12/60/34, John Fothergill to MB, 14 December 1765.

[70]   BA&H, MS 3782/12/85/207, W. Wyatt to MB, 8 September 1765.

In acquiring a water-powered site on Handsworth Heath, Boulton had the opportunity to create a landscape garden with the millpond serving the dual purpose of industrial reservoir and ornamental lake. He was not unique in doing this but Soho was a particularly fine example.[71]

The Soho Mint

The risk-taking aspect of Boulton's creativity style is well demonstrated by The Soho Mint, constructed in 1788 at a cost of between £7,000 and £8,000. The manufacture of medals was not unknown to him for he had made them for distribution on Captain Cook's second voyage in 1772, besides other medals in 1774 and 1781.[72] Moreover he had an excellent practical knowledge of metal stamping, acquired in the 'toy' trade. In 1786 he undertook to supply coinage for one of the East India Company's possessions in Sumatra by preparing the blanks at The Soho Manufactory and dispatching them to London where they were struck into coins on machines that he had erected in one of the Company's warehouses.[73] This, together with his high hopes of supplying a new national copper coinage, prompted him to invest in building his own steam-powered mint, despite having no Government contract. Boulton's first minting innovations were the perfection of his steam-driven coining machinery, besides the introduction of sharp edges to prevent counterfeiting. He then turned his attention to the duplication of dies, in better quality steel[74] with concave faces, enabling consistently produced coins to be struck with a single blow such that 'every piece becomes perfectly round, and of equal diameter, which is not the case with any other national money ever put into circulation'.[75] Boulton's minting innovations are highlighted in his unique 1802 Techniques Medallic scale medal in which each numbered circle upon the face of the medal indicates how many coins of different sizes could be struck in one minute by the Soho Mint steam powered presses.[76] Besides building the first steam-powered, most efficient, and powerful mint in the world Boulton had to be assured of a supply of copper, should he obtain the hoped-for Government order. This was achieved through his 1788 non-competing deal with Thomas Williams in

---

[71]   Phillada Ballard, Val Loggie, Shena Mason, *A Lost Landscape: Matthew Boulton's Gardens at Soho* (Chichester, 2009), p. 10.

[72]   David Symons, 'Matthew Boulton and the Royal Mint', in Dick (ed.), p. 172.

[73]   Ibid., p. 173.

[74]   Sue Tungate, 'Technology, Art and Design in the Work of Matthew Boulton: Coins, Medals and Tokens produced at the Soho Mint', in Dick (ed.), pp. 185–200, p. 193.

[75]   Ibid., p. 188.

[76]   Ibid., p. 190.

which Williams agreed not to seek a Government coinage contract but to supply the copper, thereby ensuring supplies for Boulton's other businesses.[77]

In 1789 Boulton reflected on 'Advantages of my Coining Mill' in which he listed, amongst other properties, its speed of operation, an exact record being automatically made of the number of pieces struck (a precaution against workplace theft); the machine could be easily stopped and started by a child and: 'it strikes each piece quite round, all of equal Diameter & exactly concentrick [sic] with the Edge which cannot be done quick by any other Machine ever used'.[78] Boulton was granted a patent for the 'application of motive power to stamping and coining' in 1790.[79] In 1792, the mint contained eight large coining presses (striking from 50–100 pieces per minute depending on size). Each machine was tended by a 12-year-old boy who could 'stop his press one instant, and set it going the next'.[80]

Boulton did not invent flow production but perfected it in the layout of the Soho and other mints. In the design of specific functions he adopted known methods but made numerous improvements, materials being fed from one machine to another in boxes through shoots so as to reduce handling, while the speed of each machine could be adjusted without altering the speed of the engine. Crossover innovation took place between the arrangement of the cutting-out presses in the lap room for 'toys' and the Mint, Boulton abandoning the idea of the Mint coining presses being in parallel rows and adopting, instead, a circular arrangement which was also employed in the lap house.[81] And in 1798, John Southern, head of the engine works drawing office, suggested a radical arrangement for 'working of Coining presses by means of a partial vacuum'.[82] The Mint machinery was thus radically altered and between 1798 and 1799 a new coining room was constructed. With a note of resentment Boulton compared Soho with the Royal Mint: 'they have never coin'd ¼ of the quantity I do in equal time even in their common gothic manner'.[83] After Boulton's death, William Murdock observed: 'The favourite and nearly the sole object of the last 20 years of the active part of Mr Boulton's life is in great measure to be attributed the perfection to which the art of coining has ultimately obtained'.[84]

The Soho Mint was an effective marketing showcase from which Boulton could promote the sale of complete mint packages, including steam engines, first for

---

[77] J.R. Harris, *The Copper King: A Biography of Thomas Williams of Llanidan* (Liverpool, 1964), pp. 92–93.

[78] Tait, Coinage, in Dick (ed.), p. 89.

[79] Patent No. 1757.

[80] Dickinson, *Boulton*, p. 142.

[81] Demidowicz, 'Power at Soho', in Dick (ed.), p. 126.

[82] Sue Tungate, 'Matthew Boulton's Mints: Copper to Customer' in Mason (ed.), *Matthew Boulton*, p. 83.

[83] MB to Sir Joseph Cotton, 25 November 1799, quoted in Symons, in Dick (ed.), p. 181.

[84] Dickinson, *Boulton*, pp. 147–148.

several overseas mints[85] and, shortly before his death, the Royal Mint in London. Bedridden for much of the time from 1805 onwards and using an amanuensis, Boulton designed what became a prototype for other national mints. In writing to the engineer, John Rennie, he made clear the distinction between responsibilities for the design of the productive part of a mint and its front elevation:

> The buildings in general should be plain simple and strong and all the operative buildings need not be more than two storeys high, many of them one at most, except the front which may be simply elegant in the Wyattistic style. ... Mr Wyatt may design the ornamental part but I must sketch the useful.[86]

Matthew Boulton possessed a high level of creative capacity, coupled with an innovative style of creativity and problem solving. He could tolerate ambiguity, had a low need for structure and rules, proliferated both businesses and business solutions and could think outside the existing paradigm. He was venturesome, a risk taker. He was a Producer Innovator who initiated products and processes which were intrinsically new (for example, the production system at the Soho Mint and its high quality products); he was an Adopter Innovator (for example, Sheffield Plate) and, most of all, a Crossover Innovator who applied innovations in new contexts as well as transferring them from one business to another (for example, product and process innovations transferred between his various 'toys', buckle, button and other businesses; and standardisation from steam engines to minting).

Boulton was a serial innovator who, with a low boredom threshold and an inquisitive mind, undertook the manufacture of some products which were not good business propositions and was not prepared to wind up a business solely on financial grounds until the case became compelling. But business rationality is only one element in making innovative choices. Keir wrote perceptively of his:

> ... ingenuity (which) is not sufficient to endure profitable success in trade ... Genius may plant the Hesperian tree but patient dullness more frequently reaps the golden fruit ... if Mr B did not receive from the or moulu [sic] & the other elegant branches of his manufacture the intended recompense of all commercial industry, it is certain they greatly tended to his celebrity & admiration of his various talents.[87]

Without Boulton the history of the Watt engine might have been very different. His perception of the potential market underpinned the business partnership; his energy in parliamentary lobbying for the patent extension, promoting sales in Cornwall

---

[85] For example, the Russian and Danish mints: Tann, *Factory*, pp. 92–93.

[86] BA&H, MS 3782/13/49/104, MB to Rennie, 15 November 1804. In the event the front elevation was designed by Smirke.

[87] Dickinson, *Boulton*, p. 71.

the first major market, perceiving the potential for factory engines in Lancashire and pushing Watt to develop the rotative engine were key to the success of the engine business, besides his own innovations in engine technology. Indeed the steam engine business diverted Boulton's attention from his other businesses.

But, it is suggested, Boulton had a number of business objectives of which profit was only one. His business partnerships in the arts: ormolu, silver plate, Sheffield Plate and mechanical paintings were not, on the whole, profitable but, if lifestyle and legacy were motivators for innovation, and there is evidence that they were important to him, innovation in these businesses played a significant role towards achieving those ends. Boulton, in his engagement with architecture, was 'an influential citizen in the new social order',[88] who recognised in the architecture of aristocratic England the elements of formal elegance and applied them to his factory. He was one of a small number of industrialists who engaged an architect for their factories. The boundaries between business and pleasure were finely drawn when Boulton held masques and balls at Albion Mill, attended by Sir Joseph Banks, members of the nobility and directors of the East India Company, much to Watt's displeasure.[89]

Certainly he made sufficient profit from which to engage in a lifestyle which was important to him. He was motivated by the desire for social interaction with the nobility, by reputation and, perhaps most of all, by legacy. As Murdoch perceptively put it 'The love of fame has been to him a greater stimulus than the love of gain'.[90] When Boulton considered the desirability of acquiring the freehold of his leasehold land at Soho he mused that, without the freehold, 'The monument I have raised to myself, with so much pains, & expense I shall see gliding upon the wings of time into the possession of other families'.[91] Innovation was a means to this end.

---

[88]   William Harvey Pierson Jr, 'Notes on Early Industrial Architecture in England', *Journal of the Society of Architectural Historians*, 8 (1949), pp. 1–32.

[89]   BA&H, MS 3782/12/79/61, JW to MB, 17 April 1786.

[90]   Dickinson, *Boulton*, pp. 147–148.

[91]   BA&H, MS 3782/12/111/150, MB, 'Considerations upon the Propriety of Buying Soho and the adjacent lands …'.

Chapter 4

Was Matthew Boulton a Scientist? Operating between the Abstract and the Entrepreneurial

David Philip Miller

Introduction

Many historians of science will feel some discomfort about the very question addressed in this chapter: Was Matthew Boulton a Scientist? The question seems somehow tainted, it smacks of transgressive history of science. There are a number of reasons for this, which have much to do with the way that history of science was professionalised in the years after World War II, with what we might call its 'Cold War' historiography. The first reason is the strong idealist strain in the infant field represented most succinctly by the eminent French historian Alexandre Koyré's claim that 'science is … essentially, *theoria*'.[1] Following this line of argument, a strong boundary was drawn, substantially in reaction to Marxist thought, between science proper and material practice. Even if the relationship between science and early industrialization was thought to be an interesting and important problem, and most of the time it was not, that relationship was considered by prominent historians such as A. Rupert Hall to be a weak one. In his work *The Abbey Scientists* (1966), Hall guarded against any idea that James Watt, Boulton's partner, could be appropriately commemorated in Westminster Abbey as a scientist. Faced with the awkward reality that Watt's statue was in the Abbey, Hall explained why he was out of place:

> Watt effectively ushered in the age of steam. He had a superb mechanical brain and a skilful hand: his inventions are very numerous. … He was as much interested in pure science as in technology. … In him science and technology begin to come close to each other, though Watt (rightly) never claimed that any

[1]   Alexandre Koyré, 'Commentary', in A.C. Crombie (ed.), *Scientific Change* (London, 1961, 1963), pp. 847–856, p. 856. For a discussion of these historiographical trends, especially in reaction to Marxist approaches, see Roy Porter, 'The History of Science and the History of Society', in R.C. Olby, G.N. Cantor, J.R.R. Christie and M.J.S. Hodge (eds), *Companion to the History of Modern Science* (London, 1996), pp. 32–46.

of his inventions were derived from scientific theories. The age of "applied science" was still remote.[2]

Watt's 'interest in pure science', whilst noteworthy, did not qualify him as a scientist. Only if his creative activity, his inventions, had been derived from scientific theory would Watt qualify. For Hall, Boulton, besides not being in the Abbey, could never be in the frame. He was a capitalist and entrepreneur, even more remote than his partner from the realm of science proper.[3]

In what follows, the 'Cold War' historiography and sociology of science that led to these exclusionary conclusions is outlined as are more recent trends in the field that potentially afford Boulton a 'friendlier' reception. A few examples of Boulton's activity in what I call the space between the abstract and the entrepreneurial are then examined and an assessment made of the propriety, or otherwise, of granting him the status 'scientist'. In pursuing these examples, we are concerned with a very thin slice of the rich cake of Boulton's creativity. It is nevertheless an important slice since it is what most people look to when the question of the attribution of the status of 'scientist' is in the balance.

The Historiography of Science, and of the Scientist, as Boundary Work

Koyré and Hall, among many others, were engaged in what modern historians call 'boundary work'.[4] The essential claim made through this concept is that boundaries such as those between scientist/non-scientist, between science/technology, and so on, are not natural divisions but rather are constructed, through active work, in different ways at different times and places. It becomes an important task to explain why some drawings of the line come to dominate over others, at least temporarily. The recent shifting pattern of these boundaries has created more room for Boulton the 'scientist' than Hall, Koyré and their ilk could ever grant.

---

[2]   A. Rupert Hall, *The Abbey Scientists* (London, 1966), p. 35. A fuller statement of Hall's views on this question is found in A. Rupert Hall, 'What Did the Industrial Revolution in Britain Owe to Science?', in Neil McKendrick (ed.), *Historical Perspectives: Studies in English Thought and Society in Honour of J.H. Plumb* (London, 1974), pp. 129–151.

[3]   On the history of representations of the Boulton and Watt partnership, and their consequences for perceptions of Boulton as philosopher, and of Watt as businessman, see David Philip Miller, 'Scales of Justice: Assaying Matthew Boulton's Reputation and the Partnership of Boulton & Watt', *Midland History*, 34 (2009), pp. 58–76.

[4]   The clearest articulation of the concept of 'boundary work' is Thomas F. Gieryn, 'Boundary-work and the Demarcation of Science from Non-science: Strains and Interests in Professional Ideologies of Scientists', *American Sociological Review*, 48 (1983), pp. 781–795. See also Thomas F. Gieryn, *Cultural Boundaries of Science: Credibility on the Line* (Chicago, 1999).

There were some historians in the 1960s who did not accept the sharp boundaries erected by Koyréan and Hallian historiography. Besides those maintaining an explicitly Marxist approach to the history of science, some economic historians who saw technological change as more than a 'residual' in processes of economic growth, accumulated endless instances of the *association* of scientists and industrialists. This was the approach of A.E. Musson and Eric Robinson in their mammoth study *Science and Technology in the Industrial Revolution* (1969).[5] However, even such thinkers believed that the routine mobilisation of abstract science did not occur until the late nineteenth or early twentieth century. The mixing of scientific ideas with practical purpose, even then, was widely regarded, within an ideology of pure science, as unpredictable and serendipitous. The North American sociologist, Robert Merton formulated the norms of science in the early 1940s, norms that he regarded as the basic ethic of true science.[6] He depicted the pursuit of science directly for practical, proprietorial purposes as likely to corrupt the scientific process and produce bad science. In the Mertonian schema, pursuing science in that way violated the norms of 'disinterestedness' and 'communalism'. Not surprisingly, perhaps, historians hunting down the origins and progress of the 'scientific ethos' somehow by-passed the precincts of Soho, Birmingham and similar industrial and commercial sites.

Even among the counter-currents of the history of science as it developed in the 1960s and 1970s there were developments that discouraged the asking of a question such as 'Was Matthew Boulton a Scientist?' The work of Thomas Kuhn,[7] with its model of science as paradigm-driven, its denial of a universal scientific method, indeed its denial of the very existence of 'Science with a capital S' and its advocacy of the study of the manifold 'sciences', encouraged a radical contextualism that was very wary of ideas, concepts and labels that were out of historical place. Those who celebrated Kuhnian approaches to the history of science made much of Sir Herbert Butterfield's concept of 'Whig History'.[8] They criticised old-style, positivist history of science as guilty of perpetuating Whig history because it portrayed the history of science as a history of 'good guys' versus 'bad guys', an

---

[5]   A.E. Musson and Eric Robinson, *Science and Technology in the Industrial Revolution* (Manchester, 1969).

[6]   The classic statement of the norms is in Robert K. Merton, 'A Note on Science and Democracy', *Journal of Legal and Political Sociology*, 1 (1942), pp. 115–126, reprinted as 'The Normative Structure of Science' in Robert K. Merton, *The Sociology of Science: Theoretical and Empirical Investigations* (Chicago, 1973). On the political context of Merton's work see Stephen Turner, 'Merton's "Norms" in Political and Intellectual Context', *Journal of Classical Sociology*, 7 (2007), pp. 161–178.

[7]   Thomas S. Kuhn, *The Structure of Scientific Revolutions* (Chicago, 1962).

[8]   Herbert Butterfield, *The Whig Interpretation of History* (London, 1931). For an important recent critical paper on Whig history of science exhibiting shifting attitudes in the field see Hasok Chang, 'We Have Never Been Whiggish (About Phlogiston)', *Centaurus*, 51 (2009), pp. 239–264.

heroic story of right, truly scientific, ideas (as judged in present terms) struggling to the light through the miasma of pseudo-scientific nonsense (as judged from the same perspective). Against this old style of work, Kuhnian contextualism mandated a non-Whiggish approach, which avoided imposing current scientific ideas as a framework for selecting what was interesting and important in the past. Instead, past ideas that appeared wrong or just plain crazy to us were to be treated sympathetically, as a challenge to our capacity for historical empathy. Ideas about nature were to be respected and understood as products of their times. Hence the considerable interest taken in alchemy, even, or especially, when it appeared in the thought of Isaac Newton. Hence new treatments of phlogiston theory that tried to grasp its strength and appeal rather than dismissing it out of hand as hopeless nonsense to be inevitably swept away in a 'Chemical Revolution'. Hence attempts to understand the scientific integrity of those who resisted Copernicanism and who tried and imprisoned Galileo.[9] Within this broader demand not to treat the past only in terms of lines of progress culminating in present ideas, it was also argued that scientific roles, and especially the role of 'scientist' itself, should be treated with careful attention to context. This is where my own initial discomfort with my question lay. Surely we should *avoid* speaking of Boulton as 'scientist'? To speak that way is Whiggish because it imposes upon the historical period in which Boulton lived, a concept and a role that is quite alien to it.

In 1962, Sydney Ross, published his famous article 'Scientist: The Story of a Word'.[10] Ross's key revelation was that the term 'scientist' did not exist prior to the early 1830s and only entered common usage much later. According to Ross, the halting emergence of the term 'scientist' was linked to the processes of specialisation and professionalisation of scientific enquiry. One conclusion seemed obvious: to call anyone 'scientist' before circa 1850 was anachronistic. Of course, what was being objected to here was not the use of the word as such – most recognised that such extreme nominalism was pointless – but rather the projection into the past of a role, a social status, that, it was thought, could not exist in those earlier times.

There were other features of Kuhnian history of science that seemed to militate against treating characters like Boulton as 'scientists'. Kuhnian historiography had important continuities with earlier approaches to the history of science, notably a continued emphasis upon science as *theory*, a relative neglect of scientific work

---

[9]   On Newton's alchemy see Betty Jo Teeter Dobbs, *The Foundations of Newton's Alchemy or 'The Hunting of the Greene Lyon'* (Cambridge, 1975) and Karin Figala, 'Newton's Alchemy', in I. Bernard Cohen and George E. Smith (eds), *The Cambridge Companion to Newton* (Cambridge, 2002), pp. 370–386; on approaches to the Chemical Revolution see John G. McEvoy, *The Historiography of the Chemical Revolution* (London, 2010).

[10]   Sydney Ross, 'Scientist: The Story of a Word', *Annals of Science*, 18 (1962), pp. 65–85. Ross was Professor of Colloid Science at Rensselaer Polytechnic Institute, in Troy, New York, from 1948 to 1980.

as material practice. Boulton and his ilk, even if we relaxed anti-Whig scruples, could no more be scientists within Kuhnian historiography than within Koyréan because they were men of practice and not of theory.

More recent trends have opened things up a little. Historians and sociologists such as Peter Galison and Andrew Pickering,[11] in their studies of that 'purest' of scientific ventures, high-energy physics, have emphasised scientific practice in the sense that they regard science at its core as being as much about material manipulations as about ideas. Hence derive their concepts of the 'mangle of practice' and of 'trading zones', which import manufacturing and commercial metaphors into the heart of science. This encourages us to transgress previous boundaries. In related moves, the historical relationship of science and technology in the eighteenth century has been reconceived, the boundary redrawn, notably in the work of Margaret Jacob and Larry Stewart.[12] Also we have gained much knowledge in recent years of the nineteenth-century practitioners of a hybrid 'engineering science', characters such as W.J.M. Rankine and Sir William Thomson (Lord Kelvin) whose activities defy attempts to draw boundaries between science and practice.[13] Rather than a simple linear picture of the emergence of 'science-based industry' in the later nineteenth century, historians find 'industry-based science' (a term coined by Wolfgang König[14]) to be a live tradition. In some cases, only in industry could the materials of scientific production be found. Someone like Irving Langmuir, the first American industrial scientist to win the Nobel Prize (in 1932), moved from academe to the nascent laboratories of General Electric in the early twentieth century and stayed there, because he judged that at GE he would have readier access to the cutting-edge technology for producing the physical effects that he studied. When Langmuir participated in the Solvay Congress of 1927 he was known, and loved, as the 'gadget man' from GE.[15] But Langmuir's gadgets were not trivial from a scientific point of view, as the other participants I

---

[11]   Peter Galison, *Image & Logic. A Material Culture of Microphysics* (Chicago, 1997); Andrew Pickering, *The Mangle of Practice: Time, Agency and Science* (Chicago, 1995).

[12]   Margaret Jacob and Larry Stewart, *Practical Matter: Newton's Science in the Service of Industry and Empire, 1687–1851* (Cambridge, MA, 2004).

[13]   Crosbie Smith and M. Norton Wise, *Energy and Empire: A Biographical Study of Lord Kelvin* (Cambridge, 1999), and on Rankine, Ben Marsden, 'Engineering Science in Glasgow: Economy, Efficiency and Measurement as Prime Movers in the Differentiation of an Academic Discipline', *The British Journal for the History of Science*, 25 (1992), pp. 319–346.

[14]   Wolfgang König, 'Science-Based Industry or Industry-Based Science? Electrical Engineering in Germany before World War I', *Technology and Culture*, 37 (1996), pp. 70–101. See also John V. Pickstone, *Ways of Knowing: A New History of Science, Technology and Medicine* (Manchester, 2000), pp. 162–188.

[15]   Movie footage of the Solvay Congress participants shot on Langmuir's camera is reproduced in the DVD by Roger R. Summerhayes, *Langmuir's World* (1998), available via http://langmuirsworld.not/. On Langmuir, see David Philip Miller, 'The Political Economy

think realised. Rather they were crucial to the world of scientific production that culminated in the massive technological paraphernalia of high-energy physics and of synthetic biology, in which technical and scientific accomplishment blur beyond recognition. Historians of science have been more inclined recently not just to pursue the obvious scientific heroes but also to seek out the 'invisible technician' and the 'invisible industrialist' as an important part of science.[16]

Ideas about scientific roles are moving in accord with these developments in the history of science. In *The Scientific Life* (2008), Steven Shapin argues that the Cold War historiography and sociology of the 1950s enshrined a notion of the scientist, which very imperfectly, and partially, depicts the role in the modern world.[17] The importance of 'pure science' and the 'pure scientist' were asserted with a fervour and urgency fuelled by reaction to the growing dominance of the scientific enterprise by industrial and military institutions. It was in this atmosphere of denial that Koyréan and Hallian historiography took deep root. The idea thus promulgated – that the scientist proper has stood apart from the world of industry, commerce and the making of money – is, according to Shapin, a serious distortion of the historical picture. Now, given all these developments in how we view the history of science, it begins to make more sense, to appear more legitimate, to ask whether Matthew Boulton was a scientist. The encouragement, and even the preferential reward, of the commercially-active scientist in our own times helps to normalise that role historically and inclines historians to bring people within the ambit of science who previously were excluded. In this way, our own 'boundary work' responds to the zeitgeist.

Finally, after these long preliminaries, I come to Boulton, his activities and his times. There are a number of ways in which Boulton's science, or natural philosophy, can be discussed. I want to take for granted his *avocational* interest in natural philosophy, his participation in the affairs of the Lunar Society and in the wider Republic of Letters for its own sake.[18] Boulton was undoubtedly in some degree a man of 'curiosity', a creature of what, thanks to Joel Mokyr (generally) and Peter Jones (so far as Boulton and the West Midlands are concerned), we now understand as the 'Industrial Enlightenment'.[19] However, I want to explore

---

of Discovery Stories: The Case of Dr Irving Langmuir and General Electric', *Annals of Science*, 68 (2011), pp. 27–60, and Pickstone, *Ways of Knowing*, p. 173.

[16]   Steven Shapin, 'The Invisible Technician', *American Scientist*, 77 (1989), pp. 554–563; Jean-Paul Gaudillière and Ilana Lowy (eds), *The Invisible Industrialist: Manufactures and the Production of Scientific Knowledge* (Houndmills, 1998).

[17]   Steven Shapin, *The Scientific Life* (Chicago, 2008).

[18]   An excellent survey of Boulton's part in the Lunar Society and of his natural philosophical interests is given in Jenny Uglow, 'Matthew Boulton and the Lunar Society', in Shena Mason (ed.), *Matthew Boulton: Selling what all the world desires* (New Haven and London, 2009), pp. 7–13.

[19]   Joel Mokyr, *The Gifts of Athena: Historical Origins of the Knowledge Economy* (Princeton, 2002); Peter M. Jones, *Industrial Enlightenment: Science, Technology and*

the 'space' within which Boulton deployed natural philosophy, putting it to use. Jacob and Stewart in their book *Practical Matter* (2004) have identified what we might call a space between the abstract and the *practical* within which a great deal of activity mediating between natural philosophy and the technological practice of the mechanic and artisan occurred. This activity is crucial to the real role of science in industrialisation. I want to suggest the importance of an analogous space between the abstract and the *entrepreneurial* within which Boulton spent a good deal of his time and energy. Sometimes work in this area aimed towards a definite material product. At other times it was directed to marketing. The work could also be 'defensive' in nature, designed to obviate a challenge or a threat. Work in this space might be defined as follows: *the pursuit and use of natural philosophy in a tactical and strategic fashion with a view to the maintenance and expansion of an enterprise that was bigger than any particular project within it.* Once we liberate ourselves from the narrow interpretation of 'scientist' bequeathed to us by the advocates of 'pure science' of the late nineteenth century and by 1950s 'Cold War' historiography and normative sociology of science, there is every reason to call Boulton a scientist in this larger, but specific, sense.

There is one more preliminary consideration before I present examples – publication, or rather the lack of it. Unlike his fellow Lunar Society entrepreneur, Josiah Wedgwood, Boulton did not publish investigations under his name in the *Philosophical Transactions* of the Royal Society, or elsewhere for that matter. The fact that Wedgwood did, made it relatively easy to depict him as scientist, though it misled in not highlighting the activities of the 'invisible technician', Alexander Chisholm, who did a lot of the work under Wedgwood's direction.[20] However, the failure to publish cannot be taken too seriously as a criterion of non-membership of the scientific community since Sir Joseph Banks himself (who was President of the Royal Society through the heyday of Boulton's career) published virtually nothing, exerting his influence, on the science of natural history primarily, as a botanical impresario and entrepreneur. Banks undoubtedly had a thorough command of the botanical science of his time, and his collections and record-keeping systems enabled him to mobilise that science with unrivalled facility. But Banks concerned himself little with trying to contribute to the abstract science of botany through publication. Rather he mobilised that abstract knowledge in the

Culture in Birmingham and the West Midlands, 1760–1820 (Manchester, 2008). It is important here, however, to acknowledge the argument that 'Industrial Enlightenment' is a misleading, or incomplete, account of how industrialisation was actually achieved and of the forces that drove it. A strong statement of this view is made in William J. Ashworth, 'The Ghost of Rostow: Science, Culture and the British Industrial Revolution', *History of Science*, 46 (2008), pp. 249–274.

[20]   John A. Chaldecott, 'Josiah Wedgwood (1730–95) – Scientist', *The British Journal for the History of Science*, 8 (1975), pp. 1–16; Larry Stewart, 'Assistants to Enlightenment: William Lewis, Alexander Chisholm and Invisible Technicians in the Industrial Revolution', *Notes and Records of the Royal Society*, 62 (2008), pp. 17–29.

service of a larger botanical enterprise of collection, exchange and use.[21] Thus much of Banks's scientific work was done in the same 'space' where I claim to find a good deal of Boulton's, that is, between the abstract and the entrepreneurial. Banks and Boulton came to know each other. They were in important respects kindred spirits, albeit unlikely ones, because they operated in a similar kind of space.

Boulton between the Abstract and the Entrepreneurial

A few examples will illustrate concretely what was involved in Boulton's operation between the abstract and the entrepreneurial. One example relates to Boulton's exhibition of certain attitudes that under the 'Cold War' historiography and sociology of science were regarded as antithetical to scientific behaviour, but that under the new, more capacious view of scientific life are typical rather than aberrant. Other examples relate to particular projects that were contemplated and sometimes prosecuted within the abstract-entrepreneurial space.

First, let us take an attitudinal example. In 1776, 10 years before their election to the Fellowship of the Royal Society of London, Boulton responded to Watt's contemplated publication of some of his steam investigations as follows:

> The Curve of Boiling points under diff^t pressures will do you honour if you think it prudent to publish it. I w^d explain ye Engine & things but little further than most Philosophers may do by inspecting an Engine. Intimate that great Mechanical difficulties have occurd [sic] but that We have now conquered them & renderd the Engine less liable to be out of order than a Com^n one. I think the best & most reputable advertisement w^d be a Paper in the Phlo^l [sic] Transactions.[22]

Note that here Boulton not only treats a projected paper in the *Philosophical Transactions* as an advertisement for the partners' steam engines but also advocates making sure not to communicate the details of their innovations too fully, that is, consciously to restrict the content of the paper for commercial reasons. Boulton and Watt, in fact, used this sort of rhetoric in other public accounts of

---

[21]   On Banks see John Gascoigne, *Joseph Banks and the English Enlightenment: Useful Knowledge and Polite Culture* (Cambridge, 1994), and on the way in which Banks exercised his influence through detailed command of botanical science and its possibilities see David Philip Miller, 'Joseph Banks, Empire and "Centers of Calculation" in late Hanoverian London', in David Philip Miller and Peter Hanns Reill (eds), *Visions of Empire: Voyages, Botany and Representations of Nature* (Cambridge, 1996), pp. 21–37.

[22]   Birmingham Archives and Heritage (BA&H), MS3782/12/76/113, Matthew Boulton (MB) to James Watt (JW), 10 July 1776.

their engine innovations in the 1770s and 1780s designed to sell engines.[23] The interesting point here is that Boulton urges the publication of such an account in *Philosophical Transactions*. For him there is clearly no meaningful division to be made *in practice* between the world of science and the world of marketing and business. Under 1950s historiography and sociology of science this is behaviour sufficient in itself to banish Boulton from the scientific realm. But in more recent conceptions of the 'scientific life' it is standard fare. In modern sociology of science, the 'scientific norms' are not regarded as straightforward guides or as real constraints on behaviour but rather as an ideology selectively deployed to publicly justify a great diversity of actual practice.[24] Marketing may be denied publicly as a legitimate feature of the scientific role, but it is undoubtedly central to scientific practice. In fact, many now believe that marketing reaches to the very centre of the production of 'abstract' knowledge, including the writing of the scientific paper. Recording devices switched on in modern scientific laboratories turn up similar conversations to those we find in the Archives of Soho.[25]

Turning now from work between the abstract and the entrepreneurial on the marketing front to that on the technical/scientific, I would like to re-emphasise and bring into my argument what Jim Andrew has taught us recently about Boulton as a steam engineer.[26] Even before Watt's removal to Birmingham, Boulton and William Small acted independently in working on Watt's 'Steam Wheel' – a rotary design that ultimately fell by the wayside. In the mid 1770s, Boulton dealt in detail, in Watt's absence, with the development testing of the steam engine in Birmingham. Moreover, he was able and happy to get his hands dirty, demonstrating that he could translate the abstract into the practical and back again. Boulton frequently made suggestions for improvement of the engine's design. Above all, of course, it was Boulton who, with a keen sense of the potential market for rotary power, pushed Watt towards rotative engines based on the beam engine design. In the course of that crucial innovation, Boulton took a full part in technical negotiations. Indeed, in January 1781, Boulton suggested a simplification of the sun and

---

[23]   David Philip Miller, *James Watt, Chemist: Understanding the Origins of the Steam Age* (London, 2009), pp. 37–43.

[24]   Michael Mulkay, 'Interpretation and the Use of Rules: The Case of the Norms of Science', *Transactions of the New York Academy of Sciences*, Series 2, 9 (1980), pp. 111–125.

[25]   A number of studies of the very process of writing the scientific paper have identified and stressed that making knowledge claims and marketing them are part of the same process. See, for an early example, John Law and R.J. Williams, 'Putting Facts Together: A Study of Scientific Persuasion', *Social Studies of Science*, 12 (1982), pp. 535–558. See also J. Paul Peter and Jerry C. Olson, 'Is Science Marketing?' *Journal of Marketing*, 47 (1983), pp. 111–125.

[26]   Jim Andrew, 'Was Matthew Boulton a Steam Engineer?', in Malcolm Dick (ed.), *Matthew Boulton. A Revolutionary Player* (Studley, 2009), pp. 107–115; Jim Andrew, 'The Soho Steam Engine Business', in Mason (ed.), pp. 63–70.

planet gear system, and Watt duly substituted Boulton's suggestion for his own
less satisfactory arrangement in the patent of 1782, so that Boulton's innovation
became standard on Boulton & Watt rotative engines.

Jim Andrew has also explained how the imposition upon the eighteenth
century of modern ideas about steam engines led to misunderstanding of the
regimes for testing their performance, and recording and calculating it. This had
led people to disbelieve calculations that Boulton supplied to Watt and to dismiss
him as incompetent when, in fact, the calculations were correct and just what
Watt needed. The important point to take from this for our current concern is that
Boulton could not only handle the technical aspects of steam engines but also
demonstrated technological knowledge that enabled him to see whether, and how,
marketable technologies could be constructed.

A critic might argue – in concert with Rupert Hall's ghost – that these
technicalities are not abstract science. Boulton's work with steam engines was
between the technical and the entrepreneurial. Let's accept the scientific/technical
distinction for the moment, much against my inclinations, and pursue Boulton into
spaces between abstract *science* and the entrepreneurial. Boulton could and did
take his entrepreneurial gaze deeper into the science of engines.

An early example of this concerned the theory and practice of boiler design.
Even before Watt came on the scene, Boulton, Benjamin Franklin and Erasmus
Darwin were discussing steam engines. As the story goes, Boulton's worries about
the possible lack of water power to sustain his manufactory led him to serious
investigation of steam engines from the mid 1760s. He pursued his own steam
project until he met Watt in 1768. Boulton had built a model steam engine which,
early in 1765, he had transported to London for Franklin, and others, to examine.[27]
Writing to Boulton in December 1765, Darwin expressed curiosity about how
Boulton's steam engine had been received by Dr Franklin and what Franklin's
observations on it had been. Darwin, Boulton and Dr William Small had evidently
been discussing boilers and their optimal design. This question involved them
in considering philosophically the phenomenon of evaporation in general. Did
evaporation occur by solution of water in air, in which case steam could only
be produced at a surface? Or did it occur by the combination of heat and water?
Darwin wanted to know 'your final opinion and Dr Small's on the important
Question, whether Evaporation is as the Surface of boiling Water, or not? – or if
it be as the Surface of the Vessel, exposed to the Fire, which I rather suspect'. He
continued:

> For if you boil Water in a Florence-Flask, you see bubbles arising from the
> Bottom, these are Steam, as they mount they are condensed again by the cold
> supernatant Fluid, and their Sides clapping together make that Noise called
> Simmerring. 2dly When Water boils in the Æolipile or closed Tea-Kettle, what a

[27]   See Samuel Smiles, *Lives of Boulton and Watt: Principally from the Original Soho mss* (London, 1865), p. 183.

Quantity of Steam? – is this only as the upper Surface? I humbly conceive not –
but as that Surface in contact with the Fire.[28]

This natural philosophical question of *where and how* the boiling took place was
regarded as crucial to the design of the boiler of the steam engine. Darwin sought
from Boulton his, and Small's, view on 'the best way of constructing the Boiler
for the above Purposes'.[29] The question of which variable related to the quantity of
steam depended upon which theory of evaporation was adopted. Solution theory
would be compatible with the proportionality of quantity of steam to surface area
of the water to be boiled; solution, in any case, requiring solvent (air) and solute
(water) to be in contact. The proportionality of the surface of boiler in contact with
the fire would be consistent with the idea that steam is compounded from water
and heat, or the matter of fire, since this could occur, and did on that theory occur,
without contact with the air. At about the same time, but apparently independently,
Watt was also considering 'boyler' design in much the same theoretical light.[30]

When we turn to Boulton's notebooks we find him pondering these questions
in a variety of ways. In some 'Steam Engine Facts', which he noted down in 1769,
Boulton recalled further experiments to settle the theoretical question:

> The Evaporation of Water is not in ye Proportion to ye surface nor ye quantity of it
> but as ye quantity of heat that enters it. An Iron pan of water took a certain time to
> evaporate it. It was then filld to the same highth wth Water & a piece of wood put

---

[28] BA&H, MS 3782/13/53/30, Erasmus Darwin to MB, 12 December [17]65 as
transcribed in Desmond King-Hele (ed.), *The Collected Letters of Erasmus Darwin*
(Cambridge, 2007), pp. 65–66. I have omitted Dr. King-Hele's interpolation of 'at' after
'as' in his transcription of this letter. This interpolation is an understandable but I think
erroneous 'correction' or unnecessary amplification. This is so because Darwin's concern
is with the co-variation of the size of the surface at which the steam is produced and the
quantity of that steam. So throughout the letter he refers to quantity varying either 'as' the
upper surface of the boiling water or 'as' the surface of the vessel in contact with the fire.
The location of the production of the steam ('at …') is implicit in the covariation ('as …').

[29] Ibid., p. 66. In a jocular, but nevertheless telling analogy in this same letter, Darwin
compared the way the strength and activity of man depended on the 'Quantity of vital
Steam rising into the Brain from his boiling Blood' with the way that 'this Animal' [the
steam engine] depended on the design of the boiler.

[30] Watt's notebook in 1765 recorded studies of the boiler of Sisson's engine: '… I
saw no boyler so perfect … as the common tea kitchen. … Here the fuel is always in
contact with the sides of the boyler containing the water …'. This is consistent with the
view that the significant factor in producing steam is to ensure the maximum proximity of
heat (fire particles) and water. The Notebook is reproduced in Eric Robinson and Douglas
McKie (eds), *Partners in Science. Letters of James Watt and Joseph Black* (Cambridge,
MA, 1970), the section in question being at p. 435.

into it so as to reduce ye surface 9/10 wth Lead upon it to keep it down & exposed
to the same fire ye evapor^{tn} was performed in ye same time.[31]

An earlier notebook from the 1750s[32] offers an amusing insight into Boulton's
operation between the abstract and the entrepreneurial, which appears to have
been occasioned by advice from the Right Honourable the Earl of Hopetoun on
'boyling eggs'. The fact that eggs in moderately boiling water took 3½ minutes
while eggs in furiously boiling water took only 2½ was useful practical information
for Boulton, but also set him puzzling philosophically, since 'it is known from
Experience that if water boyls the Thermometer will rise to 212 & if it boyls still
more Violent the Mercury will rise no higher'.[33] His answer was that a hotter fire
produces water particles of greater velocity and a greater ebullition will mean
that 'as great a Quantity of heat passes through the water' in a shorter period of
violent boiling as in a longer period of moderate boiling! Whether boiling eggs, or
kettles, or raising steam for an engine Boulton moved back and forth between his
understanding of the scientific laws of heat and the practicalities of business – and
of breakfast! A later notebook of 1780–86, titled 'Boylers', offers more mature
observations on design, some remarkably prescient of later developments.[34]

Another fine example of Boulton's work between the abstract and the
entrepreneurial relates to the issue of engines using substances other than steam.
Whilst from our perspective, with hindsight, it is clear that steam was to be
the only serious working substance for engines through the late eighteenth and
early nineteenth century, this could not be clear in advance to Boulton and his
contemporaries. The study of new airs was a hot scientific topic in the 1770s
and 1780s.[35] We must remember also that, viewed as a chemical compound of
water and heat, steam was conceptualised as closer to other airs in character and
properties than it was to be regarded subsequently. Also the production of airs from
solid substances, and their re-absorption, raised new possibilities for generating
the vacuum and the pressure upon which the operation of engines depended.
Boulton's reaction to these possibilities was, I suggest, a typical defensive move
of the scientist-entrepreneur in the face of the threat that such alternatives to steam
might represent.

It was round about 1780 that Boulton and Watt became alert to the possibility
of using the capability of fluids, or even solids, to release and reabsorb airs to

[31]   BA&H, MS 3782/12/108/5, MB 'General Notebook 1768–1775', pp. 99–100.

[32]   BA&H, MS 3782/12/108/1, MB 'General Notebook, 1751–1759', pp. 21–22.

[33]   Ibid.

[34]   BA&H, MS 3782/12/108/23, MB 'Boylers – 1780', 1780–86, noted in Fiona Tait,
'Coinage, Commerce and Inspired Ideas: The Notebooks of Matthew Boulton', in Dick
(ed.), *Matthew Boulton*, pp. 77–91, p. 87.

[35]   See Miller, *James Watt, Chemist*. Another useful way into the literature on the topic
is provided by Trevor Levere and Gerard L'E. Turner, *Discussing Chemistry and Steam.
The Minutes of a Coffee House Philosophical Society 1780–1787* (Oxford, 2003).

produce a vacuum and derive mechanical force thereby. Boulton wrote to Watt on 21 July 1781, explaining what had impelled him to consider the question. One circumstance had been seeing in London 'a Glass Vessell filld w[i]th [marine acid] air which became as instantaneously condensed when brought into contact with [water] as steam w[oul]d have been, & the same thing happens with [volatile alkali] air...'. Boulton explained further with accompanying diagram:

> Now if you take a Glass tube hermetically seald at A & open at B – let the space between C & D be filld with mercury & the space between D&A be filled w[i]th water into which project as much [volatile alkali] air or [marine acid] air as will saturate that [water]. ... The moment you apply heat to the water, it gives out the air & ye instant you cool it that instant it is absorbed again by ye [water]. Heat it again & it gives out air so as to raise the mercury to a considerable height. ...[36]

Besides seeing the London experiment, Boulton reported that he had been stimulated by a conversation with 'Simcox' who 'mentioned to me the Idea of detaching fixable air from chalk & then in an instant reabsorbing it'.[37]

Boulton responded to his perception of these possibilities and the potential threat they represented to the steam engine enterprise (remember that in 1780 Boulton and Watt were only five years into their 25 year extension of the steam engine patent of 1769). Like any good scientist-entrepreneur Boulton responded by setting someone to work on the problem. That someone was Joseph Priestley,

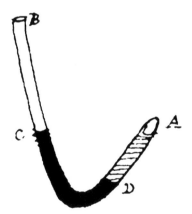

Figure 4.1   'A Glass tube', diagram in a letter from Matthew Boulton
                  to James Watt, 21 July 1781
Source: Courtesy of Birmingham Archives and Heritage, MS 3147/3/5.

---

[36]   BA&H, MS 3147/3/5, MB to JW, 21 July 1781.
[37]   Ibid.

who had just arrived in Birmingham and become part of the Lunar Society circle. Robert Schofield has noted Priestley's willingness to assist his new friends with work relevant to their industrial and commercial interests.[38] In this case, Priestley, though obviously far from invisible as a historical figure, was the 'invisible technician' to Boulton's 'scientist-entrepreneur'.

Boulton was convinced very quickly of the impracticability of deriving mechanical force by expelling and reabsorbing airs using solids. However, he remained convinced enough of the possibilities of using fluids to argue (again like the scientist-entrepreneur engaged in what became known much later as 'defensive R&D') that they should take out a patent against the eventuality. He drafted the key clause of such a patent for Watt's benefit as follows:

> ... a patent for making Engines by means of alternately emiting & reabsorbing all & severaly the different airs, Gasses, permanently elastic fluids, or by what soever other name they might be calld, seting forth that whereas there are several substances that do contain or are capable of containing several sorts of air acids &c in a fixed or solid state & that such airs &c are readily detached from, or are emitted by such substances upon being heated & again on being coold are readily reabsorbed by the s[ai]d substance & thereby a vacuum may be obtained & Mechanical force derived.[39]

No such patent was ever taken out. Indeed, Watt and Priestley rather quickly 'poured cold water' (as it were) on these alternatives to the condensing of steam, persuading Boulton that they were, after all, impracticable to the point of not representing a threat. Nevertheless, the example illustrates very nicely Boulton operating in the space between abstract science and the entrepreneurial. He knows enough of the science to perceive the potential practical importance of the experiments he has learned about. He sets a specialist to work on the problem to determine the plausibility and extent of the threat whilst acting quickly to draft a defensive patent that could be used to obviate it.

Other examples might be given with regard to steam engines and also, especially, I suspect, in relation to metallurgy and coining. However, we have already seen enough evidence for the plausibility of the position taken here to reserve for a future occasion what would be a worthwhile, and potentially very revealing, project – to try to pursue Boulton further into the spaces that it is suggested he inhabited.

---

[38]   Robert E. Schofield, *The Enlightened Joseph Priestley: A Study of His Life and Work from 1773 to 1804* (University Park, PA., 2004), pp. 160–161.

[39]   BA&H, MS 3147/3/5, MB to JW, 21 July 1781.

Conclusion

When we are tempted to say that Matthew Boulton was not a scientist, we consider excluding him not only in order to avoid an anachronistic term but also on the basis of a particular conception of the scientist's role. However, once we allow that the 'scientist' has had a more varied role than traditionally conceived and that the most usual forms of that role are not sealed off from the world of industry, money-making and entrepreneurship, then only the extreme nominalist worried about the word will deny that Matthew Boulton was a 'scientist'. In fact it is ironic that, as Sydney Ross pointed out, those who resisted adoption of the term 'scientist' in the mid nineteenth century did so because to them it 'implied making a business of science; it degraded their labours of love to a drudgery for profits and salary'.[40] Thus for some at least the word stood initially in *contrast* to the role of the disinterested scientific practitioner rather than denoting that role. They rejected the term precisely because they thought it applied to people like Matthew Boulton! Only later did 'scientist' commonly come to designate the disinterested professional.

It has been argued that, while some aspects of Boulton's scientific activities (his generalised interest in natural philosophy as a man of curiosity) simply sat side-by-side with his more familiar entrepreneurial role, other aspects so suffused his entrepreneurship that we might see him as that most modern of scientists, the scientist-entrepreneur. We might also see him as an early version of the so-called 'Invisible Industrialist' who played a part in the production of scientific knowledge and its practical use but a part that has been rendered invisible by narrow conceptions of what science is and has been, conceptions that have blinded historical enquiry to the technical and scientific Boulton.

It is entirely possible that by the time of the 300th anniversary of Boulton's death, historians will see the period during which the exclusionary ideology of 'pure science' was at play – and when historians were also in its grasp – as a mere blip in the history of science and the history of the scientist. Perhaps by that time the great Birmingham 'Hardware Man's' status as a scientist will be taken for granted.

---

[40] Ross, 'Scientist', p. 66.

Chapter 5

The Origins of the Soho Manufactory and its Layout

George Demidowicz

Matthew Boulton came to Soho in 1761 and, within a few years, the small water mill he had acquired had been transformed into one of the largest manufactories in the Birmingham area. This is even more remarkable in the knowledge that such growth was not his original intention, since first and foremost he had been in desperate need of waterpower. The Soho Mill was the seed that germinated into the Manufactory, but this could not have happened without Boulton's ambition, energy and swagger.

This paper will examine how Boulton's rapidly emerging vision for a single great manufactory was translated into specific buildings in a particular physical arrangement. Analysing the development of the Manufactory in the earliest years is crucial in understanding the function and layout of the whole complex, which reached its maximum extent in about 1805. This task is unfortunately hampered by a paucity of sources. The earliest known images of the Manufactory (a pair of perspectives of the front and rear) probably date from about 1768 and clearly demonstrate the extremely rapid growth which had taken place from 1761 (Figure 5.1).[1] After so much expensive construction work, Boulton would have been keen to show off his Manufactory to potential customers by means of such marketing images. The first known large-scale plan of the whole works is as late as 1788 and shows that the Manufactory had expanded by about another third in the intervening

---

[1]   Birmingham Archives and Heritage (BA&H) 82934, scrapbook compiled by Samuel Timmins, 1. The author has suggested that the pair of engravings was made by Edward Rooker from a drawing by William Jupp made in 1767–68, BA&H MS3782/12/23/128 William Jupp to Matthew Boulton (MB), 25 May 1769; George Demidowicz, *The Soho Industrial Buildings: Manufactory, Mint and Foundry* (forthcoming). Val Loggie has proposed an alternative artist, Thomas Feilde, BA&H, MS 3782/1/18/7, Thomas Feilde to MB, 5 July 1765; Val Loggie, 'Picturing Soho: Images of Matthew Boulton's Manufactory', in Shena Mason (ed.), *Matthew Boulton Selling what all the world desires* (New Haven and London, 2009), pp. 22–30, pp. 23–24. The Rooker authorship is more convincing as both front and rear perspectives are mentioned in Jupp's bill.

Figure 5.1   The Soho Manufactory perspectives, front and rear, c. 1768
Source: Courtesy of Birmingham Archives and Heritage.

two decades (Figure 5.2).[2] The lateness of the illustrative and cartographic record is compounded by the almost complete lack of building and machinery accounts for the earliest years. Furthermore, only a few contemporary letters describing the erection of workshops and other buildings on the site have survived.

A retrospective approach is necessary, therefore, to establish the sequence of development, which culminated in the construction of the 'principal' building between 1765 and 1767. This was to be the largest single building erected, a Palladian edifice designed to look like a mansion. Not surprisingly it dominates most of the illustrations of the Manufactory, the classic view being taken from the north. If the 'principal' building is removed from the c. 1768 pair of engravings, three courtyards of buildings remain, most of these, if not all, constructed before 1765 (Figure 5.1). Stripping away these buildings in turn should leave us with just the water mill. Its position on the engravings, however, is not obvious, as it is overshadowed by the 'principal' building in scale and architectural pretension.

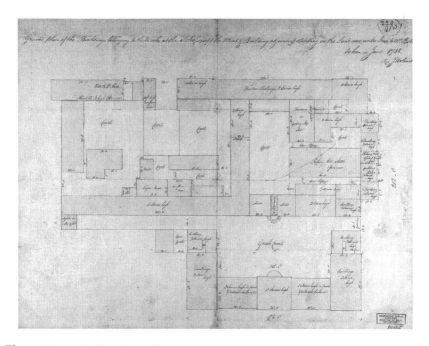

Figure 5.2   The first known large-scale plan of the Soho Manufactory
             by J.A. Smith, June 1788

Source: Courtesy of Birmingham Archives and Heritage.

---

²   BA&H, MS 3147/5/1447, Ground plan of the Building belonging to the Works at Soho (excepting the Mint and Building adjoining); standing on the Land now under lease to Mr Bolton [sic] taken in June 1788 by J.A. Smith. This plan was published by Jennifer Tann, *The Development of the Factory* (London, 1970), p. 78.

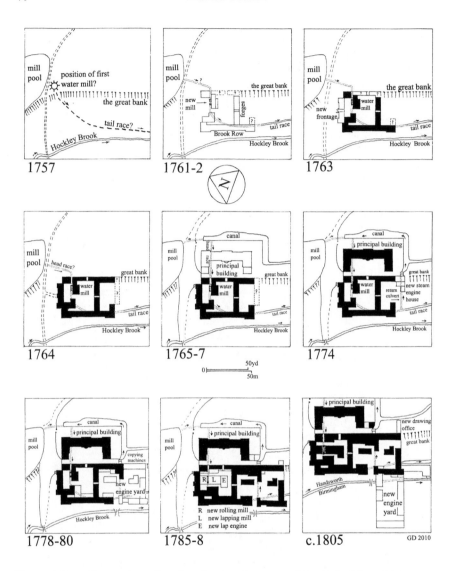

Figure 5.3   Phase plans showing the development of the Soho Manufactory
Source: Drawing by George Demidowicz.

An external water wheel can just be discerned in the front view (Batiment Vu Par Devant) in the shadow that is cast by the long elevation of the mill. A Dutch gable faces the rear of the 'principal' building (Figure 5.1).

It could be assumed that this was the original position of the water mill, but there is no evidence to support this conclusion. Its location, about 40m from the edge of the mill pond, suggests that the mill may have been moved. The mill dam was the spot normally favoured for a water mill to take advantage of the head of water created by the earthworks of the mill pool (Figure 5.3). A small-scale plan of Handsworth Heath and Soho, c. 1790, shows an old road heading directly to this location.[3] It began on the former line of the turnpike road from Birmingham to West Bromwich, passing the west front of Soho House. The road then curved sharply downhill following the edge of the Soho enclosure – later park – to reach the end of the mill pool and the likely position of the original mill. This was built in 1757 by Edward Ruston and John Eaves to roll metal and was only four years old when Matthew Boulton took over the lease in April 1761. Despite its newness, it was demolished a few months later 'as the old construction of ye Mill was ill fav^d for M B^s new plan he thought it most convenient and prudent to pull down ye old Mill wch he accordingly did & rebuilt it'.[4] Matthew Boulton was not impressed by the old-fashioned technology of the Ruston and Eaves rolling mill, but more importantly it did not accord with his 'new plan'.[5] This could be interpreted as the construction of a new manufactory for the production of Birmingham 'toys' – buttons, buckles, sword hilts and watch chains to name but a few. The scale of building was so ambitious that it was probably not feasible to crowd them around the original mill, assuming that it was situated on the dam at the edge of the pool. It would have been logical for Boulton to place his water mill conveniently at the centre of his new manufactory, and this is where it can be seen on the c. 1768 pair of engravings. His other motivation for pulling down the mill was that he wished to construct a lapping and polishing mill in preference to a rolling mill. Boulton was probably the first to mechanise this function.[6] He was aided in this innovatory scheme by John Wyatt, who allowed him to apply his invention of roller bearings to the new mill machinery.[7]

The reconstruction, and probable relocation of the water mill, took place at the same time as the new manufactory buildings were being erected (Figure 5.3). Work started around July 1761 when Boulton began '... to Build some dwellings for Workmen a Warehous wth several shops ...'.[8] The site chosen lay immediately to the south-east of the mill pool just within the area covered by Boulton's lease and between the Hockley Brook, which formed the boundary between Handsworth and Birmingham parishes, and a steep bank or terrace that ran parallel to it. This distinctive feature of the topography of the Manufactory site will be called the 'great bank' for convenience. It had been exploited and emphasised by the long lateral dam of the original mill pool and continued through the whole length of

---

[3]   BA&H MS 3147/5/1470.

[4]   H.W. Dickinson, *Matthew Boulton* (Cambridge, 1937), p. 43, quoting from 'Case between B[oulton] & F[othergill]', BA&H, MS 3782/12/60/265.

[5]   Ibid.

[6]   Dickinson, p. 203, quoting the memoir of MB by James Watt (JW), 1809.

[7]   University of Birmingham Special Collections, Wyatt Papers/20, John Wyatt (JWy) to [a friend], 29 August [1762].

[8]   Dickinson, p. 43.

Boulton's Soho land as far as Hockley Pool. The 'great bank' was approximately 16 to 20ft high (4.9 to 6.1m) and defined one side of the first Manufactory buildings (Figure 5.3). The new water mill was constructed into this bank and set in motion for the first time in August 1762.[9] Its Dutch gable, facing north-eastwards, would have been highly visible from the old road mentioned above that led down from the turnpike. The other buildings were completed about the same time. The 'dwellings for Workmen' can be identified with 'Brook Row', a terrace of houses that overlooked the Hockley Brook and which can be clearly seen in the rear view of the Manufactory (Figure 5.1). It is more difficult to be specific about which buildings comprised the 'Warehouse wth several workshops'. In the same rear view (Figure 5.1), two buildings can be seen running at right angles from 'Brook Row' and parallel to the water mill and a third, L-shaped, in line with the mill itself. The two storey workshop with six chimneys attached to an external wall was probably a forge. Not all of the buildings in view were constructed in this phase, which may have created only one courtyard of buildings so that the water mill and its wheel faced a deep forecourt defined by the 'great bank' and the projecting north end of 'Brook Row' (Figure 5.1, Batiment Vu Par Derriere). It would be no surprise to discover that the earliest buildings contained production space and the necessary casting and forging shops for the manufacture of buttons and buckles, as this was essentially Boulton's main line of business at the time (Figure 5.3).

Building work appears to have taken place in 1763. In July of that year Boulton had mortgaged his Soho lease to borrow £1,300, which may have been used to finance this phase.[10] A letter from William Wyatt to Matthew Boulton in September of that year refers to difficulties in the construction and to lack of supervision by William and his brother, Samuel, who 'drew the plan'.[11] The Wyatt brothers would not have been engaged to design ordinary workshops so buildings of some architectural ostentation must have justified their employment. These can be found in 'Pool Row' or 'Rolling Mill Row', as it was much later to be known, the range of buildings that completed the north-west side of the Manufactory (Figure 5.3).[12] They fronted the road which descended steeply from the probable site of the original mill to the Hockley Brook and then crossed into Birmingham Heath. It is clear that Boulton wanted to provide a more dignified frontage to his new Manufactory and this was achieved by a combination of symmetry and architectural detail. A central gabled bay with a Diocletian window accommodated a high vehicular arched entrance.[13] It was flanked by two-storey wings, which terminated in taller rectangular pavilions with pyramidal roofs,

---

[9]   Dickinson, p. 43; JWy, letter [to a friend], 29 August [1762].

[10]   BA&H, MS 2254; MB borrows £1300 from Henry Carver and Girton Peake on 26 July 1763; recited in mortgage dated 13 December 1766.

[11]   BA&H, MS 3782/12/85/204, William Wyatt to MB, 16 September 1763.

[12]   BA&H, MS 3782/13/126; MS3782/6/182; the buildings faced the Mill Pool and later became attached to the reconstructed Rolling Mill (see below).

[13]   A Diocletian window is semi circular divided into three lights by two mullions.

some of which supported hipped dormers. The pavilion on the 'great bank' side was constructed with an extra storey so that it did not appear buried in the slope, thereby compromising the symmetry a little. 'Pool/Rolling Mill Row' projected slightly forward for further effect and created a second courtyard out of the short-lived forecourt to the water mill. For a few years, on entering the Manufactory through the giant archway, the visitor would have been confronted with a 17ft diameter water wheel turning on the opposite side of the courtyard. The frontage of 'Pool/Rolling Mill Row' is unfortunately obscured in most of the later views of the Manufactory, by the lie of the land and specifically the higher level of the mill pool.

The Manufactory grew during the summer of 1764 in a further season of work adding some more large button shops and a stamping shop.[14] The result may have been the third courtyard visible on the right-hand side of the rear view, but is not certain (figures 5.1 and 5.3). Some of the 'great bank' may have also been colonised, a difficult site requiring the excavation of a vertical wall in order to create workshops with a single-storey at the top of the bank and having three-storeys facing inwards into the courtyards.[15] It must be emphasised that the two perspectives of the Manufactory c. 1768, though an extremely important source, are defective in detail and give no impression of the great change in level, which was such a distinctive feature of the Soho site.

By the end of 1764, Matthew Boulton had constructed the largest 'toy' manufactory in the region, if not in the country, and many an entrepreneur in their mid 30s would have been satisfied with this achievement. This was not Boulton's way, however, and the construction of his Manufactory had perhaps stimulated his desire to build on an even more grandiose scale. He was not only motivated to obtain high and unprecedented levels of production, but was also committed to raising the quality of the goods he produced. In the process he hoped to change the image of Birmingham as a place where only the tawdry and the cheap were made. What better way than producing silver ware and Sheffield Plate , highly fashionable objects hitherto largely unknown to the Birmingham manufacturers? At the same time he could widen his customer base by attracting the aristocracy and even more of the aspiring middle ranks of society.

Was unfettered ambition the only reason for the construction of the 'principal' building at Soho between the years 1765 and 1767? Boulton's restless seeking out of new projects certainly caused him many financial difficulties throughout his business career, not least being the budget-blowing costs of his new Sheffield Plate and silver manufactory masquerading as a mansion house. According to Boulton, initial building estimates of £2,000 provided by William Wyatt turned out on completion to be nearly five times higher, 'a sore grievance to an infant

---

[14]   BA&H, MS 3782/1/37, B&F to John Perchard, 18 August 1764; BA&H, MS 3782/1/34/235.

[15]   BA&H Stone Collection, 14/33/C; this photograph of the Manufactory under demolition in the early 1850s shows these 'great bank' workshops in cross-section.

undertaking …'[16] Unfortunately we see the affair only through Boulton's eyes. In a memorandum he takes no account of personal ambition and justifies the unification and expansion of production at Soho as a solution to the 'great inconveniences' of operating on two sites, the other being at Snow Hill.[17] All the 'Clarks [sic] and assistants … cryed out for a Building equal to the Whole Manufactory'.[18] It is unlikely, however, that the nagging of his staff to operate from a single site would alone have convinced him to have invested in what turned out to be an extremely showy plated works, warehouse, showroom, office and accommodation block.

Construction of the 'principal' building probably commenced in the spring or early summer of 1765.[19] The fact that Boulton's brother in law, Luke Robinson, died in 1764 so that the remainder of the Robinson family fortune probably passed to Anne, Boulton's wife, may have had some bearing on the timing of this event.[20] The site lay above the 'great bank' at the end of the mill pool and, therefore, overlooked his existing Manufactory, creating a two-level site (Figure 5.3). The multi-purpose building created a new principal front to the Manufactory facing north-eastwards up the slope towards Soho House and the turnpike road. It would have dominated the view from the curving access road and totally obscured the water mill and its Dutch gable (Figure 5.5). The assumed shifting of the mill a few years earlier would have required a channel to bring water the extra distance from the pool and this probably had to be diverted for the construction of the 'principal' building. The feed from the pool passed under the road as before, but filled a new canal constructed below a terrace or forecourt to the main frontage. A culvert then took water off the canal at right angles and passed under the 'principal' building into the new rear yard. After running below the latter, it flowed into a new timber trough or 'foreboy' constructed over the 'great bank' to disgorge its water onto the external water wheel of the mill below (Figure 5.3).[21] There was a height difference of approximately 9ft (2.9m) between the forecourt of the new building and its back yard, which produced three storeys to the front and four storeys to the rear. Furthermore, as if to emphasise this difference, vertically sliding sash windows graced the Palladian architecture of the frontage in contrast to the large cast-iron windows of the workshops facing the earlier Manufactory buildings.

The walls of the 'principal' building were completed in December 1765 and good weather in December allowed progress to be made on the roof.[22] The completion of the building and its fitting out, however, dragged on into 1767 when

---

[16]   BA&H, MS 3782/12/20.

[17]   Ibid.

[18]   Ibid.

[19]   Ibid.; BA&H, MS 3782/1/40 f.107, MB to Samuel Wyatt, 26 April 1765.

[20]   Dickinson, p. 36; Shena Mason, *The Hardware Man's Daughter* (Chichester, 2005), p. 7.

[21]   BA&H, MS 3782/12/60/19, John Fothergill (JF) to MB, 2 November 1765.

[22]   BA&H, MS 32782/12/81, John Scale to MB, 11 November 1765; MS 3278/12/60/34, JF to MB 14 December 1765; MS 3782/12/60/37, JF to MB, 19 December 1765.

a further £1,500 was spent on 'finishing of our Building & in new erections'.[23] Not surprisingly Boulton blamed William Wyatt, the architect, considering that he had been 'egregiously deceived by Wyatts estimates'.[24] On 29 December 1766 Wyatt replied to another letter of complaint: 'I am very sorry you have been misled by bad Estimates or had your Money misapplied. This I can with truth say I had no intention of either and assure you I would lay out Mr Boultons [sic] Money with as much frugality as I would my own'.[25]

Wyatt tried to pass some of the blame to Mr Newbold, the builder on site, for his 'great compassion' towards his workforce, presumably for slowing down the pace of work.[26] Despite Boulton's anger, Wyatt could not promise a visit for another eight to ten days 'having been very much engaged of late'.[27] Since we observe this affair solely through the Boulton lens, it is difficult to judge the real causes of the massive cost overrun. Spiralling costs on large building projects are not unusual and as frequent today as two and a half centuries ago. What is certain is that the unplanned capital outlay hindered the financial viability of Boulton's Soho businesses for many years to come.[28]

Anger over the costs of the 'principal' building must have been tempered by the immediate favourable reaction of customers and visitors to Soho: 'during this summer [1767] scarcely a Day hath passd without having one two or 3 Companys of foreigners or Strangers to wait upon, as well as many of our Nobility who are much delighted w'th the extension and regularity of our Manufactory ...'.[29] At least on this front, all was working to plan – as long as admiration ultimately turned into orders. Visitors and customers were attracted to Soho as a novelty, an industrial building designed by an architect in the fashionable Palladian style. Before Soho, factory buildings rarely reflected standards of architectural taste, except for occasional minor embellishment. It is no wonder that Jabez Maud Fisher, that most perspicacious and eloquent of Boulton's visitors, remarked that 'his great and wonderful manufactory' was 'like the stately Palace of some Duke'.[30] Although cloaked in the architecture of a gentleman's mansion, this extraordinary building was anything but inside. The structural, architectural and functional contrast between the front and back has already been noted. The central part of the frontage contained the warehouses and showrooms, whilst

---

[23]   BA&H, MS 3782/12/1 f.66, MB to J.H. Ebbinghaus, 2 March 1768.

[24]   BA&H, MS 3782/12/20.

[25]   BA&H, MS 3782/12/85/211, William Wyatt to MB, 29 December 1766.

[26]   Ibid.

[27]   Ibid.

[28]   Nicholas Goodison, *Matthew Boulton: Ormolu* (London, 2002), p. 28; Rita McLean, 'Introduction: Matthew Boulton, 1728–1809', in Mason (ed.), *Matthew Boulton*, pp. 1–6, 3–4.

[29]   BA&H, MS 3782/12/1 f.30, MB to J.H. Ebbinghaus, 28 October 1767.

[30]   Kenneth Morgan (ed.), *An American Quaker in the British Isles The Travel Journals of Jabez Maud Fisher, 1775–1779* (Oxford, 1992), p. 253.

the Sheffield Plate and silver ware workshops were located on the far side of longitudinal corridors that ran along each floor. The main office or counting house was situated in a semi circular bay in the centre of the rear elevation. The deeper wings at each end of the building, which produced the flattened U-shaped plan, were divided into accommodation for the more senior members of the staff working at Soho (Figure 5.5).[31]

Despite its novelty, this extraordinary hybrid building was born of an age where the division between the arts, crafts and industry were not so marked as at the present time. Objects could be created and appreciated for both their practical and aesthetic qualities and the manner of their production admired. Fisher perceived the essence of Soho – in its buildings, landscape and philosophy. He recognised that the canal constructed along the front of the 'principal' building was 'nothing more than his [Boulton's] Mill Damb [sic] and his Races, which he has ingeniously constructed to answer the *Dulce* as well as the Utile'.[32] The expensive experiment that was the 'principal' building was a success in terms of image or brand creation.[33] It has endured through the survival of engravings and aquatints so that, in most people's perception, the Soho Manufactory *is* the 'principal' building. It is for this reason that the present author drew a perspective of the Manufactory from the rear to demonstrate that the Soho buildings were far more extensive than could be discerned from Boulton's propaganda and marketing images (Figure 5.4).[34]

The pair of much earlier engraved perspectives, dated to c. 1768 (Figure 5.1), records the Manufactory at the end of a phenomenally rapid programme of building. The land on which the Manufactory was built belonged to Matthew Boulton, forming a small part of his lease of Soho from the lords of Handsworth, the Birch family. The buildings at this time belonged to the partnership of Boulton and Fothergill (established in 1762), which engaged in two basic branches of business – the 'toy' trade in the earliest part of the Manufactory complex (1761–1764) and the silver ware and Sheffield Plate trade in the new 'principal' building (1765–1767).

At this time the only power available was generated in the water mill, devoted mainly to the polishing or lapping of finished goods but perhaps with room made for a little metal rolling capacity. For the latter Boulton relied on his lease of Holford mill located on the river Tame in Handsworth (1765–1781).[35]

In view of the debt that the Boulton and Fothergill partnership carried as a result of the heavy investment in buildings in such a short period of time, it is hardly surprising that no significant construction took place in the years that immediately

---

[31]   George Demidowicz, 'A Walking Tour of the Three Sohos', in Mason (ed.), *Matthew Boulton*, pp. 99–107, 99–100 and Figs 79–81.

[32]   Morgan, p. 253.

[33]   Loggie, 'Picturing Soho', p. 20–30

[34]   Demidowicz, 'A Walking Tour', Fig. 79.

[35]   BA&H, MS 3782/1/8; MS 3782/12/25/13.

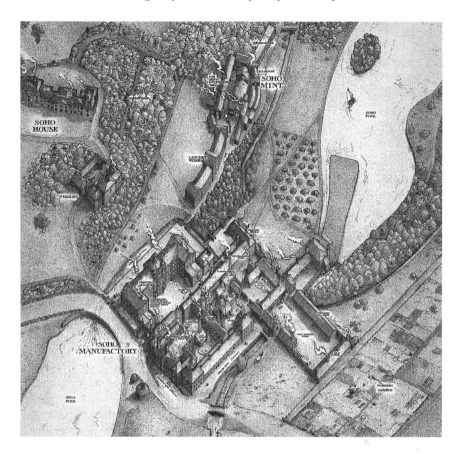

Figure 5.4   Axonometric projection of the Soho Manufactory c. 1805
Source: Courtesy of Bremner and Orr, based on a drawing by George Demidowicz.

followed. It was not until 1774 that another phase of building commenced. This time it was not to expand productive capacity, but to enhance the unreliable source of power provided by the water mill. Lesser men would have rested and consolidated, but not Boulton, for he had heard about an improved steam engine, invented by a Scotsman, James Watt. Soho certainly suffered from water-supply problems; there are ample Soho records of a horse gin being applied in times of drought to keep his water wheel in motion.[36] Six to ten horses were harnessed to the gin costing '5 to 6 guineas a week besides Injuring the Mill'.[37] A steam engine, even a more efficient one, could only work in a vertical direction at this time and was usually used for pumping water out of mines. Boulton could only apply

[36]   BA&H, MS 3782/1/34 ff. 129, 183, 208, 263; MS 3782/1/8 f. 23.
[37]   BA&H, MS 3782/12/20.

it to circulating water back to the water wheel, but he saw beyond solving his local difficulty. By attracting Watt to Soho and entering into a new partnership, he envisaged selling an improved steam engine to customers throughout the country eager to save on their coal bills. Never reticent in making use of hyperbole, Boulton boasted that he could provide 'what all the world desires to have – POWER'.[38]

The steam engine project provided the necessary new stimulus that Boulton's personality required. Fisher's incisive analysis of his character and motivation is relevant in this context:

> He is scheming and changeable, ever some new matter on the Anvil to divert his attention from the steady pursuit of some grand object. He is always inventing, and by the time he has brought his Scheme to perfection, some new Affair offers itself. He deserts the old, follows the new, of which he is weary by the time he has arrived at it.[39]

Fisher had recognised an essential Boulton trait, his constant need for the stimulus of new projects and ideas. The Mint (1788) and Latchet works (1794–1745), constructed on separate sites, are obvious examples of how his restlessness would be manifested in the future. Much of the subsequent expansion of the main Manufactory complex, however, is explained by the development of the steam engine business by the new firm of Boulton and Watt. But first Watt's engine had to be brought down from Scotland and put to work. It arrived in 1773 and building works were completed by the autumn of 1774.[40] The tail race water from the mill was intercepted by a new culvert that led back to the 'great bank'. Here the water was pumped up a vertical tube about 24ft (7.3m) high into a new section of channel that connected to the existing canal on the front of the 'principal' building. The returned water could then enter the culvert that led under this building to the water mill and be re-used to turn the wheel. A tall engine house was constructed against the 'great bank' (Figure 5.4). This added relatively little to the total area of buildings, but must have entailed considerable effort in excavation work and expense in installing the first working James Watt steam engine in the world. The position of the steam engine house is known from the 1788 plan (Figure 5.2). If this is compared to the c. 1768 view of the Manufactory from the rear, it appears that the long narrow building with three chimneys on the extreme right-hand side of the engraving and shown touching the corner of the three-storey end terrace of the 'Brook Row' did not exist in 1788. There are two possible explanations for this: first that the building was demolished before 1788. If this did take place, then the reason for its removal is unknown, since it was not in the way of the engine house and the culvert leading to it. Second, it is possible that it never existed and

---

[38]   George Birkbeck Hill (ed.), *Boswell's Life of Doctor Johnson* (Oxford, 1984), 2, p. 459.

[39]   Morgan, p. 255.

[40]   BA&H, MS 3782/1/23/13; MS 3782/16/3 f. 151.

only reflected Boulton's intention to build in this location. From an aesthetic point of view, it neatly enclosed this section of the perspective.

The pumping engine house came at the beginning of a new era of expansion at the Manufactory, which nearly doubled the area of its buildings over the next 30 years. South–eastwards, between the 'great bank' and the Hockley Brook, was the most obvious and practical direction for any extension. The premises of Boulton and Watt were initially modest, a forging shop being constructed in August and September of 1775, a few months after the partnership was created on 1 June.[41] Initially, the firm did not manufacture steam engines but sold licences to customers, allowing them to erect the engine using their own suppliers for parts. Boulton and Watt provided drawings and engineers to help assemble the engine on site and ensure that it worked. The workshop was needed to provide the more intricate specialist parts.

The earliest complete inventory of rooms in the Manufactory survives from this period and includes the Boulton and Watt forging shop. In November 1777, 128 rooms were listed having a total area on all floors of 32,039¾ sq. ft (2,976.5 sq. m).[42] Although this was one of the largest manufactories in the Birmingham region, it would have been dwarfed by the vast production sheds of factories today. Despite the existence of this inventory, it is frustrating that the location of most of the rooms and workshops cannot be identified. There is no large-scale plan of all floors which could help with the interpretation of the inventory, deepening our understanding of how the Manufactory was organised internally and the functional connections between the various workshops. We know that most of the original 'toy' trade workshops (e. g., buttons, buckles, watch chains and boxes) were located in the original lower Manufactory buildings and plated and silverware shops in the 'principal' building which overlooked them. But this division is complicated as the 'principal' building was also designed to accommodate warehousing and offices for all of the trades.[43]

The Boulton and Fothergill partnership was dissolved a few days after Fothergill's death in June 1782. It had ended acrimoniously with Fothergill insolvent and embittered.[44] One important consequence of these events was that the Manufactory buildings were transferred into the personal estate of Matthew Boulton. From now on he charged each of the Soho businesses rent for their buildings and was responsible for their maintenance. He also collected the rent from the houses on the Manufactory site.[45]

---

[41]   BA&H, MS 3782/12/76/110.

[42]   BA&H, MS 3782/13/126.

[43]   BA&H, MS 378/12/102/6, Particulars of the Houses & Workshops & Mills of Soho; Room no. 15 – 'Toy' Room, Room nos 2–3 – Button Warehouse, Room nos 4–5 – Nelson's Warehouse, Room no. 6 – Mr Scale's Room.

[44]   Dickinson, pp. 110–111; BA&H, MS 3782/12/5/304, MB to [?], 21 June 1782.

[45]   BA&H, MS 3782/12/102/5 Dimensions of the Soho Shops and Buildings; MS 3782/1/5 f. 73; MS 3782/13/126.

After the Boulton and Fothergill partnership was dissolved, Boulton divided the firm into two, entering into partnership with John Scale to manage the former 'toy' business giving it the name Matthew Boulton & Scale usually shortened to MB & Scale. The silver and Sheffield Plate business was named Matthew Boulton & Plate Co (MB & Plate Co). This branch also produced considerable quantities of ormolu from about 1768, although it suffered a terminal decline by 1782.[46] These two firms continued to occupy their respective parts of the Manufactory, following the basic chronology of their establishment at the Soho site. An inventory of about 1789 lists not only the rooms and their sizes but also the firms to which they belonged, using the abbreviations 'B. Co' (the Button Company) and 'P. Co' (the Plate Company) as they were conveniently called.[47] The former possessed the larger area with just over 14,000 sq. ft (1,300 sq. m) in 55 rooms whilst the latter's workshops measured just over 10,500 sq. ft (975.5 sq. m) in 31 rooms, altogether 24,500 sq. ft (2,275.5 sq. m). The total Manufactory area of just over 32,000 sq. ft (2.973 sq. m), recorded in 1777,[48] was made up by 5,400 sq. ft (502 sq. m) of Boulton and Watt premises and the balance, just over 2,000 sq. ft (185.8 sq. m) provided by the lapping and rolling mill and pumping engine house.

In 1785, the water mill had been completely rebuilt and reorganised to take on a significant metal rolling capacity. The old lapping mill was demolished and a larger building set out at right angles to the original plan, attached at one end to the rear of 'Pool/Rolling Mill Row' (figures 5.3 and 5.4). The water wheel remained in its original 1761 position, although it was replaced with one of larger diameter (20ft/6.1m). In the new arrangement it powered metal rolls on one side and laps on the other side of the mill, all under one roof. The position of the new mill entirely altered the original pattern of yards and blocked the original high arched entrance into them from the exterior. This can best be seen by comparing the c. 1768 engravings with the 1788 plan (figures 5.1 and 5.2).

Three years later in 1788 Matthew Boulton took up another grand and innovative scheme – the powered minting of coins. He added a rotary engine to the lapping end of the mill to free his water wheel to flatten copper alone. The lap engine house projected into the large yard between the mill and a two-storey range of forging shops that stretched from the 'great bank' to the rear of 'Brook Row'. Only a small passage through this building provided access into the next yard where the main entrance had been established from the Birmingham side across a bridge crossing the Hockley Brook. The yard was dominated by the Watt pumping engine, by this time much modified and known affectionately as 'Old Bess'.[49] Sometime in the late 1780s a steel or hardening furnace with a tall, tapering chimney was constructed in the same yard. It can be seen in two later views of the

---

[46]   Goodison, pp. 61–62.

[47]   BA&H, MS 3782/12/102/6.

[48]   BA&H, MS 3782/13/126.

[49]   H.W. Dickinson and Rhys Jenkins, *James Watt and the Steam Engine* (1927), p. 123 and Fig. XXII. The engine is on display at the Science Museum, London.

rear and in the author's perspective (Figure 5.4).[50] The next court south-eastwards, also reached via the Hockley Brook bridge, formed the main focus of the rapidly developing Boulton and Watt businesses.

By 1790, the firm of Boulton and Watt had considerably extended their operations at Soho, by building a fitting shop (1781) and an engine house (1782) where rotative engines were tested for the first time. From 1780 they had also diversified into the production of letter copying machines which Watt had invented in 1779.[51] The result was the creation of another enclosed yard. It was bounded by the copying machine workshops built into the 'great bank', the original forging shop on the south-east and the fitting shop and engine house lying parallel to the Hockley Brook (Figure 5.3).[52] The layout of the engine works premises is much better understood owing to the survival of a considerable number of plans in the separate Boulton and Watt archive.[53] The company's policy was to keep copies of its engine drawings using the Watt copying device and a drawing office constructed in about 1790 aided their preservation.[54]

It is beyond the scope of this article to provide a detailed analysis of the further expansion of the engine works, but its most rapid phase took place between the years 1801 and 1804 under the direction of Matthew Robinson Boulton.[55] By this large extension the layout of the Manufactory broke free of the constraint of the Hockley Brook for the first and only time. The new engine works was established in the parish of Birmingham on former heath land that had been enclosed and purchased by Boulton (Figure 5.3). This was achieved by culverting the brook through the new yard. From the time that the two perspectives were drawn (1768), the Manufactory had grown by nearly twice in area, mostly through the initiative of the Boulton and Watt firm, the most commercially successful of the Soho enterprises, judging by the longevity of the engine business.[56] With the completion of the 'Birmingham' engine yard the Manufactory reached its maximum size and no significant change to the footprint of the buildings subsequently took place.[57]

---

[50]   Birmingham Museum and Art Gallery, John Phillp Album, ' View of the SOHO MANUFACTORY taken from Birmingham Heath, February 1796; BA&H, MS 3069, 'View of BROOK ROW from the MANUFACTORY [1801]; Demidowicz, 'A Walking Tour', Fig. 79.

[51]   Dickinson, pp. 107–109.

[52]   BA&H, MS 3782/1/4 f. 467.

[53]   BA&H, MS 3147/5/1447.

[54]   BA&H, MS 3147/9/8.

[55]   George Demidowicz, 'Power at the Soho Manufactory and Mint', in Malcolm Dick (ed.), *Matthew Boulton: a Revolutionary Player* (Studley, 2009), pp. 116–131, 124–125; Demidowicz, 'A Walking Tour', pp. 103–104.

[56]   George Demidowicz, *The Soho Foundry Smethwick West Midlands: A Documentary and Archaeological Study* (2002), pp. 121–123 for Sandwell Metropolitan District Council; Laurence Ince, 'The Soho Engine Works 1796–1895', *Stationary Power*, 16 (2000).

[57]   The present author's perspective was drawn to show the buildings in about 1805.

Mention must be made of two separate buildings – the Soho Mint and the Crescent or Latchet Building. It is again beyond the scope of this article to deal with them in any detail as they were constructed many years after the Manufactory was founded (figures 5.3 and 5.4).[58] The Soho Mint was established in 1788 as the first steam engine powered mint in the world, whilst the Latchet building was constructed in 1794–1795 and extended in 1798–1799 to manufacture a patent flexible buckle. Both buildings reflect Boulton's unceasing quest for new schemes. The mint presented the perfect opportunity to couple the *dulce* with the *utile*: beautiful metal discs impressed by dies designed by artists and disgorged from presses at rapid rates powered by steam engines. The latchet, however, appears to have been one of Boulton's blunders – an uninspiring shoe fastener in a style that was to rapidly go out of fashion so that by 1806 production had entirely ceased.

This chapter has concentrated on the main Manufactory complex, attempting to explain its development and basic layout in the hope that those who study the extant plans can more easily interpret the intricate and confusing patterns of buildings and courtyards. Not surprisingly, Matthew Boulton's personality has

Figure 5.5   Engraving of the Soho Manufactory, *Monthly Magazine*,
                No. 17 (1796)

Source: Courtesy of Birmingham Archives and Heritage.

---

[58]   See also Demidowicz, 'Power at the Soho Manufactory' (2009), 123–124, 129–131; Demidowicz, 'A Walking Tour', pp. 104–105; Richard Doty, *The Soho Mint and the Industrialisation of Money* (London, 1998).

figured prominently in the analysis. His many skills and talents are recognised and lauded, but his unquenchable desire to erect manufactories of all sorts, equip them and staff them is often overlooked. Boulton was a builder *par excellence* and without his drive, intellectual curiosity and pursuit of novelty, the Soho Manufactory would never have been founded and evolved into the fascinating complex of buildings so tragically demolished between 1853 and 1863.

Chapter 6

Boulton, Watt and Wilkinson: The Birth of the Improved Steam Engine

Jim Andrew

On 20 April 1778, *Aris's Birmingham Gazette* published a letter, to the Birmingham Canal Company, in which it was stated 'affords irrefragable Proof of the great Utility of the new-invented Steam Engine, lately erected on the said canal, under the immediate Direction of Mess. Boulton and Watt, the Patentees'.[1] This beam engine, which like all early steam engines was for pumping water, was ordered after the canal company committee requested its chairman 'to have some conversation with Mr. Boulton about the erecting of a fire engine ... to raise the water ... into the summit'.[2] The newspaper passage refers to a test carried out on a small canal pumping engine by John Smeaton, the famous eighteenth-century engineer.[3] Smeaton had himself spent much time improving the earlier type of engine, invented by Thomas Newcomen, but without achieving the performance of even the earliest of the Boulton and Watt engines.[4]

Few history books covering the eighteenth century, and none concerning the industrial history of that century, fail to mention James Watt's improvement to the steam engine.[5] It is usually stated that the engine gave three times the efficiency, meaning it used a third of the fuel, of the earlier engine in doing the same task, and the inefficiency of the Newcomen engine is usually explained as energy lost in heating and cooling the steam cylinder every working stroke.[6]

---

[1]   H.W. Dickinson, and Rhys Jenkins, *James Watt and the Steam Engine* (Oxford, 1927), p. 136. *The Concise Oxford Dictionary* defines 'irrefragable' as indisputable or unanswerable.

[2]   The National Archives, Public Record Office, Kew, Rail 810/4, Birmingham Canal Navigations Company, Minute Book 1775 to 1784, 23 August 1776, p. 6.

[3]   The results of this test were reproduced in John Farey, *A Treatise on the Steam Engine* (London, 1827, reprinted Newton Abbot, 1971), p. 338, but with a transcription error in the performance quoted which should be 13,976,220 lbs raised 1 foot.

[4]   L.T.C. Rolt and J.S. Allen, *The Steam Engine of Thomas Newcomen* (Hartington, 1977), pp. 113–117, and R.L. Hills, *Power from Steam* (Cambridge, 1989), p. 5.

[5]   R.J. Forbes and E.J. Dijksterhuis, *A History of Science and Technology* (Harmondsworth, 1963), p. 351, and R.A. Buchanan, *Industrial Archaeology in Britain* (Harmondsworth, 1972), pp. 253–254, are good examples.

[6]   Ibid.

Any further explanation often mentions condensing steam to produce a vacuum in these engines but the popular perception is that dropping the temperature of steam below 100°C creates something like a perfect vacuum, zero pressure.[7] In fact only a partial vacuum was created in these early engines and a very partial one in the case of Newcomen engines. There is seldom any mention of Watt's engine giving considerably more power than the earlier design, but due credit is given to Matthew Boulton for the commercial success of the engine and to John Wilkinson for producing accurately bored steam cylinders for the engines.[8]

In this chapter, I will explain just why Boulton and Watt's engines were such an improvement over the earlier design, and why they offered so much more scope for development when other aspects of steam technology, such as boiler materials and design, were improved. I will argue that the Boulton and Watt engine paved the way for the exploitation of steam power through the nineteenth century and well on into the twentieth century. Details of some of Matthew Boulton's many engineering contributions to the development of the engine's design from experimental engine to fully developed water pumping engine will also be explored. The science of thermodynamics, developed a century after Boulton's time, will be used in a very simple form to look at the efficiency of these two types of engine. This and other explanations help us to appreciate and understand the brilliance of these engineers who achieved such significant improvements despite their limited understanding of how power was produced.

The Two Designs of Engine

The fundamental difference between the Newcomen and Watt engines was where the steam was condensed to form the 'vacuum' which created their power. In the Newcomen engine, a water spray condensed the steam in the steam cylinder while in the Watt engine the steam was condensed, by a similar water spray, in a separate vessel connected to the steam cylinder, which could then remain hot all the time. The popular explanation of the improved efficiency of Watt's design, that the Newcomen engine wasted heat in heating and cooling its steam cylinder every stroke, does not stand up to critical examination.[9] The steam cylinders of these engines were large and weighed between two and six tons each, while a

---

[7]   In some 30 years of speaking to the public about how early steam engines worked, the author has regularly met people who were convinced that condensing steam meant a perfect vacuum.

    [8]   Hills, *Power*, pp. 53–59.

    [9]   The author recalls this popular explanation being disputed by Donald Payne at a meeting of the Newcomen Society in Birmingham in February 2006. Payne helped assemble and then operate the Black Country Living Museum's replica Newcomen engine and had taken measurements of temperatures and quantities of injection water as well as closely observing the engine while it was running.

typical engine worked at eight to twelve strokes per minute – it does seem unlikely that this weight of metal could be heated and cooled that many times a minute, although it would clearly have fluctuated in temperature to some extent, so there must be another explanation.

The first point to consider is the relationship between the temperature and pressure of the steam within these engines, and here one is talking of what is termed 'wet steam' because both the Newcomen engine cylinder and Watt's separate condenser always contained some condensed water within those spaces. This water would be at much the same temperature as the steam within the cylinder/condenser. Figure 6.1 is a graph of the boiling point of water at various pressures with some of those pressures identified; for example, the boiling point

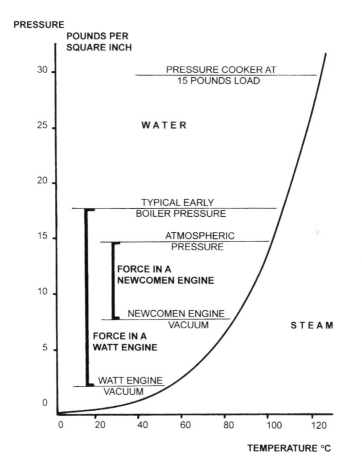

Figure 6.1   The relationship between water and steam with varying temperatures and pressures

Source: Drawing by Jim Andrew.

of water at atmospheric pressure is 100°C, and a pressure cooker with a 15 pound load (meaning that it blows off at a pressure of 15 pounds per square inch above atmospheric pressure) cooks at about 122°C. With wet steam, the pressure and temperature are always on this curved line because just cooling the steam will condense some of it and lower the pressure to match the new temperature while just lowering the pressure will let some of the hot water boil, absorbing latent heat and so lowering the temperature.[10] Similarly, raising the temperature will boil some water and raise the pressure while raising the pressure causes some steam to condense which releases latent heat and raises the temperature.

The pressures found in early steam engines are also marked on Figure 6.1, and these were controlled by the temperatures in the engine, but understanding these means looking more closely at how these engines worked.[11] Both Newcomen's and Watt's engines which we are considering were 'single acting' mine pumping engines. In this layout of engine, the steam piston was driven down by pressure, above the piston, and then lifted up because the pump piston and rods, at the other end of the beam, were weighted to return the steam piston to the top of its stroke. The steam piston rod and the pump's rods were connected to the engine's beam by chains.

The Newcomen Engine

Early steam engines used crude boilers which were only safe at a little above atmospheric pressure, in other words they were little more than large kettles.[12] Letting steam at just above atmospheric pressure from the boiler through the steam valve into a Newcomen engine's cylinder (Figure 6.2), would fill the cylinder as the piston was drawn up while blowing most of the condensed water out down the eduction or exhaust pipe. Closing the steam valve and spraying water into the cylinder condensed some of the steam to produce a partial vacuum and atmospheric pressure acting above the piston drove it down, raising the pump rods and doing work. When the piston reached the bottom of its stroke, the valve gear opened the steam valve again. Steam from the boiler then rushed into the partial vacuum and the weight of the pump rods raised the piston back up again to the top of its stroke while steam filled the cylinder. The steam would once again blow water out through the eduction pipe which ended in a non-return valve immersed in a tank of water.

---

[10]   *Chambers Technical Dictionary* defines 'latent heat' as 'the heat which is required to change the state of a substance from … liquid to gas without change in temperature'.

[11]   Contemporary pressure gauges could not monitor the fluctuating pressures in the engines but many records of condensate temperatures from both Newcomen and Watt engines exist in the notes of engine tests carried out by Boulton and Watt, held in the Archives of Soho, Birmingham Archives and Heritage (BA&H).

[12]   Dickinson and Jenkins, p. 240.

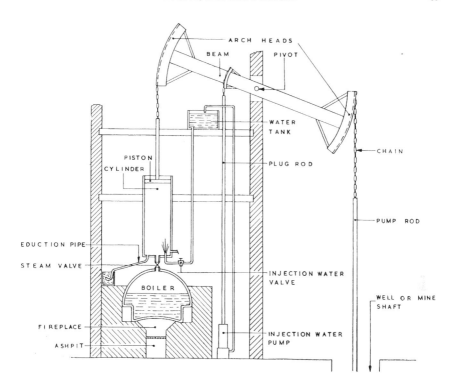

Figure 6.2   A typical Newcomen steam engine

Note: The valve gear is not shown.

Source: Drawing by Jim Andrew based on an original by W.K.V. Gale, in W.K.V. Gale and W.A. Searby, *Boulton, Watt and the Soho Undertaking* (Birmingham, 1952).

Engineers soon found that, in most Newcomen engines, cooling the steam in the cylinder below about 80°C took too long, and the engine then needed far more steam to refill the cylinder. This slower running of the engine produced less work so the engine's characteristics were set to suit the partial pressure obtained at 80°C.[13] Just cooling the steam to 80°C meant that the pressure in the cylinder only fell to about 7½ pounds per square inch and, as atmospheric pressure is about 14½ pounds per square inch, there was only some seven pounds per square inch difference in pressure between above and below the piston. The partial vacuum would be seriously affected by any air in the cylinder, either from leakage or released from the water in the boiler. This air was blown out through a non-return valve in the side of the bottom of the steam cylinder, using the slight pulse of

---

[13]   Power is the product of the forces in the engine and its speed of operation so slowing up an engine can easily reduce power even if the forces are rather higher.

pressure as the steam valve was opened, before the piston began to be drawn up. This valve made a snorting noise and was known as the snifting valve.[14]

The Watt Engine

About 50 years after the completion of the first recorded Newcomen Engine, a young instrument maker at Glasgow University was asked to repair a model Newcomen engine used in demonstrations.[15] The model performed rather badly and the instrument maker, Watt, was puzzled that it used so much steam when it did not even do any real work.[16] We now know that reducing the size of a Newcomen engine would exaggerate its inefficiency to the point where it could hardly run. If Watt had been studying a full-size engine, he might never have queried its inefficiency, although engineers had been worried about this for many years. After months of puzzling over this problem, Watt had the inspirational thought that the steam cylinder needed to be kept hot all the time while the steam needed to be condensed in something cold and this could be a separate vessel.[17] The two could be connected by a valve which opened at the right time to let the steam rush into the cold vessel where it would then be condensed, thus maintaining the vacuum. This was his famous separate condenser.

It took 10 years and a move to join Matthew Boulton in Birmingham to progress Watt's design from a simple little test rig to a fully working and full size engine.[18] Along the way, Watt found that better machining would be needed to make a seal between the piston and cylinder, something which was to be achieved by John Wilkinson in the Midlands. At this time, most steam engines used a packing of loose twisted hemp rope wrapped round the piston, soaked in some form of grease, to seal against the sides of the cylinder.[19] In the Newcomen engine, it was possible to overcome the poor machining of the piston and cylinder by having water lying on top of the piston which then kept the packing flexible and, by leaking through the packing, prevented air from entering the cylinder. This would not work at the higher temperature of the piston and cylinder of the Watt engine. Indeed, to keep the cylinder as hot as possible, Watt enclosed its top and used steam from the boiler to force the piston down, and he needed to stop this steam leaking past the piston and reducing the working vacuum. In many cases, he proposed surrounding

---

[14]   A full explanation of the operation of the Newcomen engine can be found in Rolt and Allen, pp. 89–95.

[15]   R.L. Hills, *James Watt, vol. 1: His Time in Scotland, 1736–1774* (Ashbourne, 2002), pp. 312–315.

[16]   Ibid., p. 315.

[17]   Hills, *Watt, vol. 1*, pp. 329–340.

[18]   R.L. Hills, *James Watt, vol. 2: The Years of Toil, 1775–1785* (Ashbourne, 2005), p. 56.

[19]   Dickinson and Jenkins, p. 389.

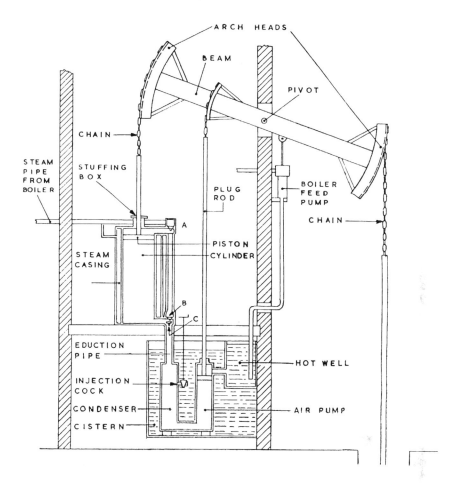

Figure 6.3   A typical Watt steam engine
Note: The valve gear is not shown.
Source: Drawing by Jim Andrew based on an original by W.K.V. Gale, in W.K.V. Gale and
W.A. Searby, *Boulton, Watt and the Soho Undertaking* (Birmingham, 1952).

the cylinder with a steam-filled jacket to ensure it remained as hot as possible.[20]
The vacuum maintained in the condenser meant that condensed water and air now
had to be pumped out rather than run off as in the Newcomen engine.

In its developed form (Figure 6.3), the Watt pumping engine let steam from the
boiler into the space above the piston, in the steam cylinder, and the steam pressure
forced the piston down against the vacuum already existing below the piston. As
the piston approached the bottom of its stroke, the steam admission valve and the

---

[20]   Hills, *Watt, vol. 1*, p. 364.

exhaust valve (A and C in Figure 6.3) were closed and the 'equilibrium valve' (B in Figure 6.3) opened which allowed the steam to be transferred from above to below the piston as it rose under the weight of the pump piston and pump rods. As the piston approached the top of its stroke, the equilibrium valve (B in Figure 6.3) closed, the steam admission valve (A in Figure 6.3) opened, and the exhaust valve (C in Figure 6.3) opened to allow the steam below the piston out into the cold condenser where it was condensed by a water spray and thus a vacuum was created below the piston. The air pump removed air and condensate from the condenser and some of this condensate was fed as hot water to the boiler.[21]

Watt's engines still used steam at only just above atmospheric pressure because boiler technology had not progressed very far, but applying this pressure to the top of the piston, in place of the air at atmospheric pressure in the Newcomen design, gave a small increase in the pressure on the piston. The vacuum, however, was now far better because the separate condenser, sitting in a tank of cold water and with a water spray, could usually operate at something below 40°C, which gave a pressure of about a pound per square inch so that the pressure difference between above and below the piston was now some 15 pounds per square inch, over twice the pressure difference in the Newcomen engine. Since both types of engine worked at about the same speed, this meant that the Watt engine immediately gave more than twice the power of a Newcomen engine with the same size cylinder. A better explanation is that a Watt engine only needed a cylinder of two-thirds the diameter of a Newcomen engine's cylinder to produce the same power: quite a saving in the cost of construction.

Watt was soon able to carry out tests using a model engine which demonstrated that his design used much less steam than the earlier one and was, therefore, more efficient, but he began having difficulties in scaling his design up from a model to a full size engine.[22] The major problem was that, at that time, the machining of the pistons, cylinders and valve gear of steam engines was very crude compared with the accuracy of the much smaller items produced by instrument makers.[23] This poor quality machining was not too serious in the Newcomen design with its water seal helping the piston's rope packing and with the lower pressure difference across the piston, but in the Watt design, this simply introduced too many problems for satisfactory running.[24]

Initially, Watt's development work was funded by John Roebuck whose business empire in Scotland included the Carron ironfoundry and collieries. Roebuck was interested in pumping engines for his collieries and for sale.[25] The Carron ironfoundry was casting and machining cylinders for Newcomen engines using a boring machine designed by John Smeaton where a long boring bar was

[21]   Dickinson and Jenkins, pp. 393–395.
[22]   Hills, *Watt, vol. 1*, p. 378.
[23]   L.T.C. Rolt, *Tools for the Job* (London, 1965), p. 52.
[24]   Ibid., pp. 50–53.
[25]   Dickinson and Jenkins, pp. 24–34.

rotated by a waterwheel, and the cylinder was mounted on a trolley running on rails so that it was drawn over the rotating cutter on the end of the boring bar.[26] The cutter tended to follow imperfections in the casting and, because of its length, often drooped giving a bore which might be round but was certainly not cylindrical.[27] When the Carron ironfoundry produced an 18-inch diameter cylinder for Watt's first experimental engine, they were unable to machine it accurately enough to allow proper sealing between the piston and the cylinder.[28] Watt tried many different flexible compounds to try to seal the piston in the cylinder but none was satisfactory.[29] Watt even considered mimicking the liquid seal of the Newcomen engine, but as water would simply have boiled away, he tried various oils without success.[30] He then considered using mercury or one of the low melting-point alloys, such as Newton's metal[31] which melts at 95°C, with a small pump to recirculate it from the bottom of the cylinder back to above the piston.[32] There is no sign that he tried this approach. He then fabricated a cylinder from sheet tin which gave a smooth surface but it could not be made sufficiently cylindrical to work for any length of time.[33] Eventually his other activities slowed development of the engine until Roebuck's financial failure allowed Matthew Boulton to take over funding the engine development in 1774.[34]

Improving the Machining of Steam Cylinders

It was Watt's move to Birmingham, where Boulton had both the contacts with industrialists, such as Wilkinson, and the determination to bring the design to satisfactory operation, which ended almost 10 years of unsatisfactory progress. Watt had brought the parts of his experimental engine with the unsatisfactory 18-inch cylinder to Birmingham with him.[35] Wilkinson had a Newcomen cylinder boring machine, similar to the one at the Carron ironfoundry, but he was also producing naval cannon using his patented cannon boring machine which allowed much more accurate machining of the cannon's bore.[36] In this machine, the cannon was rotated in bearings while a substantial non-rotating boring bar, set in suitably

---

[26]   Rolt, pp. 50–53.

[27]   Ibid., p. 51.

[28]   Hills, *Watt, vol. 2*, p. 36 gives the 'out of roundness' as ⅜ inch in 18-inch diameter.

[29]   Dickinson and Jenkins, p. 106.

[30]   Hills, *Watt, vol. 1*, p. 395.

[31]   Newton's metal is an alloy of 18¾ per cent tin, 31¼ per cent lead and 50 per cent bismuth giving a melting point of 95°C.

[32]   Hills, *Watt, vol. 1*, p. 395.

[33]   Ibid.

[34]   Ibid., pp. 446–450.

[35]   Hills, *Watt, vol. 2*, p. 36.

[36]   Rolt, p. 53.

robust mounts, was fed into the cannon to produce a truly cylindrical bore. Watt's experimental engine was of 18-inch bore and five feet stroke so the overall length and flange diameter of the cast cylinder would be within the dimensions of a large cannon's barrel length and its width across the trunnions (the supporting projections on the side of cannons).[37] It is arguable that, with suitably large bearings, the new cast-iron cylinder could be rotated in Wilkinson's cannon boring machine to give a much more accurately bored cylinder. Watt settled in Birmingham in June 1774, and it seems that a new 18-inch cylinder was ordered from Wilkinson well before the end of the year.[38] Early in 1775, Boulton wrote to Watt that he had no news of the new cylinder but would chase Wilkinson for news of it.[39]

One of the first moves, on establishing their partnership in 1775, was for Boulton and Watt to seek an extension to the patent, which Watt had obtained in 1769.[40] This meant seeking an Act of Parliament and, before there was any real progress on perfecting the engine design, Watt had to spend months in London dealing with Parliamentary matters and also time in Scotland winding up his affairs there and, a year later, in marrying his second wife.[41] Thus it was left to Boulton to make progress with the engine in Birmingham. Boulton showed considerable technical skill in his many other ventures yet many history books have overlooked the same skill being contributed to the development of the steam engine.[42] Boulton employed some skilled craftsmen who were able to make any new smaller components needed for the experimental engine. In April 1775, Boulton wrote to Watt about the fine new cylinder provided by John Wilkinson. He reported that it was tolerably true but of slightly larger diameter than the existing tin one so he had instructed Joseph Harrison, a fitter, to make a brass hoop to modify the piston.[43] In May, Boulton wrote that the engine 'goes marvellously bad' and he was testing

---

[37]   Watt's experimental engine was 18-inch bore by five feet stroke, giving a diameter of the end flanges of about 26 inches, to which can be added another three inches for the steam/exhaust connection. The length would be somewhat under 7 feet. Large naval cannon of that time were regularly 10 feet long but the width over the trunnions, the cylindrical pivots projecting from each side, is seldom recorded. Using data in Peter Padfield, *Guns at Sea* (London, 1973) and then scaling from drawings in that book, the width over the trunnions was about 37 inches.

[38]   Hills, *Watt, vol. 2*, p. 36.

[39]   Ibid., p. 38 refers to a letter of 26 January 1775, cited in J.P. Muirhead, *The Origin and Progress of The Mechanical Inventions of James Watt* (3 vols, London, 1854), vol. 2, p. 80, MB to Watt, 26 January 1775.

[40]   Hills, *Watt, vol. 2*, pp. 38–47.

[41]   Ibid., p. 77.

[42]   It is not unusual to find the myth of Watt portrayed as a hopeless businessman and MB as the sharp entrepreneur but virtually technically inept. Taken from a comment by Jennifer Tann, 'Boulton and Watt's Organisation of Steam Engine Production before the opening of Soho Foundry', *Transactions of the Newcomen Society*, 49 (1977–78), pp. 41–53, note 1, p. 51. Neither description stands examination.

[43]   Letter from MB to Watt, 24 April 1775, cited in Muirhead, p. 84.

parts of the engine, trying them for friction and directing his staff to take action to improve things.[44] Later that month, he wrote again of problems with components not fitting properly and what he had done to correct them, which had resulted in good sealing, but his tests still showed excessive friction between the piston and cylinder.[45] Further in this same letter, he commented on his tests of other components including the water pump and his efforts to reduce friction between all the engine's components.

The following month he was able to report that the engine was running very well and he sent details of temperatures and pressures found while working the engine.[46] He was able to measure the boiler and condenser pressures and was wondering if they could in any way measure the fluctuating pressure in the cylinder but it would be two decades before Boulton & Watt developed an instrument to record these pressures.[47] Within days this was followed by the first information on coal consumption for the experimental engine which was now running at about nine strokes per minute with no obvious leaks.[48] The testing of the engine now showed great promise, an Act of Parliament was granted extending the patent to 1800 and the partners could move to supplying designs of full size engines for their customers.[49]

In March 1776, the first full-size engine was set to work at Bloomfield Colliery near Tipton.[50] It had a 50 inch diameter cylinder which Watt ordered from Wilkinson in June 1775.[51] This first commercial engine was soon followed by one with a 38 inch cylinder for John Wilkinson's own works.[52] Clearly, Wilkinson had moved on from using his cannon boring machine and it seems he had perfected a much more accurate way of machining full-size cylinders. Details of his boring machine were slow to emerge although other later designs of accurate boring machines were well recorded. In 1919, a drawing was found among the Boulton and Watt papers in Birmingham.[53] This *Drawing of the Bersham Boring Mill* by one John Gilpin, undated but evidently late eighteenth century, shows how Wilkinson probably redesigned his Newcomen engine cylinder boring machine, similar to the Smeaton machine at the Carron ironfoundry. The machine shown was really four separate

[44] Letter from MB to Watt, May 1775, cited in Muirhead, p. 86.

[45] A second letter from MB to Watt in May 1775, cited in Muirhead, pp. 87–89.

[46] Letter from MB Watt, June 1775, cited in Muirhead, p. 90.

[47] This was the engine indicator, see R.L. Hills, *James Watt, vol. 3: Triumph through Adversity* (Ashbourne, 2006), pp. 83–87.

[48] A second letter from MB to Watt in June 1775, cited in Muirhead, p. 91.

[49] Hills, *Watt, vol. 2*, pp. 51–60.

[50] Dickinson and Jenkins, pp. 111–115. Reprints a report, on p. 113, from *Aris's Birmingham Gazette* for 11 March 1776.

[51] Letter from Watt to Wilkinson, 27 June 1775, cited in Muirhead, p. 113.

[52] Hills, *Watt, vol. 2*, pp. 60–62.

[53] E.A. Forward, 'The Early History of the Cylinder Boring Machine', *Transactions of the Newcomen Society*, V (1924–1925), pp. 24–38, p. 26 and Plate V.

machine tools driven by a single waterwheel. One boring machine was still of the old Smeaton design but there were two different sizes of the new more accurate boring machines, together with a large lathe which could well have been needed to machine the components of the boring bar.

This 'Smeaton' machine was altered because the large Watt engine cylinders would be too large to rotate, as in the cannon boring machine, so they were strapped down on the machine bed. While the cannon barrels were only open at the muzzle, engine cylinders were open at both ends and a substantial tubular boring bar could be passed right through the cylinder and be supported in bearings at each end. There was a slot along the length of the boring bar which allowed a cutter holder, machined to be a sliding fit, to be drawn along the bar. The cutter holder carried a number of carefully adjusted cutting tools which produced the accurately machined cylinder bore. These were not small changes to the existing machine but would have taken some time to effect. The speed of Wilkinson's move from the cannon boring machine to this accurate cylinder boring machine shows his grasp of the commercial advantage of working with Boulton and Watt.

Why the Improved Efficiency?

Before looking at the further improvements which Watt was to make to his design of engine, the reason for the initial improvement in efficiency, over the Newcomen design, needs to be considered. It is relatively easy to see why the new engine design was so much more powerful than the earlier design. Power is produced by the pressures in the cylinder and the engine's speed of operation. Thus the greater pressure difference in Watt's design, from condensing the steam to a lower temperature and hence giving lower pressure, immediately increased the power even though the engines worked at similar speed. The degree of efficiency gain is less obvious and much less easy to quantify, but it is possible to consider this in two different ways. One is a plausible popular explanation and the other relies on thermodynamics. Many engineers had wondered if they could reduce the amount of steam used in the Newcomen engine. It is quite possible that some of them realised that in addition to filling the steam cylinder with steam of a high enough pressure to allow the piston to be pulled up to the top of its stroke, a great deal of steam was condensed in the cylinder. The cool walls of the cylinder, which had been exposed to the cold air above the piston, the cool underside of the piston, the condensate lying in the cylinder bottom and the water running down the cylinder walls from the water seal above the piston, all had to be heated by condensing steam, which also provided some heat to the fluctuating temperature of the cylinder. Knowing that lots of steam was being used in this way and causing the inefficiency, however, did not really solve the problem. That would take Watt's innovative design.

Smeaton managed to improve the machining of the cylinder bore, which would reduce the amount of water running down the cylinder walls, applied some insulation to areas such as the underside of the piston and worked to ensure that pipes and valves were of such a size that they did not restrict steam flow in the engine.[54] It must be remembered that, at this time, there were no suitably fast-acting pressure gauges to record the fluctuating pressures within the engine so it was difficult to even monitor what was happening inside the engine. Smeaton's attention to detail allowed him to double the efficiency of the Newcomen engine, but this seems to have been as far as the design could be improved, and few of his customers understood the need to maintain this optimum level of care if they wanted efficient operation.[55]

These many different causes of steam being wasted are closer to the truth than the old idea of just the heating and cooling of the cylinder but do not give a clear indication of the actual improvement which might be expected. Fortunately, the efficiency of these two engine designs can be compared by using the science of thermodynamics which was not developed until a century after Watt's time. In its simplest form this predicts that the absolute maximum efficiency of a theoretical heat engine was related to the difference between the higher and lower temperatures in the engine system, divided by the higher temperature, on an absolute temperature scale, although the complete theory is a great deal more complicated than this.[56]

This would not seem to be much use when looking at these early steam engines until it is remembered that Newcomen and early Watt engines were built to similar engineering standards, using similar types of boilers with similar steam temperatures. Thus from a thermodynamic viewpoint, the only really significant difference between them was the temperature range over which they operated. Boulton and Watt recorded the temperatures of the condensate from both types of engine in a variety of tests to determine the saving from using Watt's design.[57] The

---

[54]   Many of the records of Smeaton's work on steam engines were destroyed in a fire at John Farey's house but some copies can be found in BA&H, while a published source, Farey, *A Treatise*, contain sections on the results of his investigations. These indicate that he gained power and efficiency from cooling the steam in the cylinder well beyond that of most Newcomen engines but by methods which were difficult to maintain when used by customers.

[55]   This was a problem which Watt also faced. Engine owners would often pay little attention to maintaining the optimum operating conditions for an engine but complain when its performance deteriorated.

[56]   Many books on thermodynamics give the theoretical maximum efficiency of heat engines. For example: D.B. Spalding and E.H. Cole, *Engineering Thermodynamics* (London, 1958), p. 204.

[57]   Boulton's data on his early tests of the engine, see Muirhead, vol. 2, p. 84, gives one set of useful temperatures but Boulton & Watt's later test results, recorded in notebooks, letters and on data sheets in BA&H, B&WP, show condensate temperatures on Newcomen engines of 70°C to 80°C and Watt engines of 30°C to 40°C. No doubt the steam temperature within the cylinder would be higher than the condensate temperature, so the upper of these figures has been used in this analysis.

boiler temperature can be assumed to be just above 100°C for both engines. The condensate temperatures, the cold aspects of the thermodynamics, were typically found to be about 80°C and 40°C respectively so the two temperature ranges are slightly over 20°C and 60°C. This places their efficiencies in the ratio of one to three, and suggests the Watt engine used about a third of the coal required by a Newcomen engine doing the same work, the accepted figure for the improvement of the early Watt engines. This is all very approximate and anything but an exact interpretation of the thermodynamic theory. Nevertheless it does offer some explanation of the relative efficiencies of these two designs of engine.

Later Developments with Watt's Design

An appreciation of the effect of the temperature range through an engine also helps us to understand the improvements in efficiency to be gained by the steady development of the Watt engine during the life of the patent and then through the nineteenth century. There was another significant gain from the increased efficiency of the Watt engine compared with the Newcomen design. As the Watt engine only used a third of the amount of steam for the same power it could be operated from a smaller boiler, or even fewer boilers when looking at more powerful engines. In many cases the saving in the cost of the steam engine cylinder, piston and boiler, compared to an equivalent Newcomen engine, would cover the additional cost of the separate condenser, air pump and more complicated valve gear. This meant that a Watt engine might not cost any more than a similar power Newcomen engine, built to the same engineering standard. The fee demanded by Boulton and Watt for using the patent was paid out of the saving in coal consumed by the engine, enhancing the advantages of adopting Watt's design of engine.[58] This analysis is based on data for Midland built engines but Anthony Burton has reported one case in Cornwall where the Watt engine which replaced two Newcomen engines only cost the same as one of those Newcomen engines.[59] It may well be that the economic case for the Watt engine was even stronger in areas such as Cornwall where operating and construction costs were different from those in the Midlands.

For several years after entering the engine market, it was Boulton and Watt's practice for one to supervise the business in Birmingham while the other spent time in Cornwall, one of their major markets.[60] Regular letters between them included the results of their efforts to commission engines, maintain existing

---

[58]   J.H. Andrew, 'The Costs of Eighteenth Century Steam Engines', *Transactions of the Newcomen Society*, 66 (1994–1995), pp. 77–91, p. 83.

[59]   A report by a Mr. Dalton in 1779, cited in A. Burton, *Richard Trevithick, Giant of Steam* (London, 2000), p. 22.

[60]   Hills, *Watt, vol. 2*, pp. 96–105.

ones and consider improvements in design.[61] These letters contain many further examples of Boulton's continuing technical contribution to the engine and of his dealing first-hand with the rectification of faults.[62]

Despite the initial success of Watt's production engines, he continued to seek improvements both in the mechanical components such as the valve gear and in the extraction of more energy from the steam. The early working cycle of Watt's engines had the boiler steam pressure applied to the piston almost to the bottom of its stroke but he realised that the steam pressure and the inertia of the moving parts would allow the engine to complete its stroke if the steam valve closed partway down the stroke, so letting less steam into the cylinder. This was termed 'expansive working' and tests on a second Birmingham Canal Company engine in 1779 showed a 14 per cent improvement in efficiency if the steam valve was closed at that point which just allowed the engine to complete its stroke.[63] Such progress could not be achieved with the Newcomen design, therefore this was a further benefit from Watt's separate condenser.

Although James Watt became very concerned about the safety of steam at much above atmospheric pressure, other engineers were more adventurous and sought to exploit the benefits of working engines at higher pressures. Sadly, this development did introduce the potential for boiler explosions; indeed, many lives were to be lost through them, but with improved boiler technology, insurance inspection and legislation, more powerful and efficient engines were built. Increasing the working pressure increased the engines' power; just as the Watt engine was more powerful than the Newcomen design but, in addition, the higher pressure was obtained by using a higher boiler temperature which, under the thermodynamic principles outlined above, meant greater potential efficiency. Many high powered engines continued to use separate condensers to maximise the extracted power from the fuel burnt, and the best of the beam engines in the nineteenth century reached efficiencies five times that of Watt's own designs.

The Legacy

The long term contribution made by Boulton and Watt to steam power is much more widespread than the basic detail given above. There were improvements in design, manufacture, machining and metallurgy as well as the technologies of lubrication, instrumentation and boiler operation. For almost 20 years, Wilkinson was their

---

[61]   For example the many letters from JW to MB in 1781–82, see Muirhead, pp. 129–144.

[62]   For example, in September 1778, MB was assisting with packing grease round the piston of the Chacewater engine; see Hills, *Watt, vol. 2*, p. 106.

[63]   BA&H, B&WP, MS 3147/4/1/ Box 53, JW's Blotting and Calculation book, volume for 1779–1781, p. 9. MS 3147/3/81, Letter Book No. 3, p. 219, Watt to Logan Henderson, 28 June 1779.

preferred supplier of steam cylinders and many other components. It is difficult to know how much longer Boulton and Watt would have relied on Wilkinson for accurate steam cylinders but a dispute between John and William Wilkinson in the early 1790s meant that Boulton and Watt had to resort to other suppliers until they could establish their own steam engine works, the Soho Foundry.[64]

An insight into why the Watt engine was such an improvement on the earlier design and one with massive potential for further improvement gives us all a better understanding of the brilliance of these two engineers in inventing and developing their improved steam engine. The realisation that Boulton made significant contributions to the technology of the engine, while Watt was himself no mean businessman, helps explain the great success of their partnership. The rapid development by Wilkinson of a boring machine, which would provide the accuracy needed for Watt's design to succeed, has long been accepted, but often without understanding that his rapid change from the cannon boring system to the new cylinder boring arrangements was fundamental to the development of the improved steam engine.

---

[64]   Hills, *Watt, vol. 3*, p. 103.

Chapter 7

Matthew Boulton's Copper[1]

Peter Northover, Nick Wilcox

When Matthew Boulton secured his contract for the 1797 regal coinage on 9 June, he was faced with the need to secure very rapidly sufficient copper to coin 480 tons of 1d coins and 20 tons of 2d coins at a specific price if he was going to both fulfil his contract and make any profit.[2] In fact, the requirement was for rather more than 500 tons of copper. The problems he encountered, exacerbated by his falling out with Thomas Williams, have been well and entertainingly described by George Selgin.[3] The origins of the copper he obtained lay mainly with Cornish producers. We have only limited contemporary accounts of the quality and composition of that copper. Later, the United States Mint had cause for complaint with some of the planchets shipped from Birmingham for the production of cents and half cents.[4]

Nevertheless, this sort of information is most important if we are to understand the operation of the Soho Mint. For example, a communication from Boulton recorded in the Minutes of the Privy Council for 28 March 1797, sets out his requirements for the copper and how he would process it:

> I offer to receive the Copper from the Smelting Furnace in its first state which is called <u>Tough Cake Copper</u>: to Roll it hot, to scale and clean it, and then to Roll it cold in polished Rollers exactly to the size required; After which I will cut out the Dubs and Coin them in such Dies as I will produce and shall first be approved by Your Lordships and for the better Preservation of the Coin in Carriage as well as the more easy Distribution of it, I will wrap them in Paper packets of one or two shillings worth in each Packet, and then pack them in

[1] The authors would like to thank Mr Chris Salter for assistance with the microprobe analysis, Prof Ronald Bude of the University of Michigan for the provision of the United States coins, Dr Derek Jenkins for the Boulton copper coins, Mr Tony Morfett for access to the Manx 1733 coinage, and Prof Hubert Emmerig of the University of Vienna for providing the Austro-Hungarian coins. We also thank the Hampshire and Wight Trust for Maritime Archaeology and Mr Kevin Camidge for making copper sheathing and bolts available for analysis.

[2] George Selgin, *Good Money: Birmingham Button makers, the Royal Mint and the Beginnings of Modern Coinage, 1775–1821* (Michigan, 2008), p. 163.

[3] Ibid., pp. 178–184.

[4] Richard Doty, 'Early United States Copper Coinage: the English Connection', *British Numismatic Journal*, 57 (1987), pp. 54–76.

small Casks to the amount of about three hundred Weight each ... I will remelt
the Scissel and stand to all other Expenses incurred by Manufacturing Cake
Copper and Scissel into Coin.

The same Pricy Council minutes also record that the Committee on Coin '... are
therefore of the Opinion that the Coin should be made of good malleable copper
and not of any mixed metals, as is practised in many other Countries of Europe'.[5]

Other problems were undoubtedly caused by the design of the machinery,
a design that was changed fairly soon afterwards in 1801.[6] Yet others were
caused by the stress of striking such large pieces of copper, especially the 60g
two pence pieces. A small amount of metallurgical data has been published on
the 1797 coinage[7] but the further analysis of coins was undertaken to obtain a
more complete picture, and to determine whether the copper hastily acquired in
1797 differed in any way from that used in subsequent contracts. Further 1797
coins were analysed, together with examples from the 1801 and 1806/1807 UK
coinages, and Liberty head cents and half cents from the United States coinage
from 1799 to 1837, when Soho's contracts to supply the United States Mint were
terminated. The coins obtained included one from 1824 when Belles and Harrold
had been supplying the planchets for the cents.[8] To provide a comparison and
contrast to the British copper in Boulton's production, Austro-Hungarian one
Kreuzer coins struck from copper mined in that empire were recently sampled and
analysed. These analyses were undertaken by one of the authors (Nick Wilcox) for
a dissertation for the degree of MEng in the Department of Minerals, University
of Oxford, and the results are published here for the first time.[9]

Sampling and Analysis

2d, 1d and ½d coins from the United Kingdom coinage and 1c and ½c coins
from the United States coinage were purchased for sectioning, while the Austro-
Hungarian 1 Kreuzer coins were provided by an Austrian colleague. In all cases,
a section through the thickness of the coin was cut and then hot-mounted in a
carbon-filled, thermosetting resin, ground and polished to a 1mm diamond finish.
Some of the United Kingdom coins were sampled parallel with the surface as well.
Compositional analysis was by electron probe microanalysis with wavelength

---

[5]   National Archives, Kew (NA), PC1/37/114.

[6]   Richard Doty, *The Soho Mint and the Industrialisation of Money* (London, 1998).

[7]   Peter Northover, 'Copper in the Industrial Age', in Christian Degrigny (ed.), *Metal 07, Vol. 1: When Archaeometry and Conservation Meet* (Amsterdam, 2007), pp. 83–90.

[8]   Richard Doty, 'Early United States Copper Coinage', *British Numismatic Journal*, 57, pp. 54–76.

[9]   Nicholas Wilcox, 'Copper in the Industrial Revolution' (University of Oxford, unpublished MEng thesis, 2009).

dispersive X-ray spectrometry. Five to ten analyses were made on each sample over 30x50mm rasters for 17 elements with detection limits of 100ppm. For a small number of the 1797 coins point analyses were made to examine the partition of key elements between the solid solution copper matrix and oxide inclusions. Microstructures were studied by optical metallography in both polished and etched states, and by scanning electron microscopy with electron backscatter diffraction (EBSD). The metallographic etches used were ammoniacal hydrogen peroxide, an acidified aqueous solution of ferric chloride further diluted with ethanol, and an alkaline solution of cuprammonium chloride. The metallography was complemented by microhardness testing and also tensile testing of contemporary and closely similar copper from wooden warships.

British Copper

The mintage of all the United Kingdom regal copper produced at Soho is listed in Table 7.1, using data published by Richard Doty.[10]

Table 7.1   Mintage of United Kingdom regal coinage struck at Soho

Date	1/4d	1/2d	1d	2d
1797–1799			43,969,204	722,180
1799–1800	4,224,000	42,480,000		
1806	4.833,768	87,893,526	19,355,480	
1807	1,075,200	41,394,384	11,290,168	

Clearly, a statistical picture is virtually impossible to establish as a result of the analysis of only a handful of coins out of so many millions issued. Nonetheless, the results of the United Kingdom coins over the decade from 1797 present a consistent and interesting picture. Table 7.2 presents the mean compositions of the different issues and also the mean compositions of copper bolts and sheathing of Royal Navy vessels, from *HMS Colossus* built in 1787 and lost in 1798, and *HMS Pomone*, built in 1805 and lost in 1811.

All the copper has the same broad characteristics: the principal impurities are arsenic, bismuth, lead, silver and, usually, nickel. As can be seen from Table 7.2, nickel, antimony, and silver contents are rather uniform, while arsenic, bismuth and lead are more variable. The reason for this lies in the smelting process used in British copper works at the end of the eighteenth century, as will be explained in a little more detail below. Of particular note is the bismuth content with maximum

---

[10]   Doty, *Soho Mint*, pp. 318–331.

Table 7.2   A comparison of mean compositions of coinage and ships' copper

Analyses	Date	Object	Ni	Cu	As	Ag	Bi	Pb
5	1797	1d	0.05	98.19	1.10	0.09	0.40	0.08
5	1797	2d	0.05	98.09	1.13	0.09	0.46	0.09
6	1799	1/4d, 1/2d	0.06	98.78	0.74	0.10	0.18	0.09
7	1806	1/2d, 1d	0.05	98.98	0.56	0.09	0.23	0.03
5	1807	1/2d	0.06	98.02	1.28	0.09	0.44	0.03
7	1787–1811	Bolts	0.05	98.48	0.78	0.09	0.21	0.27
5	1787–1811	Sheathing	0.09	99.17	0.42	0.06	0.05	0.08

mean values of 0.40–0.46 per cent in the coinage with individual values almost reaching 1 per cent.

Bismuth is well-known as the cause of embrittlement in copper through the formation of bismuth films in grain boundaries,[11] and bismuth concentrations such as these should leave the copper unusable. However, careful metallography and microanalysis using point analyses has shown that, within the limits of detection of electron probe microanalysis, the bismuth is entirely present as oxide inclusions and so does not affect the mechanical properties of the copper. This was demonstrated when sections of copper bolts from HMS Impregnable (built in 1786 and wrecked in 1799) were tensile tested. Associated with bismuth in the oxide inclusions in the majority of the samples were some of the arsenic and, probably, some of the lead, although this last could not be proved.

Next we must consider the variation of the different elements in the different issues of the regal coinage. In the 1797 dated 1d and 2d coins, the mean arsenic contents are effectively the same (1.1 per cent). This observation has several implications: the 1d and 2d coins dated 1797 were minted from 1797 through 1798 and into 1799, although in very different quantities, and this argues for some consistency of composition in the copper supply. A similar level of arsenic was observed in the 1807 ½d coins but much lower levels (0.74 per cent for the 1799 ¼d and 0.56 per cent for the 1806 ½d and 1d). For a possible interpretation of these phenomena we must turn to the naval copper.

The mean arsenic content of the copper bolts studied was 0.78 per cent and that of the sheathing 0.42 per cent. It is presumed, unless there is an extensive history of repair in a vessel, that the copper bolts used in the hull were part of its

---

[11]   D. Hanson and G.W. Ford, 'Investigation into the Effects of Impurities on Copper, Part 5: The Effect of Bismuth on Copper', *Journal of the Institute of Metals*, 37 (1927), pp. 169–180. B.D. Powell and H. Mykura, 'The segregation of bismuth to grain boundaries in copper-bismuth alloys', *Acta Metallurgica*, 21 (1973), pp. 1151–1156.

original structure when built, and were likely to have been supplied new from the Navy's copper contractors. In contrast, sheathing was renewed regularly, perhaps every three or four years and the dockyards became very efficient at recycling it. When copper of this type is re-melted, unless the conditions are strongly reducing, elements such as arsenic and lead will be lost to both the vapour and oxide phases, while elements already present as oxides, arsenic and lead again, as well as bismuth, will be skimmed off as dross. When the copper solidifies and is processed into new products, such as replacement sheathing, the arsenic lead and bismuth contents will all be decreased and this is seen in the ships' copper with the lead (0.27 per cent vs 0.08 per cent) and bismuth (0.21 per cent vs 0.05 per cent) as well as the arsenic. On this basis, it is reasonable to suggest that the lower arsenic and bismuth contents in the 1799 ¼d and ½d and 1806 ½d and 1d coins could be due to a higher level of remelted scissel being used. For the 1799 coinage this might well be from the 1797–1799 production of 1d and 2d coins. Given the pressure on the Mint to get, at least, the first 500 tons into circulation there may not have been much time for the careful housekeeping Boulton had promised. Also, since he started production of the ½d coins without a contract he would perhaps be more likely to use metal stocks to hand rather than risk money on further new purchases of copper.

The United States Copper

Twenty-two worn United States copper cents, of which 19 had legible dates, and three half-cents, were sectioned and analysed; the date range was 1803–1837. The analyses are presented in Table 7.3. For the majority of the coins, the compositions are very similar to those of the British coins analysed, and there is no clear trend with time; no doubt a consequence of the analysis of so small a sample from so large a production. Particular features are the generally low level of lead, and the continued significance of bismuth as an impurity. This extension of the time-line to 1837 amply demonstrates the consistency of the copper supplies that Soho could achieve. There are a small number of exceptions; for example, increased lead contents in 1803 and 1827–28. Above we proposed that there was a tendency for Anglesey copper to have a higher lead content than the Cornish. By 1827–28 though, Anglesey had been in severe decline for a long time as estimated production figures show.[12] A clue that some of this copper might still be from the south-west can be seen in the 0.15 per cent tin in the 1828 cent. Meanwhile, the 1828 half cent analysed also had a slightly elevated tin concentration at 0.04 per cent, overall a measure of the periodic fluctuations in the copper supply.

---

[12]   John C. Symons, 'The mining and smelting of copper in England and Wales, 1760–1820' (Coventry University in association with University College Worcester, unpublished PhD thesis, 2003), Appendix 2.

Matthew Boulton

Table 7.3   Composition of United States copper coinage, 1803–1837

Coin	Ni	Cu	As	Ag	Bi	Pb
1803 cent	0.03	97.91	0.98	0.07	0.58	0.29
1803 half cent	0.06	99.08	0.42	0.07	0.25	0.01
1807 cent	0.04	98.88	0.58	0.08	0.27	0.04
1808 cent	0.04	98.73	0.20	0.10	0.29	0.03
1816 cent	0.07	98.76	0.51	0.11	0.43	0.03
1816 cent	0.04	98.51	0.83	0.07	0.40	0.03
1816 cent	0.05	98.92	0.47	0.08	0.29	0.10
1818 cent	0.05	99.05	0.54	0.12	0.14	0.02
1818 cent	0.05	98.58	0.86	0.07	0.28	0.02
1818 cent	0.06	98.70	0.51	0.07	0.52	0.02
1819 cent	0.09	99.16	0.27	0.07	0.18	0.17
1820 cent	0.03	97.96	1.28	0.11	0.50	0.04
1820 cent	0.03	98.90	0.63	0.07	0.28	0.02
1821 cent	0.10	98.95	0.53	0.06	0.23	0.03
1824 cent	0.02	98.58	0.91	0.12	0.28	0.03
1825 half cent	0.06	99.04	0.40	0.10	0.29	0.03
1827 cent	0.11	98.60	0.33	0.11	0.42	0.34
1828 cent	0.07	96.97	1.03	0.09	1.09	0.51
1828 half cent	0.02	99.12	0.47	0.08	0.18	0.03
1837 cent	0.02	99.15	0.47	0.09	0.03	0.03
1837 cent	0.06	98.59	0.37	0.10	0.74	0.04
1837 cent	0.03	99.54	0.18	0.08	0.03	0.01

The end of Birmingham's dominance in the supply of copper to the United States mint in Philadelphia in 1837 is exemplified by the three cents analysed. One has a typically British composition while the other two have trace levels only of bismuth and lower arsenic concentrations, of which one also contains 0.14 per cent antimony, never a feature of the British copper. These three analyses represent three different sources of the supply of copper to the United States' authorities.

Austro-Hungarian Copper

To provide an example of how British copper could differ from that used in other countries, and to re-emphasise the consistency of its composition, 13 1 Kreuzer coins from both Austrian and Hungarian mints and dating from 1800 to 1881 were analysed. As with the British copper, the composition was very consistent across the whole date range but the impurity pattern was very different with lower arsenic, increased lead, and trace bismuth, but with a very high antimony content, the mean being 0.87 per cent. Metallography showed that much of the antimony, like the bismuth in the British copper, was present as oxide inclusions.

Table 7.4   Comparison of coinage copper

Coinage	Ni	Cu	As	Sb	Ag	Bi	Pb
Mean, UK coinage	0.05	98.41	0.96	0.03	0.09	0.34	0.06
Mean, US coinage	0.05	98.70	0.62	0.02	0.09	0.35	0.08
Mean, Austro-Hungarian coinage	0.06	98.42	0.19	0.87	0.07	0.23	0.03

Structure and Properties of the Copper

All the coins examined share the same microstructure, with a fully recrystallised grain structure with annealing twins, with parallel rows of oxide inclusions. Microanalysis has shown these to be of cuprite (cuprous oxide) surrounded by bismuth and arsenic oxides and, possibly, some lead oxide.[13] The grain size is variable with some signs of secondary grain growth in many of the coins studied, a function of prolonged annealing, or the copper cooling slowly after hot rolling. There are no visible signs of cold work, which is to be expected since the blanks were close to fully softened before striking. The final cold rolling in polished rollers was to ensure that the correct gauge of strip was produced and that it had a good surface finish, the overall reduction in thickness in this process being quite small. This is confirmed by microhardness measurements on the United States cents, where the mean hardness from 21 coins was HV79.8 with a standard deviation of ±5.

Even the 1797 2d coin is not big enough to produce suitable tensile test specimens, so tensile testing was carried out on material machined from ships' bolts. These differ from the coinage copper in being finished by cold rolling, so their yield stress and ultimate tensile stress will be greater than for the coinage copper, and the strain at failure lower. Mean values from 10 tests from two bolts

---

[13]   Wilcox, *Copper*, pp. 33–65.

was a yield stress of 252MPa (2600 kg-force/sq. cm), an ultimate tensile strength of 305MPa (3110 kg-force/sq. cm), and a strain at failure of 11 per cent. The two bolts used came from ships built several years apart and the results again reflect the consistency that was possible in the copper supply.

Besides grain and deformation structures, the etching of the copper also reveals the segregation of individual elements as parallel dark bands. These bands result from the microsegregation of arsenic inherent in the cast structure of the ingots being elongated by rolling and not being homogenised because the annealing times and temperatures were not sufficient to achieve this (it makes no observable difference to the working properties of the copper). The banded structures can be divided into coarse and fine, the dark bands in the coarse structure being up to 250μm, and the fine no more than a third of that. The difference is due to the different sizes of ingot being rolled down, the coarser being from smaller ingots needing less reduction in size. Although not directly relevant to the 1797 coinage or the United States cents it is recorded that in 1791–92 Soho was supplied with ingots of touch pitch copper ('tough cake') of 36–40lb, 48–50lb and 90–100lb, and it is to be presumed that the practice continued.[14]

Given the consistency with which copper could be supplied, concerns about quality most probably arose from quality control problems on the part of the supplier. The correct quality (touch pitch) was recognised at the last stage of refining by removing and casting a sample from the melt. When cut half through and then broken, it would show a 'fine close grain, a silky fracture, and a light-red colour', while an ingot forged while red hot should be soft and free from cracks at the edges.[15] Copper that still contained too much oxygen could be embrittled by cuprous oxide, and with too little could be embrittled by films of impurities in grain boundaries, as has already been described for bismuth. There was, however, a bigger margin for error than the metallurgists of the time realised because, in the copper examined here, the hard cuprous oxide particles which could initiate cracks are usually sheathed by the arsenic and bismuth oxides which are soft and deformable.

Ormolu

Beyond the coinage the only area of Boulton's output at Soho that has received any metallurgical study is the ormolu: a number of pieces were analysed for a recent publication.[16] These were produced at the end of the 1760s and through the 1770s using a distinctive alloy. The impurity pattern of the brass is very much what would be expected from re-melting British copper, i.e. it is rather like

---

[14]   Susan Tungate, PhD student at the University of Birmingham, pers. com.

[15]   C.L. Bloxam, *Metals, their properties and treatment*, new edition partly rewritten and augmented by A.K. Huntington (London, 1882), pp. 226–232.

[16]   Nicholas Goodison, *Matthew Boulton: Ormolu* (London, 2002).

that recorded for ships' sheathing. French ormolu, with which Boulton wished to compete, throughout the eighteenth century contained of the order of 20–22 per cent zinc and up to 1 per cent each of lead and tin to improve castability, while the impurities showed more antimony and nickel than were found in the British copper. Boulton's brass contained only 10 per cent zinc, perhaps because the finished colour of the gilded brass was preferred, being somewhat warmer. Although he might have had the alloy made to order, it has become apparent that this could well have been a commercially available formulation much earlier in the eighteenth century. Current research on the Manx 1733 copper alloy coinage, which was produced on the Isle of Man using metal bought from the mainland, used an alloy of a very similar composition.[17]

Conclusion

The analysis of the copper used at Soho in the production of the 1797 and subsequent United Kingdom regal coinages, and of the planchets supplied to the United States Mint in Philadelphia, showed that the copper used over a period of four decades was of a consistent quality. The properties of this copper were determined by the use of Cornish and, to a lesser extent, Welsh and possibly Devonian, ores, and the standard smelting and refining technology of the late eighteenth and early nineteenth centuries. The copper has a distinctive pattern of impurities and oxide inclusions that has been found to characterise British copper and enable it to be distinguished from the output of, at least, some other countries. There was probably a reasonable margin of error in the refiners' quality control which would still allow the shipping of copper that would fit Boulton's requirements but, clearly, there were lapses. The same quality of copper was bought by the Admiralty East India Company for the bolts and sheathing of their ships and, indeed, a suggestion was made that the copper recovered from warships when dry-docked should be used for the coinage. A copy of a letter from the Commissioners of the Navy to the Copper Coin Committee in 1797 summarises the prices the Navy paid for finished copper: approximately £116 per ton for sheathing and £135 per ton for bolts, and received for copper recovered: £75 per ton, compared with the £122 per ton which the Copper Coin Committee were considering paying.[18] The Commissioners of the Navy also emphasised the cost of melting and re-refining the recovered copper, but that was largely to protect their own stocks of copper. These sources and the results of the analyses demonstrate that Boulton was using the standard industrial copper of his day with no additional refining or other modification of its composition. The expertise of the Soho Mint lay in the conversion of this copper to a mass-produced coinage. Metallurgical study has allowed us to confirm the documentary record for the

---

[17] Anthony Morfett, numismatist, pers. com.

[18] NA, PC1/37/114.

supply of ingots of tough cake copper and their conversion to coin. The same compositional patterns enable us to identify British copper in the alloy chosen by Boulton for his ormolu.

Chapter 8

The Mechanical Paintings of
Matthew Boulton and Francis Eginton

Barbara Fogarty

The term mechanical painting, which was used as early as May 1778, is an intriguing one.[1] How far the process was machine-based and how much it relied on hand-finishing is explored in this chapter. Mechanical paintings were the first-known attempt to reproduce the look and feel of oil paintings and, although comments on their quality varied, at their best they were said to be indistinguishable from oil paintings.[2] The process was invented by Francis Eginton and applied commercially by Boulton from about 1777 to 1781. Eginton was working for Boulton in 1764 and, by 1771, was described as the 'chief designer' at Soho.[3] He was in partnership with Boulton and Fothergill from at least 1776 to 1778 managing the 'Silver, Plated and Ormolu Goods' department which included japanned ware and mechanical paintings. From 1778 to 1780, he was in another partnership with Boulton and Fothergill solely for the production of mechanical paintings and japanned ware.[4] Matthew Boulton marketed the ingenuity and novelty of the process as much as its quality to his eighteenth-century customers. Using archival evidence, this chapter explores the origins of Boulton's so-called mechanical paintings and the features of the mechanical painting process. It also considers the findings and significance of recent scientific investigation of some of the alleged mechanical paintings by using Infrared Reflectogram, paint cross-section and other techniques.

Brendan Flynn and Martin Ellis described the production of mechanical paintings as one of the 'most intriguing enterprises' of the Soho Manufactory.[5]

---

[1]   Birmingham Archives and Heritage (BA&H), MS 3782/12/63/5, John Hodges (JH) to Matthew Boulton (MB), 10 May 1778.

[2]   William T. Whitley, *Artists and their Friends in England 1700–1799* (2 vols, London, 1928), 2, p. 28.

[3]   Kenneth Quickenden, 'Boulton and Fothergill's Silversmiths', *The Silver Society Journal*, No. 7 (Autumn 1995), pp. 342–356, p. 351.

[4]   Ibid., p. 352. The mechanical paintings included copies of Old Master oil paintings and contemporary works by Angelica Kauffman, P.J. de Loutherbourg and Benjamin West. This chapter considers copies of two paintings by de Loutherbourg entitled *Summer* and *Winter*.

[5]   'Mechanical Paintings', in Shena Mason (ed.), *Matthew Boulton: Selling what all the world desires* (New Haven and London, 2009), p. 220.

Although their manufacture was unprofitable and short-lived, mechanical paintings have long fascinated scholars, such as Eric Robinson and Keith R. Thompson and Antony Griffiths, not least because the process has yet to be precisely understood.[6] The account is further complicated by a later, similar polygraphic process practised by Joseph Booth in London from about 1784.[7] Since the method of producing mechanical paintings remained a secret, there are many conflicting theories. In 1862, the 'Sun Pictures', a collection of seven paper prints, now in the Science Museum, were discovered in the library at Soho. The Photographic Society of London in 1863 suggested that they were created using a photographic method.[8] However, George Wallis proposed instead that they were aquatints and part of an experimental transfer process for mechanical paintings.[9] Wallis had never seen a mechanical painting and doubted whether any had survived. Robinson and Thompson identified a number of mechanical paintings including various copies of Philippe Jacques de Loutherbourg's *Winter* and *Summer*, as illustrated in Figure 8.1. Mechanical paintings have often been conflated with Joseph Booth's polygraphic copies beginning with James Watt's memoir of 1809 in which he described Eginton's mechanical paintings as 'what are now called polygraphic pictures'.[10]

New scientific evidence from David Saunders and Antony Griffiths at the British Museum (BM), proved that examples of *Winter* and *Summer* previously described as Eginton's mechanical paintings show no underlying printed image and appear to be of Booth's manufacture, thus leaving no substantiated example of the mechanical painting process.[11] Their findings suggest that the tested paintings were made by a different technique to that described in the archives.

---

[6]   Eric Robinson and Keith R. Thompson, 'Matthew Boulton's Mechanical Paintings', *Burlington Magazine*, 112/809 (1970), pp. 497–507. Antony Griffiths, 'The Mechanical Paintings of Boulton and Eginton', unpublished paper, *The Image Multiplied Symposium*, 16 February 1988, British Museum, London.

[7]   Joseph Booth, *A Catalogue of Pictures, Copied for Sale by a Chymical and Mechanical Process, The Invention of Mr Joseph Booth, with the Originals from which they have been taken, by the Polygraphic Society [...] Fifth Exhibition: Opened in November 1790* (London).

[8]   Minutes of the Photographic Society of London, *The Photographic Journal*, No. 134 (15 June 1863), pp. 29–305, p. 291.

[9]   George Wallis, 'The Ghost of an Art-Process practised at Soho, near Birmingham, about 1777 to 1780, Erroneously Supposed to have been Photography', *Art Journal* (August 1866), pp. 251–255. Aquatints are tonal prints which give the effect of watercolour, made by etching plates through a resin ground.

[10]   Quoted in H.W. Dickinson, *Matthew Boulton* (Cambridge, 1937), p. 107.

[11]   David Saunders and Antony Griffiths, 'Two "mechanical" oil paintings after de Loutherbourg: History and technique', in Marika Spring (ed.), *Studying Old Master Paintings: Technology and Practice* (London, 2011), pp. 186–193, p. 188 and 191.

Figure 8.1   Copy after P.J. de Loutherbourg, *Summer*, c. 1778
Source: Courtesy of Birmingham Museums and Art Gallery.

The Origins of Mechanical Painting

This chapter will situate Boulton's mechanical paintings within the eighteenth-century fascination with ingenuity. The role of ingenuity in marketing was strong enough to persuade Boulton to pass off some hand-copied pictures as mechanical paintings. Maxine Berg has looked at the process whereby imitation led to invention and the transfusion of existing technologies.[12] She proposed a link between imitation (the copying of unobtainable or highly expensive goods) and invention (the creation of a new method of doing something), arguing that the key driver for product development in the eighteenth century was imitation.[13] In mechanical paintings Boulton was trying to imitate oil paintings, as he said he had discovered a more accurate way of reproduction than hand-painting alone. Collectors and antiquarians appeared more than happy to open up their collections to manufacturers of Boulton's status. In 1779, Boulton approached Sir Watkin Williams Wynn, asking for the loan of pictures and describing his commercial aims: 'I am engaged in painting as a manufacture and that by some peculiar

---

[12]   Maxine Berg, 'From Imitation to Invention: Creating commodities in Eighteenth Century Britain', *The Economic History Review*, New Series, 55/1 (2002), pp. 1–30.
[13]   Ibid., p. 19.

contrivances, I am enabled to make better copies of good originals than can be done otherwise, without much greater expense … by multiplying these copies when once obtained'.[14] Boulton referred to the 'peculiar contrivances' of the manufacture and justified the slightly higher cost by the better quality, compared with hand-painted copies.

The idea of copying oil paintings by a mechanical process started when artists made several copies of their own paintings.[15] Prints were an established method of reproducing paintings, but until the advent of mezzotint and aquatint, they could not imitate the smooth tonal gradations of watercolour or oil paint, or the effects of glossy colour and raised texture. Jacob Christoph Le Blon was the first to try to reproduce pictures and drawings in full colour by overlaying the primary colours using three mezzotint plates.[16] William Aitken, who produced a monograph on Francis Eginton in 1871, stated that Boulton had seen Le Blon's work.[17] Peter Perez Burdett may also have suggested the idea of printing in imitation of painting as he tried to find a market for his own aquatints, two of which were shown at the Society of Arts Exhibition in 1772.[18] Burdett's ideas were shared with Boulton and his circle of friends, including Benjamin Franklin who wrote to Burdett on 21 August 1773 that 'I should be glad to be inform'd where I can see some example of the new Art you mention of printing in Imitation of Paintings. It must be a most valuable Discovery: but more likely to meet with adequate Encouragement on this Side the water than on ours [America]'.[19]

In fact, Eginton, Burdett and the painter Thomas Chubbard were all involved in early experiments to imitate the aquatint process of Jean Baptiste Le Prince and others who developed it in France in the late 1760s.[20] Eginton produced a set of aquatints (now in the British Museum) in 1774 which he described as 'Specimens of a new method of engraving in imitation of washed drawings invented at Soho Manufactory near Birmingham'.[21] Griffiths has observed that Eginton used Burdett's technique of brushing acid over an aquatint ground rather than Le Prince's method of drawing with a special solvent ink which weakened

[14]   Dickinson, p. 104.

[15]   Barbara Fogarty, 'Matthew Boulton and Francis Eginton's Mechanical Paintings: Production and Consumption 1777 to 1781' (University of Birmingham, unpublished MPhil thesis, 2011).

[16]   Susan Lambert, *The Image Multiplied: Five centuries of printed reproductions of paintings and drawings* (London, 1987), p. 87.

[17]   William Costen Aitken, *Francis Eginton: A Monograph* (Birmingham, 1871), p. 8.

[18]   Richard T. Godfrey, *Printmaking in Britain* (Oxford, 1978), p. 59.

[19]   Ibid., p. 59.

[20]   Martin Hopkinson, 'Printmaking and Print Collectors in the North West 1760–1800', in Elizabeth E. Barker and Alex Kidson (eds), *Joseph Wright of Derby in Liverpool* (New Haven and London, 2007), pp. 85–103, p. 88, 89. Antony Griffiths, 'Notes on Early Aquatint in England and France', *Print Quarterly*, IV/3 (1987), pp. 255–270.

[21]   Griffiths, 'The Mechanical Paintings of Boulton and Eginton'.

the ground so it could be wiped off and then acid used to etch the plate.[22] Eginton's friendship with Burdett, who fled England in 1774 because of debts, is confirmed by a letter to Boulton in which Burdett asked to be remembered to Francis and his brother John.[23]

However, printing directly onto a prepared canvas would crack the surface as the intaglio process of forcing paper to take the ink from the recessed areas of the plate requires intense pressure, so an intermediate transfer process from paper to canvas would also have been required. In Liverpool, John Sadler had developed transfer printing onto ceramic bodies and held a public demonstration of the speed and quality of the process notarised in 1756.[24] Wedgwood did business with Sadler and sent his Queensware there to be transfer-printed from 1760 to 1794.[25] Wedgwood himself did not carry out printing in his Etruria premises until about 1784 when Sadler retired and many of his hands came to Staffordshire.[26] Burdett was also negotiating with Wedgwood between 1771 and 1773 about the use of aquatint for ceramic transfer printing but Wedgwood ended the relationship without taking the proposition further.[27] Bernard Watney and R.J. Charleston's research into patents suggested that transfer-printing onto enamel was already established in the Midlands as John Brooks of Birmingham applied for a patent in 1751 to transfer-print enamels and china.[28] Enamelling was one of the trades which Lady Shelburne recorded on her visit to Birmingham in May 1766; she wrote: 'Mr Taylor, the principal manufacturer there, dined with us, and we went afterwards to Mr Boldens [Boulton] who trades very much in the same way'.[29] She recorded a demonstration of transfer printing onto enamel at the manufactory of John Taylor.[30] Thus, there is evidence of at least two instances of transfer printing onto a hard moulded surface to give Eginton the inspiration for transfer printing onto canvas, unless he himself was already using transfers on his japanned ware. Yvonne Jones has argued that transfers were not widely used on japanned ware until the nineteenth century, although Stephen Bedford took out a patent for transfer printing in 1759, and some transfers were finished by being painted

[22]   Griffiths, 'The Mechanical Paintings of Boulton and Eginton'.

[23]   BA&H, MS 3782/12/24/122, P.P. Burdett to MB, 15 September 1777.

[24]   Llewellyn Jewitt, 'Liverpool Pottery: A notice of the various "Delft ware" works, and of the invention of printing on china and earthenware in Liverpool', *Art Journal* (August 1865), pp. 241–244, p. 242.

[25]   Ibid., p. 243.

[26]   Ibid., p. 243.

[27]   Hopkinson, 'Printmaking and Print Collectors in the North West', pp. 88–92.

[28]   Bernard Watney and R.J. Charleston, 'Petitions for Patents concerning Porcelain, Glass and Enamels with special reference to Birmingham, "The Great Toyshop of Europe"', *Transactions of the English Ceramic Circle - English Ceramic Circle*, 6/2 (1966), pp. 61–124, p. 61.

[29]   Ibid., p. 79.

[30]   Ibid., p. 80.

over.[31] Berg has argued that this diffusion of ideas (the spread of new technology to broader usage and other contexts) was typical of commercial inventiveness in the eighteenth century.[32] The transfer from paper to hard bodies and the invention of aquatint were both adapted by Eginton to produce mechanical paintings.

Correspondence between John Hodges, Eginton and the artist, Joseph Barney, contains several observations about the mechanical painting process. The accuracy of these letters, even allowing for some exaggeration, can be relied upon as they were concerned with practical business matters and were never meant to mislead a public audience. The terms 'plate' and 'impression' were used many times in the correspondence which points to a printed process.[33]

The etching of the copper plates was expensive so Boulton was reluctant to offer alternative sizes unless he felt the market could bear the additional cost.[34] James Keir, too, called the mechanical paintings 'painted impressions' in his memoir of 1809.[35] For the larger paintings more than one plate was required as can be seen in one of the 'Sun Pictures', the *Physician [Erasistratus] discovering Antiochus' love for his mother in law Stratonice*, after Benjamin West, which has two sheets joined together. However, there is no evidence to suggest multiple plates were used to facilitate colour printing by overlaying one on top of the other, each to print a separate colour.[36] After the etching came the printing of the copper plate onto paper. The print on the paper was then transferred onto canvas, forming a tonal impression which could then be over-painted in oils like a 'painting by numbers' kit; the whole process is illustrated in Figure 8.2. In order for this to be economic, it was carried out by 'the boys' and hand-finished by more professional artists.[37]

---

[31]   Yvonne Jones, "'A New Species of Japanning lately introduced" – the early, hitherto unexplored years of the Midlands japanning industry', unpublished paper delivered at a conference *Made in (Middle) England: Design, Consumption and the Arts in the Midlands*, University of Birmingham, 21 March 2009.

[32]   Berg, p. 4.

[33]   BA&H, MS 3782/1/30/15, FE to JH, 5 July 1781 and BA&H, MS 3782/1/32/12, JB to JH, 20 September 1781.

[34]   David Alexander, 'Kauffman and the Print Market in Eighteenth-century England', in Wendy Wassyng Roworth (ed.), *Angelica Kauffman, A Continental Artist in Georgian England* (London, 1992), pp. 141–178, p. 175.

[35]   BA&H, MS 3782/13/37, James Keir, Memoir of Matthew Boulton, 3 December 1809.

[36]   Matthew Piers Watt Boulton, *Remarks concerning Certain Pictures supposed to be Photographs of Early Date* (London, 1865), p. 7.

[37]   BA&H, MS 3782/12/65/43, James Keir (JK) to MB, 2 December 1779.

etched aquatint on
copper plate

↓

printed onto paper

↓

transferred to canvas
(dead-colour)

Figure 8.2   The mechanical painting process

The materials and machines used in the mechanical painting process included a rolling press, copper plates and canvas. The rolling press was required to print the paper transfers and was an expensive item of equipment, as indicated in a letter to Eginton, when Boulton was trying to terminate the partnership in 1781:

> Respecting the Roling [sic] Press, if you do not think well of taking it at 8 Guineas which is less than half its cost, please to redeliver it. We do not wish to encume [sic] you with any thing that may not be agreeable, but desire to have our affairs *now* finally settled.[38]

The copper plates were supplied by Wittow & Large, copper plate makers of Shoe Lane, London, through Bayley and Dyott, London agents.[39] The plates could have been etched by Eginton as his early attempts in 1774 showed, but his brother John was also working as an engraver and chaser at Soho.

Canvas was the preferred support for the finished mechanical paintings and was used in great quantities. In another attempt to conclude the partnership with Boulton, Eginton had to reach some agreement about the materials left on his premises; as he wrote to Hodges on 8 March 1782: 'I have no objection to keeping all the 24 yards of canvas, having orders for pictures that will nearly work it all up'.[40] However, it appears that canvas was not the only support, as indicated in a letter sent by Barney to Hodges on 4 May 1782. It identified the painting of *Cupids blindfolded* as being on wood and the *Hebe* as being on copper.[41] The painting on wood may be an original for copying, but the Hebe on copper is described as 'dead-colour' – a technical term for under-painting which was part of the mechanical painting process. The allusion to copper as a support is rare in the Boulton correspondence, and the possibility of extant mechanical paintings on copper has not been raised by any previous writers. However, a bicentenary exhibition of Angelica Kauffman's work in 2007 in Bregenz, Austria, included four copies on copper of the same small oval picture *Penelope weeping over the Bow of Ulysses*.[42] Although mechanical painting was not mentioned in the exhibition, the appearance of four identical copies suggests that some mechanical paintings on metal may have survived. Kauffman was Boulton's most copied artist; in a catalogue of mechanical paintings sent out in 1780, 14 out of 24 pictures were copies of her work, and Boulton was known to own several original paintings by

---

[38]   BA&H, MS 3782/1/30/2, John Hodges (JH) to Francis Eginton (FE), 10 January 1781.

[39]   BA&H, MS 3782/1/11, pp. 27–28, 33, Boulton and Fothergill to Mess'rs Bayley and Dyott, 14 and 28 June 1777.

[40]   Dickinson, p. 105.

[41]   BA&H, MS 3782/1/32/16, Joseph Barney (JB) to JH, 4 May 1782.

[42]   Tobias G. Natter (ed.), *Angelica Kauffman: A Woman of Immense Talent*, exhibition catalogue, Bregenz, Vorarlberger Landesmuseum (Ostfildern, 2007), p. 247.

her including *Penelope*.[43] A letter from Boulton & Fothergill to Bayley & Dyott, about the price of 'Brass plates that are prepared for paintings', also suggests the use of other metals as a support.[44] Although the technique of painting on copper is similar to painting on canvas, the transfer of the impression onto copper might have required a different application of inks and binder.[45] The copies on copper were more expensive than the copies on canvas. [46] A *Penelopy* [sic] on glass is also mentioned in a letter from Eginton to Hodges in 1782, and Eginton was known to have made a reputable career out of painting on glass after he left Soho, which might also have involved a transfer process.[47]

Having printed from the aquatint plate onto paper, the next stage of producing the mechanical paintings was transferring the paper print onto the canvas to produce the dead-colour. Normally, it refers to underpainting but in this case it probably refers to the transferred impression on the canvas which laid down the composition and the light and shade.[48] Once completed, that stage would have been followed by the first over-painting in oils carried out by employees known as 'the boys'. Keir, in the role of acting manager, writing to Boulton at Chacewater on 2 December 1779 stated that 'He [Barney] consented to work by the day to retouch the boys' pictures, at 10/6 per day. If the painting business is to be carried on, and the boys continue to paint, certainly the value of the pictures will be enhanced more than the expence [sic] of his wages'.[49] The painted canvases were, thus, expertly hand-finished by artists with a little definition or impasto to add authenticity and value. These artists included Eginton, Barney and a Mr Wilson and Mr Simmons. According to Hodges' letter to Boulton in 1780, 'Mr. Simmons' time (being three days a week) is chiefly employ'd at landscapes, being what suits his abilities best. He has done several pieces from prints, &c. that came from Mr. Eginton's, which are judg'd likely to sell if at reasonable prices'.[50]

Hodges, writing to Boulton in 1780, asked if Wilson could be employed in finishing some mechanical paintings for speculative sale with Boulton's own agent in London, John Stuart, 'as there are many large pieces finish'd and many in the dead color'd state that may be finish'd by Mr. Wilson, &c., perhaps you would

---

[43]   Robinson and Thompson, 'Matthew Boulton's Mechanical Paintings', pp. 504–506.

[44]   BA&H, MS 3782/1/11 pp. 27–28, B&F to Bayley and Dyott, 14 June 1777.

[45]   In painting on copper the copper plate is abraded to improve the adhesion of the paint and the pigments appear slightly more saturated in appearance because they are not absorbed into the copper.

[46]   M.P.W. Boulton, p. 4.

[47]   M.P.W. Boulton, p. 6.

[48]   Erma Hermens (ed.), *Looking Through Paintings: The Study of painting Techniques and Materials in Support of Art Historical Research* (London, 1998).

[49]   BA&H, MS 3782/12/65/43, JK to MB, 2 December 1779.

[50]   BA&H, MS 3782/12/63/16, JH to MB, 12 September 1780.

deem it not amiss to send a few pieces for trial'.[51] In the same letter, Hodges goes on to report that:

> Mr. Barney, not being able to learn that you left any direction what he was to do in particular, purposes going on with some of the pieces for your own apartments. He has just been about one painting for Mrs. Montagu, which is now nearly ready to send off. Next he will proceed, with Mr. Wilson, in executing the small paintings order'd for Sir Sampson Gideon.[52]

After the manufacture of mechanical paintings ceased at Soho, Eginton and Barney were involved at different times in painting the whole of the canvas from the impression, that is without the boys. Despite the involvement of the professional artists, the quality of the finished paintings was often unacceptable either due to problems with the impression or the resemblance of the reproduction to the original. Barney reported that 'the last picture of *Trenmor* which you brought is in so indifferent a State that it cannot possibly be finished without being dead coloured again [I] am exceedingly sorry I did not examine the impression you brought before I sat down to paint at it'.[53]

Barney took such pains with one mechanical painting after an original by Benjamin West that he went down to London to check the picture, writing to Hodges 'I should take it as a favour if you will please to forward one the Picture of *Stratonice* which I am to paint for Mr Boulton as I purpose being in London in about a fortnight and taking the picture in order to finish it from the Original at Mr Wests'.[54] Although Aitken reported that Boulton's grandson, Matthew Piers Watt Boulton, remembered his father, Matthew Robinson Boulton, saying that the mechanical painting process 'copied colour mechanically, not merely chiaroscuro' and two of the 'Sun pictures' aquatints are in colour, the present author does not think it would have been necessary or economically viable to print onto canvas in colour, only to be totally painted over.[55]

Although Robinson and Thompson thought the mechanical paintings were dead-coloured and only touched up by hand, the evidence from the archives appears to support the argument that the canvases were entirely painted over and that the 'mechanical' process only referred to the production and printing of the aquatint onto the canvas to provide a tonal impression for the artists to follow.[56]

---

[51]   BA&H, MS 3782/12/63/12, JH to MB, 17 April 1780.

[52]   BA&H, MS 3782/12/63/12, JH to MB, 17 April 1780.

[53]   BA&H, MS 3782/1/32/12, JB to JH, 20 September 1781.

[54]   BA&H, MS 3782/1/32/4, JB to JH, 29 June 1781.

[55]   Aitken, p. 7.

[56]   Robinson and Thompson, 'Matthew Boulton's Mechanical Paintings', p. 502.

Scientific Evidence

The results of Saunders' technical analysis of the 'Sun Pictures' and of various versions of *Summer* and *Winter* after de Loutherbourg, supported some of the theories of production detailed above but offered no confirmation of a printed transfer process. Saunders made an infrared reflectogram of the coloured aquatint print of *Venus and Adonis* after Benjamin West, as illustrated in Figure 8.3. When the reflectogram was compared with the original it showed that the coloured inks were visible in the infrared spectrum, although not as strongly as the black-brown inks, suggesting that if the transfer print was used in the mechanical painting process, the inks would be seen in infrared reflectograms of the finished pictures. Analysis by Fourier Transform Infrared Spectroscopy (FTIR) of the medium carrying the ink confirmed that it was pure gum and not albumen or gelatin. This disproves Wallis' assertion that the medium was albumen, which strongly suggested the collodion photographic process to FP Smith of the Museum of Patents and some members of the Photographic Society.[57] Saunders and Griffiths proposed that the pigments held in the gum which is soluble in water could have been transferred by placing the paper transfer on the canvas and applying a wet cloth or heat to the back of the transfer print. The fact that the medium is soluble supports the theory that the prints were an intermediate process and not a finished product which would have been more stable.

A visual examination of the fully finished reproductions of *Summer* and *Winter* from the BM and Birmingham Museums and Art Gallery (BM&AG) revealed that the paint layer looked thin and blocky, with little tonal transition, and there were some impasto and visible brushstrokes. Saunders overlaid two images of a detail of the BM's and of BM&AG's *Summer* and found that the dimensions of the figure groups were exactly the same, suggesting that the process defined the overall composition. The overlay image highlights the differences and similarities in colour and the slight differences seen in the faces and clothing may be due to their having been hand-finished. Infrared reflectograms of three versions of *Winter* (BM, BM&AG and Brodsworth Hall, South Yorkshire) and four of *Summer* (the same locations plus the National Portrait Gallery, London) were taken. These revealed no evidence of an underlying printed image. Saunders suggested three possibilities: that the paint layers are opaque to infrared radiation (but this would be unlikely across all the areas sampled), that the inks used in the transfer print do not absorb infrared radiation (but this was not the case in the 'Sun Pictures' where the inks were visible to the infrared camera) or the more likely explanation that there is no underlying printed area as would have been expected from the archival evidence. Cross-section analysis of the layers showed no traces of gum and no unusual pigments, with oil as the only binding medium. The different versions of *Summer* and *Winter* are so similar they appear to come from the same process. Although most of them were previously attributed to Boulton and Eginton's

---

[57]   Minutes of the Photographic Society of London, *The Photographic Journal*, p. 292.

Figure 8.3   Infrared reflectogram of Francis Eginton after Benjamin West,
               Venus and Adonis, 1778–1781, Science Museum, London
Source: Courtesy of the British Museum, London and the Science Museum, London.

mechanical paintings, the Brodsworth *Winter* had an original label on the back
which identified it as a 'Polygraphic Copy'.[58] Thus, all the versions of *Summer*
and *Winter* appear to be Booth's polygraphs. This is a very significant finding in

---

[58]   The label reads 'A Polygraphic Copy Of A Landscape, representing A Winter
Scene; from an original picture, by De Loutherbourg; Which, with its companion, *a
Summer Scene*, cost, at Mons. Des Enfans' Sale One Hundred and Fifty Pounds. *Now in the
Possession of the Society.* The POLYGRAPHIC ART, of copying or multiplying Pictures in
Oil Colours, by a chymical and mechanical Process, is the original Invention of Mr. Booth,
and now carried on by the Polygraphic Society in London. N.B. This Picture, like all others

itself as BM&AG, BM and National Portrait Gallery images were all previously identified as by Eginton. Saunders and Griffiths suggested the possibility that the process was based on a screen or block print. The most curious findings were the unusually thick preparatory layers on the canvases. The upper-most layer contained pumice which has the characteristic of being very absorbent. Saunders and Griffiths argued that if the paintings were made by stencilling or block printing then it would be an advantage to speed up the process of drying before another block of colour was laid. The pumice-rich layer would account for the flat, lean appearance of the oil paint as it was absorbed into the preparatory layer.[59]

In summary, the scientific results prove that all the tested versions of de Loutherbourg's *Summer* and *Winter*, previously attributed to Francis Eginton as mechanical paintings, did not involve an underlying printed image, and may be by Booth's polygraphic method. Saunders and Griffiths' challenging results imply a need to find other examples of mechanical paintings of proven provenance and further research into Booth's polygraphic process.[60]

The Role of Ingenuity in Marketing

Underlying the promotion of mechanical paintings was the assumption that the ingenuity of manufacturing carried more prestige than hand-painting. Peter Jones explored the eighteenth-century fascination with the link between natural philosophy and technology which Boulton exploited in his attraction of 'industrial tourists' to his Soho manufactory.[61] Boulton intended mechanical paintings to compete with hand-painted copies and he emphasised the mechanical ingenuity of the reproduction method by alluding to 'peculiar contrivances' which guaranteed improved quality at only slightly higher cost. In fact, the mechanical part of the process reduced the time (and therefore cost) by almost a half, as Barney informed Hodges in September 1781 when he was asked to paint two blank canvases as if they were mechanical paintings of *Telemachus*: 'your Idea was perfectly right respecting *Telemachus* had it been mechanised, but at prisent [sic] the outline and dead colour take nearly half the time'.[62] Hodges had written 'Mr Boulton ... begs they may be good pieces and exactly alike, for they go as mechanical paintings'.[63]

---

in Oil, may hereafter want Varnish, in that Case it may be varnished in the same Manner as any other Picture in Oil'.

[59]   Saunders and Griffiths, 'Two mechanical oil paintings after de Loutherbourg', p. 191.

[60]   Robinson and Thomson, 'Matthew Boulton's Mechanical Paintings', p. 506, suggested Moses Haughton's *The Owl* (BM&AG) and some pictures at Culzean Castle.

[61]   Peter Jones, *Industrial Enlightenment* (Manchester, 2008), p. 96.

[62]   BA&H, MS 3782/1/32/10, JB to JH, 6 September 1781.

[63]   Minutes of the Photographic Society of London, *Photographic Journal*, no. 139, 16 November 1863, pp. 385–399, p. 395.

It is clear that Boulton intended to sell the hand-painted copies as mechanical paintings, because he was marketing the novelty of the products. However, this presented Barney with the problem of how to make money out of the hand-painted copies when they took more time than the mechanical paintings. Barney was a talented artist who had studied under Kauffman's future husband Zucchi, in 1777, exhibited at the Royal Academy from 1784, and worked as a drawing master at the Royal Military Academy from 1793 to 1820.[64] Nevertheless, between 1781 and 1782, he was grateful for the employment that Boulton offered him. He thought his hand-painted copies were as accomplished as the original, even if they were not as accurate as mechanical paintings 'for though the one I have now done has full as much effect it is by no means equal to it [the original of *Trenmor*] in correctness'.[65]

Marketing ingenuity was a significant strategy of Boulton's as invention was prized by Boulton's circle and visitors to Soho. Eginton's process for copying oil paintings pre-dated Erasmus Darwin's mechanical copying machine or 'bigrapher' in 1777, and may have informed James Watt's invention of a letter copying press in 1779 which used special inks and paper to take an offset copy of a hand-written letter.[66] Benjamin Franklin subscribed for three of Watt's copying presses 'as I love to encourage Ingenuity'.[67] Princess Dashkova, who had a 'deep desire for knowledge and profound respect for modern technology and science', visited Soho in 1780 and also subscribed for the 'Copying Machine' as well as buying several small [mechanical] pictures.[68] However, as Boulton was to find out with mechanical paintings, imitation, invention and ingenuity alone were not enough to guarantee that the project was economically viable.

Demise of the Enterprise

The production of mechanical paintings ceased at Soho in 1781, although it continued off-site for about another 10 years, though orders and sales still passed through Boulton.[69] The reasons for the demise were variable quality, lack of supervision by Boulton and ultimately the non-viable cost of production. The mechanical paintings never seemed to have made a profit. When Keir was managing

[64]   R.J. Cleveley, 'Barney, Joseph (1783–1829?)', *Oxford Dictionary of National Biography*, Oxford University Press, 2004; online edition, January 2008 [http://www.oxforddnb.com/view/article/1486, accessed 3 March 2011].

[65]   BA&H, MS 3782/1/32/12, JB to JH, 20 September 1781.

[66]   Jenny Uglow, The Lunar Men: Five Friends Whose Curiosity Changed the World (London, 2002), p. 306.

[67]   Olga Baird, 'Benjamin Franklin, Catherine Dashkova and James Watt's "Art of Copying"', *Benjamin Franklin and Russia: The Philosophical Age. Almanac 31* (St Petersburg, 2006), pp. 121–129, p. 125, 128, and www.ideashistory.org.uk

[68]   BA&H, MS 3782/12/63/14, JH to MB, 15 May 1780.

[69]   Minutes of the Photographic Society of London, *The Photographic Journal*, p. 390.

Soho for Boulton in 1779, he tried to reduce the losses of the mechanical painting business. He wrote to Boulton: 'we are taking some other economical steps in that business by which a great deal of money may be saved. For *saving* not *gaining* is the object in that business'.[70] The success of the various enterprises was reported annually and, in January 1780, John Scale commented to Boulton 'I am afraid the Painting and Refining trade will turn out very indifferent'.[71] Fothergill, never happy with his partner's cavalier attitude to cash-flow problems, wrote to Boulton of the 1779 Painting and Japanning Trade annual accounts that 'the losses we have sustain'd prior to the keeping a separate Account of this Article of our Business [mechanical paintings] must farr [sic] exceed £1000, you will now determine if it is prudent to continue so destructive a branch, without further delay'.[72] If Boulton's financial position had been stronger he would not have needed to curtail this enterprise, but in 1777 Boulton's overdraft was nearly £25,000 and he could not afford to sustain these losses.[73] By April 1782, the mechanical painting business had been sub-contracted to Eginton, Barney and Wilson, but there was still concern about the remaining stocks of pictures: Hodges warned Boulton that the 'painting trade (considering what sales were made last year) I doubt not will turn out well, and it is necessary this branch should be attended to in order to get off the stock on hand'.[74]

The quality of the mechanical paintings was not always up to Boulton's standards and was also an overriding concern for the firm of Clarke and Green who placed a large export order of 60 to 100 pictures.[75] They insisted that they be painted 'in a much more masterly manner than the pictures you sent as samples'.[76] Even Barney's work was criticised as he acknowledged in a letter to Hodges in 1781, writing: 'sorry I have not succeeded in my endeavours to please Mr Boulton in the last picture in respect to *Patience and Perseverance*'.[77]

A final reason for the demise of the enterprise, which Boulton himself gave in letters to John Garnett and Isaac Hawkins Browne about not being able to produce any new subjects, was that his time was 'almost wholly engaged in his steam engine business'.[78] He liked to supervise his businesses personally, as he revealed in a letter to Watt in 1769: 'I am determined never to embark on any trade that I have not the inspection of myself'.[79] During the period of the mechanical painting

---

[70]   BA&H, MS 3782/12/65/43, JK to MB, 2 December 1779.

[71]   BA&H, MS 3782/12/72/27, John Scale to MB, 15 January 1780.

[72]   BA&H, MS 3782/12/60/188, John Fothergill to MB, 5 February 1780.

[73]   J.E. Cule, *The Financial History of Matthew Boulton 1759–1800* (University of Birmingham, unpublished MComm thesis, October 1935), p. 74.

[74]   BA&H, MS 3782/12/63/28, JH to MB, 6 April 1782.

[75]   BA&H, MS 3782/1/30/10, Richard Clarke (RC) to B&F, 18 May 1781.

[76]   BA&H, MS 3782/1/30/16, RC to B&F, 10 July 1781.

[77]   BA&H, MS 3782/1/32/2, JB to JH, 17 May 1781.

[78]   M.P.W. Boulton, pp. 7–8.

[79]   Dickinson, p. 82.

enterprise, Boulton was often away in Cornwall, overseeing many of the 40 steam pumps built between 1776 and 1780, so he could not give the fledgling mechanical painting business the attention it required. Keir was managing in Boulton's absence, and production decisions on improving the quality by employing artists to finish the paintings and finding other ways to save money were made by him, not Boulton. Although Boulton dissolved the partnership in 1780, he did not immediately withdraw his capital from the business. As described earlier, he offered the 'roling [sic] press' and other equipment to Eginton at a reasonable price so that he could carry on off-site, which Eginton did until at least 1791 when he sent 24 paintings to Soho for £13.7s.6d.[80]

The mechanical painting process combined a tonally printed under-painting (dead-colour) with a totally hand-painted surface. However, the scientific evidence from pictures previously identified as mechanical paintings found no trace of a printed impression and suggested they may be attributable to Booth. The nature of the link with his polygraphic process remains to be discovered and mechanical paintings will continue to fascinate until substantiated by an example which coincides with the archival records. However, the inventiveness of Boulton and Eginton in adapting existing technologies and the role of ingenuity in both the promotion and reception of mechanical paintings are instructive in being typical of eighteenth-century enlightenment concerns.

---

[80]   Minutes of the Photographic Society of London, *The Photographic Journal*, p. 390.

Chapter 9

Samuel Garbett and Early Boulton and Fothergill Assay Silver

Kenneth Quickenden

On 19 February 1773 Samuel Garbett gave evidence to a Parliamentary Committee appointed to determine whether or not Birmingham and Sheffield should be granted assay offices.[1] Their applications were successful, despite strong opposition by London's silversmiths, who feared that the establishment of assay offices in those towns would encourage silversmithing there and generate competition. The arguments, put forward to the Committee in several documents by silversmiths from the capital and the two provincial towns, centred on the issue of whether there was a need for assay offices in Birmingham and Sheffield.[2] Much of Garbett's evidence focused on Matthew Boulton since he was the key figure in the efforts to secure The Birmingham Assay Office and since his early silversmithing was relevant to the Committee's enquiry. The purpose of this chapter is to assess the accuracy of Garbett's evidence about early silver by Boulton, and his then partner John Fothergill, and to gauge its impact upon later writings by scholars on the subject. This chapter is structured according to the questions put to Garbett: when the partners started silversmithing; their use of the Chester Assay Office before The Birmingham Assay Office opened; the range and volume of their assay silver production and the potential of their business.

Garbett (1717–1803) was born in Birmingham, and this was where in 1746 he joined John Roebuck at a laboratory for the application of chemistry to industrial problems. They also set up a refinery for gold and silver.[3] Garbett lobbied Parliament in 1759 seeking relief for firms from the necessity of taking out licences for trading in silver when using only small quantities, which was characteristic of Birmingham's firms.[4] He took a general interest in the well-being of the town, promoting canals and the General Hospital.[5] He knew Boulton from the 1740s,

---

[1] Minutes of the Committee on Sheffield and Birmingham Petitions, Goldsmiths' Company MS G, 11, 2, 5, pp. 13–21.

[2] Sally Baggott '"I am Very Desirous of Becoming a Great Silversmith" Matthew Boulton and The Birmingham Assay Office' in Shena Mason (ed.), *Matthew Boulton, Selling What all the World Desires* (New Haven and London, 2009), pp. 47–54, 48–51.

[3] R.H. Campbell, *Carron Company* (Edinburgh and London, 1961), p. 8.

[4] H.W. Dickinson, *Matthew Boulton* (Cambridge, 1937), p. 63.

[5] Campbell, pp. 7–20.

who much later referred to Garbett as his 'oldest and most respectable friend'.[6] Garbett was widely regarded as a man of judgement and integrity.[7] Garbett was therefore a credible witness, but could his evidence be reliable when he must have been dependent on Boulton for providing most of the evidence about his firm; and could he be entirely disinterested when the interests of his friends in Birmingham were challenged by the London trade?

The Introduction of Silversmithing

When asked by the Parliamentary Committee in 1773 'How long ago is it since he [Boulton] began to manufacture silver wares?' Garbett replied: 'I believe since he began to manufacture *plate* only is *within* [my emphasis] these seven years'.[8] 'Silver wares' might have been taken to mean silver products of any kind and Boulton had been making small silver items for many years. The manufacture of such items was widespread in Birmingham.[9] Silver was used at Boulton's father's Birmingham works in the 1750s.[10]

Boulton inherited that business in 1759 and started to develop the Soho Manufactory from 1761.[11] He bought silver in small quantities from Garbett from at least 1763.[12] Clearly, therefore, Garbett knew that Boulton had been using silver before the mid–1760s. However, since the Committee's question was asked in the context of Boulton's wish to obtain an assay office, Garbett turned the question into one about plate, denoting larger pieces of silver which, unlike very small items, were required to be assayed at an assay office before sale. By an Act of 1738 (Act of 12 Geo. 11 c. 26) articles such as jewellers' work, thimbles and clasps, and items which might have been damaged by the assaying process (interpreted

---

[6]   Birmingham Archives and Heritage (BA&H), MS 3782/12/36/67, Matthew Boulton (MB) to Joseph Ewart, 28 April 1791.

[7]   Campbell, pp. 19–20.

[8]   Minutes of the Committee on Sheffield and Birmingham Petitions, Goldsmiths' Company, MS G 11 2 5, p. 13.

[9]   Dickinson, p. 63.

[10]   BA&H, MS 3782/12/108/1/53.

[11]   Dickinson, pp. 25–43.

[12]   In the assay year (early July to early July) 1762/3 Boulton purchased £60.5s.2d. of precious metal under a Gold and Silver account with Garbett (Kenneth Quickenden, 'Boulton and Fothergill's Bullion Supplies for Assay Silver', *The Silver Society Journal*, 12 (Autumn, 2000), pp. 45–52, 52).

until 1789 as all such articles irrespective of weight)[13] and anything less than 10 pennyweights (half a troy ounce) were exempt from assay.[14]

H.W. Dickinson in *Matthew Boulton* (1937), misquoting Garbett, wrote that Boulton and Fothergill began '… making silver plate *about* [my emphasis] seven years prior to that date'[1773].[15] Elsewhere in the book Dickinson gave 1765 as the beginning of silver plate manufacture.[16] Later scholars have followed Dickinson's lead.[17]

This evidence undermines a claim Boulton made to an agent in 1763, stating that he had started producing silver candlesticks, saucepans and tablewares in large quantities for trade customers in London.[18] In 1763 Boulton purchased a £2 annual licence to sell silver.[19] The licence was not required for items up to and including 5 pennyweights but was required for items up to 30 ounces.[20] That was enough for products as large as candlesticks and a range of tablewares but the licence would have been needed even for some modest items such as buckles.[21] There is nothing to confirm that candlesticks or tablewares were being made at this date. Innumerable letters to customers in the early and mid 1760s listing the firm's wide range of goods, including such items as buttons, buckles and chains[22] in materials such as steel[23] and platina,[24] a silvery white alloy of copper and zinc,[25]

---

[13]   J.S. Forbes, *Hallmark: A History of the London Assay Office* (London, 1999), p. 232. A judgement at the Court of Common Pleas ruled that from 1789 the 1738 Act's exemptions be interpreted as applying only to articles weighing less than 10 dwts.

[14]   J. Paul de Castro, *The Law and Practice of Hall-marking Gold and Silver Ware*, 2nd edition (London, 1935), pp. 71–72.

[15]   Dickinson, p. 67.

[16]   Ibid., p. 53.

[17]   For example, Baggott, 'I am very desirous of Becoming a Great Silversmith', p. 47; Kenneth Quickenden 'Silver and Its Substitutes' in Malcolm Dick (ed.), *Matthew Boulton: a Revolutionary Player* (Studley, 2009), pp. 153–169, 153.

[18]   BA&H, MS 3782/1/40/95, Boulton and Fothergill (B&F) to 'Montreal' (code for an unidentified agent), 15 December 1763.

[19]   BA&H, MS 3782/1/34/170, 8 October 1763, licence to vend silver plate, dated Oct. 24 1763 £2.0s.0d.

[20]   William Chaffers, *Hall Marks on Gold and Silver Plate*, 9th edition (London, 1905), p. 70. Act 32 George 11 c. 14.

[21]   The Birmingham Assay Office (hereafter BAO) 'The Register of Plate and Silver Wares Assayed and Marked or Broke at The Birmingham Assay Office …', 1773–92, 31 August 1773 p. 1B, gives weights for articles made by B&F: 6 pairs of candlesticks 116 troy ounces (ozs) 13 pennyweights (dwts); a coffee pot 18ozs 18dwts; 30 pairs of buckle rims 38 ozs 6 (dwts) 12 grains.

[22]   BA&H, MS 3782/1/40/36–7, MB to Hagen Godhard, 11 November 1762.

[23]   BA&H, MS 3782/1/40/181, B&F to Francois du Plot, 5 November 1764.

[24]   BA&H, MS 3782/1/37/64, MB to George Trout, 18 February 1765.

[25]   G.D. Hiscox, *Formulas, Processes and Trade Secrets* (London, 1937), p. 80.

made no mention of large silver items. The firm was too preoccupied with orders of these kinds as well as dealing with the building work at Soho[26] to find time to develop silversmithing. The comments to the agent may be taken as an attempt to impress, though they also signalled an aspiration.

But was Garbett's implied claim that silver plate was being made by the mid 1760s true? By 1767 silver was advertised in a local trade directory as a material being used for making candlesticks in Birmingham.[27] This could have applied to the partners since by the mid 1760s they had developed a substantial manufacture of Sheffield Plate (silver on copper) candlesticks,[28] which were stamped with dies that could readily be used with silver.[29] Moreover, the first recorded order for silver plate was in 1766 when Benjamin Molineux of Wolverhampton ordered six pairs of candlesticks. The firm responded: 'our demand, in general, being more for plat'd [i.e. Sheffield Plate] than silver candlesticks, we rarely keep any quantity of the latter reddy [sic] made, therefore shall be somewhat hurried to get you six pair of silver ones compleat'd to the limitt'd [sic] time'.[30] There is, however, nothing to confirm that the order was supplied, and even if it was, it did not mark the beginning of continuous production. Boulton promised an agent the availability of silver candlesticks in December 1766,[31] but this parading of progress was probably linked to Boulton's wish to persuade the agent, J.H. Ebbinghaus, a merchant of Iserlohn in Germany, to become a partner. Expansion required extra capital and Ebbinghaus became a partner at the end of 1766, providing £2,000 then and £500 in the following year.[32] There is no evidence that candlesticks were sent nor is there any other evidence of assay silver production in the assay years 1766–67 or 1767–68. Although the archive is thin during this period,[33] it is very doubtful that production had begun in earnest. James Watt's list of items made at Soho when he first visited in 1767 did not include silver plate,[34] and it is unlikely that Boulton would have ordered silver plate from London in this period,[35] if he could have been supplied by his own

[26]   BA&H, MS 3782/1/37/11, B&F to John Perchard, 18 August 1764.

[27]   *Sketchley's Birmingham, Wolverhampton, and Walsall Directory; Upon an Entire, New and Improved Plan*, Birmingham, 3rd edition (1767), p. 16.

[28]   BA&H, MS 3782/1/37/50, B&F to Thomas Jeffries, 12 January 1765.

[29]   Kenneth Quickenden, '"Lyon-faced" Candlestick and Candelabra' in *The Silver Society Journal*, 11 (Autumn, 1999), pp. 196–210, 206.

[30]   BA&H, MS 3782/1/37, p. 325, B&F to Benjamin Molineux, 28 February 1766.

[31]   BA&H, MS 3782/12/63/98, MB to J.H. Ebbinghaus, 1 December 1766.

[32]   J.E. Cule, 'The Financial History of Matthew Boulton 1759–1800' (University of Birmingham, unpublished MComm thesis, 1935), p. 22.

[33]   BA&H, MS 3782. The evidence for production is confined to letters in this period whereas later there were ledgers (MS 3782/1/1) and journals (MS 3782/1/3) from 1776.

[34]   Dickinson, p. 202.

[35]   BA&H, MS 3782/12/23/5, Nathaniel Jefferys (NJ) to MB, 11 June 1760 (an order worth £50); MS 3782/6/190/206, NJ to MB, 4 June 1767 (order worth £5.18s.0d.).

firm. Moreover, a new block was felt to be necessary for the manufacture and display of large ornamental metalwork: Sheffield Plate, silver and ormolu (lavish clocks and vases with gilded mounts).[36] The block, an impressive building with a classical façade, was started in 1765 and finished in 1767.[37]

Even now the silver business developed very slowly: in the assay year 1768–9 only eight candlesticks are known to have been made.[38] In the following assay year only nine more candlesticks plus a coffee pot and stand, and a mazarine (a fish strainer) are known to have been produced.[39] Yet the Sheffield Plate business developed considerably, and included a wide range of tablewares.[40] Ormolu production rapidly developed from 1768 and peaked a few years later. It seems that Boulton only energetically developed the silver business from 1771, and this was almost certainly linked to his realisation that ormolu sales were not as large as he had hoped for.[41]

Garbett's comment in 1773 that silver plate manufacture had begun at Soho '… *within* these seven years …', if taken to mean, as scholars have done, that production of assay silver began in the mid 1760s, cannot be substantiated; only at the end of the 1760s does evidence survive. Taken literally, *within* encompasses those later years, but his remark gave a false impression about the length of continuous production. Significantly, however, the remark helped Boulton's cause for it conflicted with the insistence of London's silversmiths that silversmithing was not sufficiently established in Birmingham to justify setting up an assay office there.[42]

---

[36]   Nicholas Goodison, *Matthew Boulton: Ormolu* (London, 2002), pp. 52–53.

[37]   George Demidowicz, 'Power at the Soho Manufactory and Mint' in Dick (ed.), pp. 116–131, 118.

[38]   A set of four in a private collection; a pair at Birmingham Assay Office (BAO) (accession number 1140); an attributed (i.e. without makers' marks) pair at the Speed Art Museum, Louisville, Kentucky (accession number 1993.9.192 a–b).

[39]   Candlesticks: an attributed pair at Birmingham Museums and Art Gallery (BM&AG) (accession number 2002 M.364.1–2); one, attributed, in the Cobbe collection; an attributed pair at the Grosvenor Museum, Chester (accession number 2002.300); an attributed pair sold at Christies, London, 2 June 2009 lot 0277; a pair sold at Byrne's, Chester, lot 100, 22 September 2010; attributed coffee pot and stand BM&AG (accession number 1996 M 1); Mazarine, BAO (accession number 23). 'Attributed' indicates that pieces are without makers' marks.

[40]   BA&H, MS 3782/1/9, p. 107, B&F to John Porzelius (JP), 17 May 1771.

[41]   Goodison, *Ormolu*, pp. 44–56.

[42]   Case of the Goldsmith's, Silversmith's and Plate Workers of London and Places Adjacent, 17 February 1773, Goldsmiths' Company, MS G 11 1 5, pp. 1–2.

Chester Assay Office

The slow evolution of the silver business was partly due to the difficulties of sending silver for assaying to Chester, the nearest assay office to Birmingham, but nevertheless some 72 miles away.[43] Asked by the Parliamentary Committee what inconveniences Boulton suffered as a consequence, Garbett replied: 'It occasions delay and uncertainty. Very frequently goods are damaged in carrying backwards and forwards. The expense of carriage, which must be included'.[44] These comments were similar to those earlier circulated by Boulton as a part of his campaign: 'By these means delays are occasioned, ... their works ... are very often damaged and sometimes ruined by accidents in carriage, and careless packing and repacking at the Assayers; a great price is paid for carriage'.[45] That Boulton had to pay the expenses of carriage to and from Chester,[46] as well as his agent's expenses there[47] which were absorbed by the firm,[48] is beyond dispute. However, the other comments are questionable, though accepted by historians, with a claim by one that delays in sending pieces to Chester could be 'months',[49] and acceptance of Boulton's claims about delays and damage by another.[50]

James Folliott arranged for the assaying in Chester. Extensive correspondence survives between him and Boulton from 1771,[51] up until the foundation of The Birmingham Assay Office in 1773.[52] Yet not one letter can be used to support the claim about damage to silver, though once Folliott was given advice about handling a box.[53] There was, however, one occasion when damage to a consignment was alleged to have happened: in 1771 two pairs of candlesticks for Lord Shelburne were said by Boulton to have been so badly damaged while away at Chester that new parts had to be substituted. Boulton added, 'This is not the first time by

[43] Dickinson, p. 63.

[44] Minutes of the Committee on Sheffield and Birmingham Petitions, Goldsmiths' Company, MS G 11 2 5, p. 15.

[45] *Memorial Relative to Assaying and Marking Wrought Plate at Birmingham etc.* [1773], Goldsmiths Co. G 11 5 4, p. 1.

[46] BA&H, MS 3782/1/36/135, 28 December 1772, Carriage of a box from Chester 1s.6d.

[47] BA&H, MS 3782/1/9/435, Anthony J. Cabrit (AJC) to James Folliott (JF), 25 April 1772. Credited with expenses of 3s.9d.

[48] BA&H, MS 3782/1/9/ 61, John Wyatt (JWy) to William Matthews (WM), 2 March 1771. Account for candlesticks listed no expenses connected with Chester, or carriage to London, for a customer there.

[49] Eric Delieb and Michael Roberts, *The Great Silver Manufactory: Matthew Boulton & the Birmingham Silversmiths 1760–90* (London, 1971), p. 45.

[50] Maurice H. Ridgway, *Chester Silver 1727–1837* (Chichester, 1985), p. 61.

[51] BA&H, MS 3782/1/9/ 38, B&F to JF, 12 February 1771.

[52] BA&H, MS 3782/21/1/156, AJC to JF, 11 June 1773.

[53] BA&H, MS 3782/1/9/355, B&F to JF, 23 January 1772.

several that I have been served so. I had one parcel of cand[lesticks] quite broke by their careless packing'.[54] These unsubstantiated remarks can best be regarded as a part of Boulton's campaign, since he went on to link his remarks to a need for an assay office. It was surely not a coincidence that the only specific order for which Boulton claimed damage, should have been for an influential man who as early as 1766, had shown himself to have sympathy for the difficulty of labouring without an assay office in Birmingham.[55] Significantly, around this date, Garbett was in close touch with Shelburne and Boulton.[56]

Of course, the need to send silver to Chester caused some delay in completing orders, but remarks by Boulton and Garbett considerably exaggerated the difficulties. One consignment was sent from Soho on 13 May 1772.[57] It was returned from Chester on 20 May.[58] Exactly when the package was received back at Soho is not clear but a letter from Soho to a friend and business associate of Boulton's, put the average delay in sending pieces to Chester at only one week.[59] Boulton frequently exaggerated these delays when writing to customers, partly to disguise the time it took to make silver at Soho. Candlesticks were ordered by a customer on 20 June 1771.[60] They were not sent to Chester until 30 July.[61] When Soho received them back is not clear, but, after final modifications and polishing, they were sent to London on 13 August when their lateness was entirely blamed on the need to send them to Chester, even though the delay in manufacture was several times longer.[62] Many other letters show that the delays in sending pieces to Chester were used as an excuse for lateness.[63] These replies were orchestrated by Boulton as part of a public campaign: there is not a single letter to Folliott complaining about delays in assaying or returning silver, though once he was urged to ensure there would be no delay.[64]

The Parliamentary Committee also asked Garbett: 'Is it [Boulton and Fothergill silver] all assayed there? [Chester]' to which Garbett replied 'I believe

---

[54]   BA&H, MS 3782/1/9/1–2, MB to the Earl of Shelburne, 7 January 1771.

[55]   Dickinson, p. 64.

[56]   BA&H, MS 3782/12/23/87, Francis Garbett [Samuel's son] to MB, 16 December 1766 and 3782/12/23/96, Samuel Garbett (SG) to MB, 23 March 1767.

[57]   BA&H, MS 3782/1/9/448, AJC to JF, 13 May 1772.

[58]   BA&H, MS 3782/1/9/458, AJC to JF, 25 May 1772.

[59]   BA&H, MS 3782/1/9/ 47–8, JWy to WM, 18 February 1771.

[60]   BA&H, MS 3782/1/9/172, Charles Wyatt (CW) to Henry Morris (HM), 13 August 1771.

[61]   BA&H, MS 3782/1/9/157, CW to JF, 30 July 1771.

[62]   BA&H, MS 3782/1/9/172, CW to HM, 13 August 1771.

[63]   BA&H, e.g. MS 3782/1/9/278, B&F to Nicholas Smyth, 19 November 1771; MS 3782/1/9/162–3, MB to the Duke of Grafton, 1 August 1771.

[64]   BA&H, MS 3782/21/1/137, B&F to JF, 27 May 1773.

it is'.[65] Garbett was here following the partners who had implied or firmly stated this position in many letters to customers[66] and in 1771 Boulton and Fothergill insisted: 'We have always sent our plate to Chester to be mark'd where our mark is enter'd'.[67] This view has largely been perpetuated by later historians.[68] Yet in 1772 Lord Boston was advised to get his tea-urn assayed in London, because the partners lacked the time to send it to Chester, and offered the help of a London agent in arranging the assaying 'at any time you can spare the urn'.[69] That the partners thought, very unusually for them, of assaying the urn in London rather than Chester is of no particular significance, but they were being cavalier over discharging their legal obligation to ensure that assaying had taken place somewhere before sale.[70] In 1772, Patrick Robertson of Edinburgh was sent two pairs of drinking cups.[71] There is no Folliott correspondence that shows that they were sent to him and since at a later date (1776) Robertson, a maker and retailer, was sent at his request two pairs of unassayed candlesticks it is quite possible that this malpractice also happened in 1772.[72] In cases where Boulton failed to have pieces marked, this may have been overcome by retailers placing their so-called 'makers' marks on pieces, a widespread and seemingly acceptable practice, but it still conflicted with Garbett's evidence.[73] However, the partners in 1772 supplied Ely House in Dublin with sets of door-knobs and escutcheons, 11 of which survive today, none of which exhibit hallmarks, even though they were not on the official list of exemptions.[74] The partners directly dealt with the Earl of Ely,[75] as they did with another customer in Dublin, Henry Meredith, who was asked whether he wanted his candlesticks assayed at Chester

[65]   Minutes of the Committee on Sheffield and Birmingham Petitions, Goldsmiths' Company MS G 11 2 5 (1773).

[66]   BA&H, e.g. MS 3782/1/9/162–3, MB to the Duke of Grafton, 1 August 1771, and MS 3782/1/9/161, MB to James Stuart (JS), 1 August 1771.

[67]   BA&H, MS 3782/1/9/277, B&F to Parker and Wakelin, 16 November 1771.

[68]   Delieb and Roberts, p. 45 and Dickinson, p. 64. However, Ridgway, p. 61, did point to one case of a tea–urn for Lord Boston which may not have been assayed.

[69]   BA&H, MS 3782/1/9/ 626–7, B&F to Lord Boston, 1 November 1772.

[70]   Forbes, p. 16.

[71]   BA&H, MS 3782/1/38, p. 99, B&F to Patrick Robertson, 18 November 1772.

[72]   BA&H, MS 3782/1/10/658, B&F to Patrick Robertson, 13 July 1776.

[73]   Helen Clifford, *Silver in London: The Parker and Wakelin Partnership 1760–1776* (New Haven and London, 2004), p. 57.

[74]   Nicholas Goodison, 'The Door Furniture at Ely House', *Bulletin of the Irish Georgian Society*, X111, 2–3 (April–September 1970), pp. 45–48, 48.

[75]   BA&H, MS 3782/1/9, p. 387–389, B&F to the Earl of Ely, 6 March 1772.

and it is not clear whether the silver was assayed.[76] Boulton, wrongly, acted as if it was legal not to hallmark plate for export.[77]

Nevertheless, these were a relatively limited number of lapses, given that there are 27 occasions recorded in the Matthew Boulton Papers when silver was sent to Chester for marking from 1771 to 1773,[78] when consignments usually consisted of several items.[79]

However, lapses would have caused considerable embarrassment if they had come to the attention of the Committee. There would have been similar consternation if it had come to light that some pieces of early Boulton and Fothergill silver, contrary to the law,[80] were not stamped with their makers' marks. This applies to a pair of candlesticks in the assay year 1768–69.[81] Several others come from the assay year 1769–70 including a coffee pot and stand and at least seven candlesticks (Figure 9.1).[82] Normally makers' marks were struck after silver was returned from an assay office.[83] The failure to add marks in these cases was probably due to a lack of time, a frequent problem at Soho.[84]

Another issue raised by the Parliamentary Committee with Garbett, was the danger of patterns being copied by rival manufacturers when they were sent to assay offices considerable distances from Birmingham. The matter had earlier been raised by Boulton in the *Memorial Relative to Assaying and Marking Wrought Plate at Birmingham* (1773): '... their fresh designs, which have often cost them considerable sums of money, and always Pains and Time, are communicated to rivals before the Inventors have reaped benefit from them'.[85] Boulton had occasionally shown anxiety about the matter, as when he sent a special cup for Thetford races for the Duke of Grafton, via a London agent, which Boulton asked the agent not to allow people to see.[86] Uniquely, when this cup was sent to Chester, Boulton sent a clerk with it for hallmarking, which may suggest that he was keen that no one, apart from the Assay Office

---

[76] BA&H, MS 3782/1/9, p. 640, B&F to Henry Meredith (HMe), 18 November 1772. Since the bill was £20.1s.8d. the order was probably for a pair of candlesticks (MS 3782/1/38/311–12, B&F to HMe, 4 September 1773).

[77] Forbes, p. 192.

[78] BA&H, MS 3782/1/9 and MS 3782/21/1, totals calculated from these documents.

[79] BA&H, e.g. MS 3782/1/9, p. 219, B&F to JF, 12 October 1771. Three pairs of candlesticks.

[80] Forbes, *Hallmark*, p. 19, Statute 37 Edw. 111. c.7,1363 and p. 163 Goldsmiths' Co., order of 1698.

[81] See footnote 38.

[82] See footnote 39.

[83] Forbes, p. 19.

[84] BA&H, MS 3782/1/9, p. 172, CW to HM, 13 August 1771.

[85] Goldsmiths' Company, G 11 5 4, p. 1.

[86] BA&H, MS 3782/1/9, pp. 137–138, John Scale (JSc) to WM, 1 July 1771.

Figure 9.1   Attributed to Boulton and Fothergill, *Lyon-faced* candlestick,
                silver, 1769–70

Source: Cobbe Collection, courtesy of Alec Cobbe.

staff, would see it there.[87] However, there is no correspondence with Folliott in which Boulton complained that his designs had been plagiarised in Chester. The dangers of this happening in London were greater, because that was the main centre of silversmithing in England, as Boulton acknowledged,[88] but there was no question of Boulton, at least normally, using the Assay Office there.

Therefore, when asked by the Parliamentary Committee about the matter, Garbett's comments related mainly to other makers, especially buckle-makers in Birmingham, who Garbett maintained invented many new patterns, at risk in London because Birmingham makers sometimes used the Assay Office there. This evidence usefully supported Garbett's and Boulton's argument. When asked whether 'They [patterns] are not liable to be as much exposed at [an] assay office at Birmingham as at London [?]', Garbett responded: 'I think not'. When asked 'Why', Garbett replied: 'We should send servants to the assay office with the new patterns and endeavour to find some means to prevent impressions being taken

[87]   BA&H, MS 3782/1/9, p. 133, B&F to JF, 21 June 1771.

[88]   *Reply to the Petitioners from Birmingham and Sheffield to the Case of the Goldsmiths, Silversmiths and Plateworkers of the City of London and Places Adjacent* (1773), quoted in Arthur Westwood, *The Assay Office at Birmingham Part 1: its Foundation* (Birmingham, 1936), p. 21.

from them in the short time they were in the office' and continued we '... can control carriers in Birmingham to avoid problems which might arise with carriers in more distant places'.[89]

The argument that by sending articles considerable distances, the partners' silver was exposed to the risk of plagiarism, has been repeated by later writers.[90] However, with the possible exception of the *Lyon–faced* candlestick (Figure 9.1), a French pattern used in England by Boulton in 1768–69, and subsequently in a slightly attenuated form by firms in London, at least from 1771–72, there is no evidence that his patterns in this period were copied anywhere.[91]

The Range of Assay Silver

Garbett was asked by the Parliamentary Committee: 'What species do they [Boulton and Fothergill] manufacture?' Garbett replied: 'A great many sorts of silver wares. Many ornamental Utensils: tureens, candlesticks, vases, coffee pots. All of solid silver'.[92] This generalisation has been repeated in modern scholarship without question.[93]

Boulton had been making that range, but Garbett's response conceals the fact that candlesticks were by far the most numerous among his range of assay silver: at least 107, while only 48 other pieces, plus at least 11 sets of door furniture for Ely House, are known to have been made.[94] At least 10 candlesticks were cast, but Boulton's characteristic method was stamping and that method of manufacture

---

[89]   Minutes of the Committee on Sheffield and Birmingham Petitions, Goldsmiths' Company, MS G 11 2 5, pp. 16–22.

[90]   'Matthew Boulton' in *Birmingham Gold and Silver 1773–1973*, exhibition catalogue (Birmingham, 1973); Jennifer Tann, *Birmingham Assay Office, 1773–1993* (Birmingham, 1993), p. 20.

[91]   Quickenden, '"Lyon–faced" Candlesticks', pp. 198–200. London designs could have been based directly on French candlesticks.

[92]   Minutes of the Committee on Sheffield and Birmingham Petitions, Goldsmiths' Company, MS G 11 2 5, p. 13.

[93]   Ridgway, p. 62.

[94]   Totals calculated from BA&H, MS 3782 plus evidence from surviving pieces. These surviving pieces are as follows for 1768–69: a pair of candlesticks at Birmingham Assay Office (BAO 1140); an attributed pair of candlesticks at The Speed Art Museum (accession no. 1993.9.192a–b) and a set of four in a private collection. Surviving pieces for 1769–70 are: an attributed pair of candlesticks in BMAG (2002 M.364.1–2); Cobbe Collection: one attributed candlestick; a pair of attributed candlesticks at the Grosvenor Museum, Chester (accession no.2002.300); an attributed pair Christie's, London, 2 June 2009, lot 0277; an attributed coffee pot and stand at BMAG (1996 M 1); a mazarine (BAO 23); a pair of candlesticks for sale at Byrne's, Chester, 22 September 2010, lot 100. Ridgway, pp. 61–62, lists three Masonic candlesticks of 1768 but these have since been shown to be mainly of Sheffield Plate (Mason (ed.), p. 162).

was far more associated with candlesticks than any other large items. The large demand was due to their low price. Stamped *Lyon–faced* candlesticks (Figure 9.1) used only 38 ounces of silver per pair. They were priced at £17.2s.0d. as against the 108 ounces used for cast candlesticks, the method normally used in London, priced at £44.11s.0d.[95] London's silversmiths adopted a dismissive attitude towards stamped work and consistently characterised Birmingham's and Sheffield's work as inferior work of that kind. A petition presented to the House of Commons by London plateworkers insisted that '… most of the plate they make in those Towns is made or stamped with dies …'[96] A document by the Goldsmiths' Company argued that as a result '… there are very few real Goldsmiths, Silversmiths or Plate workers in either of those towns'.[97] Garbett was acting in Boulton's best interests by implying that his candlesticks, generally made by stamping, were no more significant in the firm's range than the other items, produced by the more demanding and traditional techniques frequently used in London.

The number of hand-wrought pieces made by the partners prior to the opening of The Birmingham Assay Office was very small. Their production may have had to wait until the recruitment of the specialised silversmith, Thomas Bunbury, from Dublin in 1770, who was primarily responsible for making the most important commission for the silver business before 1773, a tureen for the Admiralty (Figure 9.2),[98] which was probably obtained through James Stuart, the prominent neo-classical architect who designed the piece. The work required high levels of skill, especially chasing, for which Boulton had few staff,[99] and castings of dolphins that required models from Josiah Wedgwood.[100] The Admiralty had budgeted £100 but the bill came to £140.16s.0d. Boulton left the Admiralty to pay whatever it wished, putting the apparent error down to inexperience.[101] But Boulton was in fact intent on obliging. He only charged 2s.6d. per ounce for the fashioning, while reckoning that a London silversmith would have charged 5s.0d. per ounce.[102] At this point Boulton pursued a policy of very low pricing to obtain orders.[103] Even though Garbett made reference in his evidence to the production of tureens in the plural, the Admiralty tureen was the only one known to have been made by the partners before Garbett gave evidence.

---

[95]   Quickenden, '"Lyon–faced" candlesticks', pp. 204–205.

[96]   Goldsmiths' Company, MS Court Minute Book 17, 1767–1777, p. 232.

[97]   *Case of the Goldsmiths, Silversmiths and Plateworkers of London and Places Adjacent*, 25 February 1773, Goldsmiths' Co., G 11 1 5, p. 1.

[98]   Kenneth Quickenden, 'Boulton and Fothergill's Silversmiths', *The Silver Society Journal*, 7 (Autumn, 1995), pp. 342–356, 349.

[99]   BA&H, 3782/1/9, p. 161, B&F to JS, 1 August 1771.

[100] Goodison, *Ormolu*, p. 377, note 52.

[101] BA&H, MS 3782/1/9/212–3, B&F to JS, 5 October 1771.

[102] BA&H, MS 3782/1/11/738, John Hodges (JH) to MB, 4 April 1781.

[103] BA&H, MS 3782/12/2/72, MB to the Duke of Richmond (DoR), 4 December 1772.

Figure 9.2   James Stuart, design for a silver tureen for the Admiralty, inscribed
1781 but original design c. 1771, Pattern Book 1 1762–90
Source: Courtesy of Birmingham Archives and Heritage, MS 3782/21/2, p. 111.

The other articles in Garbett's list would also have been primarily hand-wrought, sometimes with castings, but the total was not very large. The reference to 'vases' was probably made with the Thetford Cup in mind.[104] However, the firm had also made three tea-urns.[105] Coffee pots were more numerous: Boulton once offered to make one for 3s.0d. per oz., half the price thought to be charged in London.[106] In total six were made: three with stands[107] and three without.[108]

Garbett's response highlighted the most impressive pieces made at Soho, but production also included at least one example of each of the following types of

[104]   BA&H, MS 3782/1/9/137–8, JSc to WM, 1 July 1771.

[105]   BA&H, MS 3782/12/60/51, John Fothergill to MB, 4 April 1771; 3782/1/9, p. 140, CW to JF, 10 July 1771; 3782/1/9/626–7, B&F to Lord Boston, 1 November 1772.

[106]   BA&H, MS 3782/12/1/154–5, B&F to Mr Udney, 12 June 1773.

[107]   BMAG (accession no. 1996 M 1), attributed, assay year 1769–70; BA&H, MS 3782/1/9/448, AJC to JF, 13 May 1772; 3782/1/9/464, AJC to JF, 30 May 1772.

[108]   BA&H, MS 3782/672, B&F to JF, 17 December 1772; 3782/21/1, p. 91, B&F to JF, 16 April 1773; 3782/21/1/111, AJC to JF, 1 May 1773.

assay silver: mazarine,[109] plate, fish trowel,[110] tea canister,[111] sugar dish, candlestick branch, cream jug,[112] cruet frame,[113] buckle,[114] and drinking cup.[115]

The Volume of Assay Silver

London's silversmiths attempted to minimise the extent of assay silver production in Sheffield and Birmingham,[116] but Boulton strained to do otherwise.[117] Garbett was asked by the Commons Committee: 'Has Mr Bolton's [sic] manufacture increased lately?' and he replied: 'Yes' adding, 'I sell him several thousand pounds worth of silver in a year'.[118] Garbett might have been further questioned over whether or not this sum applied just to supplies for assay silver, but he was not. Nor has the question been asked subsequently.[119]

It is true that Boulton paid Garbett £5,113 for bullion in the assay year 1771–72.[120] But much of this was for gold, which was especially used at this time for gilding ormolu.[121] Moreover, much silver was for Sheffield Plate, which by now had become a principal branch producing a wide range of tablewares.[122] Additional silver would have been needed for silver items exempt from assay such as filigree needle books.[123] Given Garbett's awareness of Boulton's Sheffield Plate production,[124] and Garbett's almost certain awareness of the production of small non-assay silver items at Soho (apart from their close friendship, Garbett

[109]   BAO, accession no. 23.

[110]   BA&H, MS 3782/1/9/252, B&F to JF, 2 November 1771.

[111]   BA&H, MS 3782/1/9/522, AJC to JF, 23 July 1772.

[112]   BA&H, MS 3782/21/1/91, B&F to JF, 16 April 1773.

[113]   BA&H, MS 3782/21/1/137, B&F to JF, 23 May 1773.

[114]   BA&H, MS 3782/21/1/156, AJC to JF, 11 June 1773.

[115]   BA&H, MS 3782/1/38/99, B&F to Patrick Robertson, 18 November 1772.

[116]   *Case of the Goldsmiths, Silversmiths and Plate-workers of London & Places Adjacent* (1773), Goldsmiths' Company, G 11, 1, 5, p. 2.

[117]   BA&H, MS 3782/12/2/72, MB to the DoR, 4 December 1772.

[118]   Minutes of the Committee on Sheffield and Birmingham Petitions (1773), Goldsmiths Company MS G 11 2 5, pp. 13–14.

[119]   Dickinson, p. 67.

[120]   Quickenden, 'B&F's Bullion Supplies', p. 52.

[121]   Goodison, *Ormolu*, p. 55.

[122]   BA&H, MS 3782/1/9/107, B&F to JP, 17 May 1771.

[123]   BA&H, MS 3782/1/9/322, B&F to Charles Vere, 20 December 1772.

[124]   Later in his evidence, in response to another question, Garbett mentioned in passing that MB 'deals in plated work' i.e. Sheffield Plate (Minutes of the committee on Sheffield and Birmingham Petitions, Goldsmiths' Company MS G 11 2 5, p. 16). No one on the Committee thought to question Garbett about the significance of that answer to his earlier answer about the volume of silver he sold to MB.

supplied silver before Sheffield Plate and assay silver were being produced at Soho), Garbett knew that silver was being used there for a range of purposes.[125] The total amount of silver being purchased was therefore a poor indicator of the level of assay silver production, which was the subject of the Committee's enquiry. The assay year 1771–72 is richly documented in the Matthew Boulton Papers but only £510.16s.3d. of silver can be shown to have been sent for assay in that year.[126] The silver door furniture for Ely House was, additionally, supplied in that period but the total cost for that order, including locks, was only £39.7s.0d.[127] Garbett had access in his own firm's accounts to the level of the partners' purchases and while his comment about the level of silver purchase was true it was also knowingly misleading.

The Future of Silversmithing at the Soho Manufactory

The Commons Committee, which was keen to be re-assured that the foundation of The Birmingham Assay Office would have sufficient business to justify its existence, asked Garbett: 'Why do you think Mr Boulton's Manufactory might be carried on to a greater extent?' Garbett replied:

> I believe that Mr Boulton has many thousand dies that he c[oul]d make use of in different parts of the manufactory [Soho] and which he does not now make use of in plate [ie silver] because of the inconveniences he w[oul]d be under in those articles by sending them to distant places to be assayed ...[128]

The use of dies at Soho was widespread for articles including buttons,[129] buckles,[130] for parts of tablewares such as waiter borders[131] and candlesticks. Dies were used for a variety of metals, including already to some extent silver, so that, for example, the dies used for the silver *Lyon-faced* candlestick (Figure 9.1) were also used for the same pattern in Sheffield Plate. They were priced at £7.17s.6d. which was

---

[125]   Quickenden, 'B&F's Bullion Supplies', p. 52.

[126]   Total ounces based upon BA&H, MS 3782/1/9. Sterling cost 5s.9d. per ounce (MS 3782/1/9/443, AJC to MB, 7 May 1772) and totalled approximately 663 ounces costing £190.12s.3d. plus about £100 of silver used for the Admiralty tureen (MS 3782/1/9/212–3, B&F to JS, 5 October 1771). Purer, softer silver, to pick up more detail from the die, was used for stamped candlesticks and cost 6s.0d. per oz. (3782/1/9/111–2, B&F to James Shrapnell, 29 May 1771). Approximately 734 ozs. were used, costing £ 220.4s.0d.

[127]   Goodison, 'The Door Furniture at Ely House', p. 48.

[128]   Minutes of the Committee on Sheffield and Birmingham Petitions (1773), Goldsmiths' Company MS G 11 2 5, p. 19–20.

[129]   BA&H, MS 3782/1/10/234–5, B&F to J.P. Du Roveray, 21 January 1775.

[130]   BA&H, MS 3782/1/10/112–3, B&F to W. Webb, 29 August 1772.

[131]   BA&H, MS 3782/1/9/89, B&F to WM, 22 April 1771.

£9.4s.6d. less per pair than the same pattern in silver.[132] Further materials used, or experimented with for candlesticks, included pinchbeck – an alloy of copper and zinc resembling gold, copper, enamelled base metals,[133] brass, ormolu, tutenague – a silvery-looking alloy of copper, zinc and nickel, and platina – a whitish alloy of copper and zinc.[134] Dies lasted indefinitely and Boulton maximised their use.[135] Therefore, articles such as candlesticks and buttons in a wide variety of materials were, and remained, staples in the firm's range.[136]

This approach to manufacture, and his generally low prices, provided support for Boulton's non-elitist contention in the campaign to secure The Birmingham Assay Office that, as he put it in a letter to Lord North, the Prime Minister, early in 1773, a successful outcome would be '… tending to the good of the community in general'.[137] In the *Reply of the Petitioners from Birmingham and Sheffield to the case of the Goldsmiths, Silversmiths and Plateworkers of the City of London and Places Adjacent* (1773), it was argued that silver plate by London makers was 'exorbitant' in price, which '… renders it unsaleable to all but a few rich people …' with the effect of 'diminishing the demand …' and preventing it becoming '… a considerable Article of Commerce with the neighbouring Nations …' and depriving '… workmen of bread …'.[138] By implying that Birmingham would be aiming at a wider and different market, Boulton might have been trying to allay the fears of London makers who thought: '… if the goldsmiths manufactory in Birmingham and Sheffield should be increased that in London must suffer a proportional decrease'.[139] But the Birmingham case was also calculated to appeal to politicians, keen to promote employment, prosperity and exports and who had to decide whether Birmingham should be granted an assay office.

While other Birmingham makers may well have had a wide market in mind, it is very doubtful that Boulton intended to make this the central thrust of his silver plate business. Prior to obtaining The Birmingham Assay Office, his prices for hand-wrought items had been around 50 per cent of Londoner's fashioning charges. This may have encouraged many to believe that he was aiming at a

---

[132]   Quickenden, "Lyon–faced' candlesticks', p. 206.

[133]   Kenneth Quickenden, 'Silver, "Plated" and Silvered Products from the Soho Manufactory', *The Silver Society Journal*, 10 (Autumn, 1998), pp. 73–95, 86.

[134]   Quickenden, 'Silver and its Substitutes', pp. 159–160.

[135]   BA&H, MS 3782/12/1/62, B&F to John Motteux, 23 February 1768.

[136]   In 1780, for example the firm made 4,428 silver-gilt buttons, 1,224 pierced silver-gilt buttons, 28,800 Sheffield Plate buttons on bone, 31,824 Sheffield Plate buttons on box wood, 30 silver candlesticks and 2,572 Sheffield Plate candlesticks: Quickenden, 'Silver, "Plated" and Silvered products', p. 92.

[137]   National Archive, Kew, T 1/498/161–2, MB to Lord North, 2 February 1773.

[138]   Arthur Westwood, *The Assay Office at Birmingham Part 1: Its Foundation* (Birmingham, 1963), p. 21.

[139]   *Case of the Goldsmiths, Silversmiths and Plate-workers of London and Places Adjacent* (1773), Goldsmiths' Company G 11 1 5, p. 3.

wide clientele. However, at this stage he was keen to enter the market, create a reputation and gain support for the Assay Office campaign amongst the influential. He had once offered to make tureens for Sir Robert Murray Keith without profit, for the prestige of making them,[140] and other items had been made at a loss.[141] However, immediately following the passing of the 1773 Act, Boulton pushed up prices for essentially handmade items so they became about 20 per cent less than London prices. He made plates for the Bluestocking Elizabeth Montagu for 1s.2d. per ounce, while reckoning that a London silversmith would have charged 1.6d.[142] Boulton's competitive prices attracted orders from the 'rich'. He only made full services of plate for the aristocracy, such as Lord Craven and the Earl of Malmesbury or the gentry, such as John Turton of Sugnall Hall near Stafford, as well as Mrs Montagu. There were instances of purchases by middle-class customers, but they were overwhelmingly confined to articles such as plain teapots at £10.0s. 0d. or buckles, at 12s.4d. a pair.[143] While the annual incomes of the aristocracy exceeded £10,000 and those of the gentry reached £8,000, they could afford services of over £1,000, but such items were beyond the middle classes where incomes were usually less than £600.[144] When making services of plate proved to be unprofitable, silversmithing generally was run down from the late 1770s.[145] However, Sheffield Plate, which was sold for under half the price of silver, with trade discounts that were refused for silver, at home and abroad, was successfully promoted for a wide market.[146] During the Boulton and Fothergill partnership between 1762 and 1782, silver production was elitist in ambition.[147] This perspective conflicts with the traditional tendency to characterise Boulton's silver as being non-elitist in character.[148]

---

[140]   BA&H, 3782/1/9/554, B&F to Sir Robert Murray Keith, 23 August 1772.

[141]   BA&H, 3782/1/11/738, JH to MB, 4 April 1781.

[142]   BA&H, MS 3782/1/10/830, B&F to Elizabeth Montagu, 15 February 1777.

[143]   Kenneth Quickenden and Arthur J. Kover, 'Did Boulton and Fothergill Sell Silver Plate to the Middle Class? A Quantitative Study of Luxury Marketing in Late Eighteenth-Century Britain', *Journal of Macromarketing*, 27/1 (March, 2007), pp. 51–64, 56–57.

[144]   Ibid., p. 56.

[145]   Kenneth Quickenden, 'Boulton and Fothergill Silver; Business Plans and Miscalculations', *Art History*, 3/3 (September, 1980), pp. 274–294, 288.

[146]   Ibid., pp. 281–282.

[147]   Quickenden and Kover, 'Did Boulton Sell Silver to the Middle Class?', pp. 58–59. Purchases by the aristocracy and gentry combined for assayed silver as shown by the firm's papers were six times greater than those of the middle classes even though the population of the latter group was more than 36 times greater, though is should be noted that documentation for wealthy customers is more substantial than that for the middle classes.

[148]   Ridgway, p. 60 wrote that Boulton's '... original aim seems to have been to supply the middle class with mass-produced items ...'.

The claims about the possibility of extending exports in *Reply of the Petitioners* ... have not been questioned in some modern scholarship.[149] Garbett's evidence, which held that the export of buckles by Birmingham manufacturers was already considerable, contended that export could be extended to more pieces of silver plate.[150] There had been several instances of Boulton exporting buckles,[151] but few larger pieces had been made for export, an unspecified number of candlesticks for a London merchant in 1771 being an exception.[152] The largest order was for Sir Robert Murray Keith, British Ambassador to Vienna, which was worth £268.19s.0d. but, not untypically for export orders, there was a delay in payment, which may have inhibited the partners' ambitions to export plate.[153] Other orders included the door furniture for Ely House and candlesticks for Henry Meredith, both in Dublin. Apart from the physical proximity to Ireland, a key factor was Ireland's use of the same sterling standard as England.[154] That did not generally apply on the Continent but Boulton and probably Garbett were unaware of that in 1773 since Boulton resolved in 1775 to find out about the intrinsic standard required in Paris, Hamburg and Amsterdam.[155] Even later it was an issue before the Birmingham Commercial Committee, of which Boulton was a member, which conducted research showing that whereas the standard in England was 11/12ths of fine silver, standards abroad were only 9/12ths or 10/12ths.[156] There were few export orders after 1773; major exceptions were those for the Duke of Holstein-Gottorp,[157] and Alexei Semyonovich Moussin-Pushkin, the Russian Ambassador.[158] Silver plate was entirely omitted from a long list of Boulton's exports to France.[159] Boulton's apparent enthusiasm for exporting silver in 1773 cannot have been strong given his difficulties in receiving payments from abroad and his lack of research over intrinsic standards in different countries. Significantly, in 1781, a

---

[149]   Westwood, *The Assay Office at Birmingham*, p. 21; A.H. Westwood, 'The Birmingham Assay Office' in *Birmingham Gold and Silver, 1773–1973* (Birmingham, 1973).

[150]   Minutes of the Committee on Sheffield and Birmingham Petitions (1773), Goldsmiths' Company, MS G 11 2 5, pp. 15–16.

[151]   BA&H, MS 3782/21/1/115–6, B&F to WM, 4 May 1773 and MS 3782/1/18/8, 15 July 1769.

[152]   BA&H, MS 3782/1/9/172, B&F to WM, 14 August 1771.

[153]   BA&H, MS 3782/12/60/102, Account of debts, 1 April 1773, £7,744.15s.3d. of which £5,054.3s.4d. was for debts from abroad.

[154]   Douglas Bennett, *Irish Georgian Silver* (London, 1972), pp. 20–21. See footnotes 75 and 76.

[155]   BA&H, MS 3782/12/107/10, p. 20.

[156]   BA&H, MS 3782/12/87/76, 25 January 1785.

[157]   BA&H, MS 3782/1/10/253, B&F to WM, 6 February 1775.

[158]   BA&H, MS 3782/1/4/58, note of bill, 29 December 1778.

[159]   BA&H, MS 3782/1/11/16, B&F to J.C. Preidel, 7 June 1777.

Swiss agent of many years standing expressed surprise that Boulton had started silver plate production many years before.[160]

Conclusions

Garbett's evidence to the Parliamentary Committee was favourable to Boulton and though plausible it was also misleading. Garbett said that Boulton's production of assay silver was more firmly established, larger and more impressive than it actually was. He had also implied that the problems of assaying silver at Chester – the delays, the damage, the plagiarism – were greater than they really were, and his assertion that silver had always been assayed before sale was untrue. His comments exaggerated the partners' prospects of widening the market for silver, through targeting less elitist markets and exporting.

Though Garbett's evidence was misleading, it seems to have been entirely acceptable to the Committee. Garbett's reputation for integrity and his knowledge of the trade may have protected him from more rigorous questioning. He was perhaps fortunate too that the Committee was chaired by a relatively local MP to Birmingham as well as the Soho Manufactory and Boulton – Thomas Skipwith, MP for Warwickshire.[161] Garbett's reputation remained intact: in 1785 the Prime Minister, William Pitt wished to consult Garbett, as a Birmingham representative, over his plans for free trade with Ireland.[162]

Damage to Garbett's reputation over the deliberations of 1773 would largely have been unjust since his evidence must have been heavily dependent on Boulton. There is a lack of surviving correspondence between the men over the issues raised by the Parliamentary Committee, prior to the meeting, though there is surviving correspondence about other issues in the early 1770s.[163] But Garbett's witness statements were in perfect concert with the arguments advanced earlier by Boulton in his campaign, which Garbett must have been aware of. While on most issues – the range of assay silver, the use of the Chester Assay Office, widening the market – Garbett was probably heavily reliant on Boulton for information, in the matter of the volume of silver supply, where Garbett had access to his own firm's accounts, he had independent information. Though he did not say so, Garbett was also aware that Boulton required silver for more than just assay silver. In implying that the volume of silver used for assay silver by the partners prior to 1773 was greater than it was, Garbett was betraying, though not to the Committee, that he was doing what he could to aid Boulton's campaign.

---

[160]   BA&H, MS 3782/12/26/116, S. Clais to MB, 10 December 1781.

[161]   Baggott, 'I am Very Desirous of Becoming a Great Silversmith', p. 49.

[162]   Campbell, p. 20.

[163]   For example, canal development: BA&H, MS 3782/12/2/40–43, MB to SG, 12 February 1771.

The unified picture presented by Garbett and Boulton was persuasive not only to the Parliamentary Committee but also to later scholars' accounts of the partners' early assay silver. But as this analysis has attempted to show, the validity of that picture has been undermined by information which Boulton left behind in the Matthew Boulton Papers.

The report of Skipwith's Committee and the findings of special meetings to enquire into the conduct of Assay Offices in England, which found that Boulton and Fothergill's silver had always been of the required sterling standard, ensured that the Act which brought into existence assay offices at Birmingham and Sheffield received Royal Assent on 28 May 1773.[164]

---

[164]   Baggott, 'I am Very Desirous of Becoming a Great Silversmith' p. 51.

Chapter 10

Hegemony and Hallmarking:
Matthew Boulton and the Battle for
The Birmingham Assay Office

Sally Baggott

Matthew Boulton succeeded in gaining an Act of Parliament for the establishment of an assay office in Birmingham in May 1773, overcoming in the process, powerful objections from the Worshipful Company of Goldsmiths and the London trade. Now the largest assay office in the world, The Birmingham Assay Office is the sole surviving enterprise in which Boulton had a hand, and it serves as a tangible example of his legacy in Birmingham and beyond.[1] Yet, Boulton's campaign has been largely obscured by the sheer weight of historical significance that his life and works represent, and, where his campaign has been attended to, it has been treated only as one episode in the context of greater narratives. The story of Boulton and The Birmingham Assay Office, however, is worth re-telling for what it reveals about the wider cultural context of the late eighteenth century. What emerges is a culture beset by conflict between London and the provinces and between craftsmen and emergent industrialists and, in order to make sense of this, I want to situate the events of 1773 within Raymond Williams' concept of culture. In particular, his interpretation of Antonio Gramsci's notion of hegemony provides a productive framework for understanding the forces at work during Boulton's campaign.[2]

Existing accounts of Boulton's battle for an assay office in Birmingham take a variety of approaches. In 1936, Arthur Westwood, who served as Junior Assay Master at Birmingham from 1893–1911 and as Assay Master from 1911–1951, published the most focused and comprehensive commentary to date.[3] In his

---

[1]   See Jennifer Tann, *Birmingham Assay Office 1773–1993* (Birmingham, 1993). At the start of 2011, The Birmingham Assay Office remained the largest in world terms and had the largest market share in the UK. For publication of quarterly hallmarking figures for the four UK Assay Offices see *The Anchor* published by the Birmingham Assay Office, http://www.theassayoffice.co.uk/anchor.html

[2]   See Raymond Williams, *Marxism and Literature* (Oxford, 1977), pp. 108–127 and Antonio Gramsci, *A Selection from the Prison Notebooks*, ed. and trans., Quintin Hoare and Geoffrey Nowell-Smith (London, 1971).

[3]   Arthur Westwood, *The Assay Office at Birmingham* (Birmingham, 1936).

biography of Boulton from the following year, H.W. Dickinson recounts the events of the 1773 campaign within the wider narrative of Boulton's life and acknowledges Westwood's contribution.[4] In an article from 1964, Eric Robinson views the events of 1773 as significantly formative of Boulton's skill as a Parliamentary lobbyist; and Jennifer Tann begins her history of The Birmingham Assay Office, published in 1993, with Boulton's role in its establishment.[5] Kenneth Quickenden's extensive work has provided useful context to the 1773 campaign in terms of the aesthetic, technical and economic dimensions of the manufacture of silver, Sheffield Plate and other metallic materials at Soho.[6] His more recent research included here, tests the veracity of claims made during the campaign for The Birmingham Assay Office and thus provides a corrective to previous interpretations.[7] All drawing extensively on the documentary evidence, now housed in the Archives of Soho, Birmingham Archives and Heritage, these accounts contain useful archival detail; projecting an image of Boulton as an ambitious and astute though not always successful businessman, endowed with considerable skill in argument, rhetoric and public relations. His campaign to establish The Birmingham Assay Office, however, warrants a different kind of approach; set within a theoretical framework, Boulton and his battle with London are indicative of significant tensions in the wider cultural context of the late eighteenth century. If we are to grasp the full significance of Boulton's campaign, we need to understand the exact nature of those tensions in terms of how the London faction sought to legitimate and maintain their dominance, and how Boulton and the Birmingham trade challenged London's position. It is also worth considering the extent to which Boulton, in winning the battle for The Birmingham Assay Office, was successful in that challenge.

It is here that I want to turn to Williams' reading of the Gramscian concept of hegemony. For Williams, Gramsci's work represents 'one of the major turning-points in Marxist cultural theory'; certainly Gramsci has been influential in Marxist theory and in British Cultural Studies more specifically since his concept of hegemony overcomes one of the foremost limitations of traditional Marxist

---

[4]   H.W. Dickinson, *Matthew Boulton* (Cambridge, 1936), pp. 64–70.

[5]   Eric Robinson, 'Matthew Boulton and the Art of Parliamentary Lobbying' in *The Historical Journal*, VII, 2 (Cambridge: 1964), pp. 209–229 and Jennifer Tann, *Birmingham Assay Office 1773–1993* (Birmingham, 1993).

[6]   See Kenneth Quickenden, 'Matthew Boulton's Silver and Sheffield Plate' in Shena Mason (ed.), *Matthew Boulton: Selling What All The World Desires* (New Haven and London, 2009), pp. 41–46; 'Silver and its Substitutes' in Malcolm Dick (ed.), *Matthew Boulton: A Revolutionary Player* (Studley, 2009), pp. 153–169; *Boulton Silver and Sheffield Plate: Seven Essays by Kenneth Quickenden* (London, 2009).

[7]   Quickenden, 'Samuel Garbett and Early Boulton and Fothergill Assay Silver' in this volume.

theory.[8] For Marx, all cultural formations are reduced to their economic origin, and it is ideology, which serves only the interests of the bourgeoisie and in which the proletariat are passively incorporated, that functions to maintain the status quo. Confined to prison by Benito Mussolini's Fascist regime, Gramsci was concerned to explain why the proletarian revolution that Marx thought inevitable had still not taken place in early twentieth-century Europe. Gramsci reasoned that the answer lay in the deep entrenchment of capitalism in Western culture and, particularly, in the concept of hegemony which provided the mechanism by which the proletariat actively colluded in the bourgeois ideology even though it did not serve their interests as a class. Thus far, however, hegemony does not seem that different to ideology, but when Williams invokes the concept of hegemony, he warns against its conflation with the traditional Marxist notion of ideology, as domination. With the implied notion of counter-hegemony in mind, he argues instead that hegemony 'has continually to be renewed, recreated, defended, and modified. It is also continually resisted, limited, altered, challenged by pressures not all its own'.[9] As an instrument of cultural analysis, then, the concept of hegemony is useful precisely because it problematises culture, opening it up as a site of conflict.

The context in which Boulton's campaign took place was certainly one of regional conflict. The motivation for Boulton was the privilege of the London trade, afforded by their access to facilities for assaying and hallmarking at the Goldsmiths' Hall, and, moreover, the resulting disadvantage suffered by the Birmingham trade due to the lack of an assay office in Birmingham within a system that nevertheless required precious metals to be assayed and hallmarked by law. Indeed, in Britain, it has been a legal requirement since 1300 that precious metal articles must be assayed or tested to determine the extent of their precious metal content, and hallmarked to guarantee that the metal is of a legally required standard if they are to be offered for sale.[10] In order for their products to be legally saleable, therefore, silver manufacturers in Birmingham, including Boulton and his partner, John Fothergill were obliged to send their goods for assay and hallmarking to an assay office, the nearest being Chester. The return journey from Birmingham to Chester of 144 miles increased costs, and there were also the added

---

[8] Williams, *Marxism and Literature*, p. 108. On Gramsci's contribution to Marxist theory see, for example, Walter L. Adamson, *Hegemony and Revolution: Antonio Gramsci's Political and Cultural Theory* (California, 1983); on Gramsci's contribution to British Cultural Studies see, for example, Ben Agger, *Cultural Studies As Critical Theory* (London, 1992).

[9] Williams, p. 112.

[10] It has been a legal requirement since 1300 that articles above a certain weight made in gold and silver must be assayed and hallmarked for them to be legally saleable. At first, Birmingham and Sheffield sought permission to assay and hallmark only silver. Birmingham has assayed and hallmarked gold from 1824, and Sheffield from 1903. Platinum was brought under the hallmarking legislation in 1975 when the Hallmarking Act (1973) came into force.

risks of damage or robbery, considerable delays and designs being copied whilst pieces were away from Soho.[11] The partners made much of these risks during the Parliamentary campaign; nevertheless, as early as 1766 Boulton had begun to consider obtaining an assay office in Birmingham and he continued to do so intermittently.[12]

Eventually, however, the matter became so pressing that he travelled to London on 22 January 1773, intent on bringing a petition before Parliament.[13] Having joined forces with the Sheffield Cutlers Company, who wished to establish an assay office in their own town, Boulton put the Birmingham petition before Parliament on February 2, the day after the Sheffield petition was presented, and the Birmingham trade presented one further petition on 25 February.[14] During the whole campaign, these were the only petitions submitted to Parliament by the Birmingham and Sheffield trades, with Boulton circulating three supporting documents: the *Memorial relative to Assaying and marking wrought plate in Birmingham*, distributed at the same time as the first Birmingham petition; *the Reply of the Petitioners from Birmingham and Sheffield to the case of the Goldsmiths, Silversmiths and Plateworkers of London and Places Adjacent*; and *Observations Relative to The Standard of Wrought Plate*, these last circulated in response to the various counter-petitions from London.[15] The Worshipful Company of Goldsmiths presented two counter-petitions, and the London trade, a further four.[16] Taken together with Boulton's private correspondence, these public documents present a remarkable index to the conflict between the dominant force of the Worshipful Company of Goldsmiths and the London trade and the emergent trades in Birmingham and Sheffield in the context of late eighteenth-century Enlightenment culture.

---

[11]   On the apparent difficulties caused to the partners and on Boulton's objections to this, see Sally Baggott, 'I Am Very Desirous of Becoming a Great Silversmith: Matthew Boulton and The Birmingham Assay Office' in Mason (ed.), *Matthew Boulton*, pp. 47–54.

[12]   Sally Baggott, 'Real Knowledge and Occult Misteries: Matthew Boulton and The Birmingham Assay Office' in Dick (ed.), *Matthew Boulton*, pp. 201–216.

[13]   In his diary for 1773, Boulton noted the expenses for his journey to London on this date. Cited in Westwood, note p. 10.

[14]   Westwood, note pp. 12 and 15.

[15]   The term 'plate' is used to refer to articles made in solid silver as opposed to plated wares made from base metals with a covering of silver. Birmingham Archives and Heritage (BA&H) MS 3782/12/89/23 and cited in Westwood, pp. 12–14; BA&H, MS 3782/12/89/13 and cited in Westwood, pp. 21–23; BA&H, MS 3782/12/89/6 and cited in Westwood, pp. 23–24.

[16]   The first petition of the Worshipful Company of Goldsmiths submitted 17 February 1773 BA&H, MS 3782/12/88/10; the second submitted 6 May, cited in Westwood, p. 18. The first petition of the London trade submitted 18 February, cited in Westwood, p. 14; the second submitted on 25 February, cited in Westwood, p. 15; the third submitted 8 March, cited in Westwood, p. 17; the fourth submitted 6 May, cited in Westwood, p. 18.

London: Dominance, Tradition, Institution and Formation

Williams is clear that 'The complexity of culture is to be found not only in its variable processes and their social definitions – traditions, institutions and formations – but also in the dynamic interrelations, at every point in the process, of historically varied and variable elements'.[17] Identifying these as the dominant, the residual and the emergent, he argues that '… definitions of the emergent, as of the residual, can be made only in relation to a full sense of the dominant'.[18] Taking Williams' lead, I want to begin by considering how the London faction sought to maintain and legitimate their dominant position in their efforts to thwart Boulton in 1773. Throughout the series of six counter-petitions submitted to Parliament, the Worshipful Company of Goldsmiths insisted upon their significance as overseers of the law relating to assaying and hallmarking, whilst the London trade emphasised their ability to be all things to all people in terms of the national market for silver. The trade also demonstrated their loyalty to their guild on which they depended, amongst other things, for the assaying and hallmarking of their work, consistently underlining the importance of the duties the Goldsmiths' Company carried out under their royal charter. Examined more closely, these documents provide notable examples of the metropolis flexing its hegemonic muscles in its efforts to maintain dominance over the provinces.

Williams' delineation of the ways in which hegemony operates involves distinguishing three specific aspects of culture: tradition, institution and formation.[19] For Williams, '… tradition is in practice the most evident expression of the dominant and hegemonic pressures and limits', and this is especially apposite in relation to the Worshipful Company of Goldsmiths since throughout their representations to Parliament they stressed their traditional credentials.[20] However, this is not just a straightforward common sense notion of tradition that Williams is talking about; he states, '… what we have to see is not just a 'tradition' but a *selective tradition*: an intentionally selective version of a shaping past and a pre-shaped present'.[21] The emphasis is Williams', and there is evidence in the sources that the Worshipful Company of Goldsmiths was involved in exactly the kind of selection to which he refers. Submitting their first petition on 17 February, as well as reminding Parliament that the duty of assaying and hallmarking with which they were charged by the Crown was of national importance, they stated that they had been 'a Guild or Corporation since time out of Mind'.[22] In order to defend their dominant position, they lengthened considerably the actual 446 years since they had received their first Royal Charter in 1327, interpreting it instead

---

[17] Williams, p. 121.

[18] Ibid., p. 123.

[19] Ibid., pp. 115–120.

[20] Ibid., p. 115.

[21] Ibid., p. 115.

[22] BA&H, MS 3782/12/88/10.

as 'time out of Mind'.[23] Furthermore, when they stated that their duty was 'a very great and important trust' that they had executed with 'the utmost care and fidelity', they chose not to select events in 1478 when their Wardens elected to be no longer answerable for any substandard wares that were hallmarked in error, necessitating the appointment of a salaried assayer at London and the introduction of the date letter to the hallmarking system.[24]

For Williams, then, hegemony relies on traditions which are fundamentally selective and constructed. In turn, however, he writes 'It is true that the effective establishment of selective traditions can be said to depend on identifiable institutions'.[25] The separation between the categories of tradition and institution is not so straightforward, though, in relation to the Worshipful Company of Goldsmiths; at pains to demonstrate their own traditional standing, as the guild of goldsmiths in London and, therefore, the guardians of their craft, they were undoubtedly an institution in their own right. In addition, their traditional and, therefore, dominant status rested on other, much greater institutions. In their first petition, put before Parliament on 17 February, they stated that their privileges had been '... confirmed and enlarged from time to time by several Charters from His Majestys [sic] Royal predecessors Kings and Queens of this Realm'.[26] Having made their associations with the Crown clear, they quickly moved to the correlate institution of nation, going on to argue that the legal standards for precious metals were 'the same as those appointed for the Gold and Silver Monies of this Kingdom' and were, therefore 'for the Honour and Riches of the Realm'.[27] They littered the petition with the words 'realm', 'nation' and 'kingdom', as they proceeded to assert that the issue at hand was a matter of national security, that by failure to uphold the standard of precious metal 'his Majesty's Subjects will be defrauded, the fair trader will be injured and the Wealth the Honour the Credit and the Commerce of this Kingdom will be diminished'.[28] The London plateworkers, in a petition presented the following day, took things a step further; the power of assaying gold and silver, they said, was 'sacred'.[29] The weight of the Williams' argument about hegemony hardly needs pointing out; the dominant status of the Worshipful Company of Goldsmiths inhered in a constructed tradition that further depended on identifiable institutions. Indeed,

---

[23]   J.S. Forbes, *Hallmark: A History of the London Assay Office* (London, 1998), pp. 18–19.

[24]   BA&H, MS 3782/12/88/10. On the introduction of the date letter to the hallmarking system and the reasons for it, see J. Paul de Castro, *Law and Practice of Hallmarking Gold and Silver Wares*, 2nd edition (London, 1935), pp. 50–54.

[25]   Williams, p. 117.

[26]   BA&H, MS 3782/12/88/10.

[27]   BA&H, MS 3782/12/88/10.

[28]   BA&H, MS 3782/12/88/10.

[29]   BA&H, MS 3782/12/88/11.

the Goldsmiths' Company was an institution in itself, sanctioned by the greater institutions of the crown, the nation and by God.

The final element in Williams' tripartite structure, formation, is pertinent to the actual workings of the precious metals trade during the period since he describes formation as '... a mode of specialised practice'.[30] In their successive representations to Parliament, the London contingent reinforced their position by implying that they were the only ones sufficiently trustworthy and knowledgeable to carry out the specialised practice of assaying and hallmarking. The Worshipful Company of Goldsmiths stated in their first petition that '... the power of trying touching assaying and marking of gold and silver plate is a very great and important trust and ought to be committed to such persons and in such places only where the same is likely to be executed and discharged with the greatest care and fidelity'.[31] It was left to the London trade to spell out the subtext in their petition submitted the following day; here, in support of the Goldsmiths' Company, they alleged that '... in the Towns and neighbourhoods of Sheffield and Birmingham there are few persons conversant with or skilled in the Gold or Silver Plate Manufactory' and that '... such an Establishment [an assay office] in those places might open a door to deceit and uncertainty in respect to the Standards of Gold and Silver plate'.[32] Furthermore, according to the London trade, London's dominance inhered in manufacturing skill. They stated, 'That the Manufactory of such plate is now carried on in the City of London by the most exquisite and skilful Artists in all its variety of useful and curious branches, in much greater perfection than in any other place in these Kingdoms', and, as a result, their work could boast '... deserved and established credit for every Quality which can contribute to Elegance Utility and true intrinsic value'.[33] Not only was their skill in gold and silversmithing unsurpassed, they were also quite able to satisfy the market since the London trade had developed '... to an extent sufficient to supply the greatest demands'.[34] The dominance of the Londoners was justifiable, then, as they were the sole possessors of the necessary knowledge, skill and abilities within their particular specialised mode of practice. The Panglossian turn of their argument was not lost on Dickinson, who wrote 'The gist of this petition was that everything was for the best in the best of all places'.[35]

London's position depended on the three particular elements of dominant cultural forms identified by Williams; on a constructed tradition, on institutional power and on the ways in which they operated in the particular cultural formation of assaying, hallmarking, silversmithing and the market that constituted the trade in the eighteenth century. For Williams, though, the complexity of culture

---

[30] Williams, p. 119.

[31] BA&H, MS 3782/12/88/10.

[32] BA&H, MS 3782/12/88/11.

[33] BA&H, MS 3782/12/88/11.

[34] BA&H, MS 3782/12/88/11.

[35] Dickinson, p. 67.

is also found in the dynamic interrelations of its different elements. Besides the dominant, he argues, residual and emergent elements of the cultural process must be recognised. The residual, he says, '… has been effectively formed in the past, but it is still active in the cultural process. … It is in the incorporation of the residual – by re-interpretation, dilution, projection, discriminating inclusion or exclusion – that the work of the selective tradition is especially evident'.[36] Processes of re-interpretation and discriminating inclusion or exclusion were an integral aspect of the tradition constructed by the Worshipful Company of Goldsmiths, but the concept of the residual is equally relevant. The Goldsmiths' Company's status as a guild with a history of some 446 years, which went under the archaic title of 'the Wardens and Assistants of the Company or Mistery of Goldsmiths of the City of London', coupled with their assertion that they had been so since 'time out of Mind' immediately marks them out as residual in the historic sense.[37] Furthermore, the vigour with which they opposed the petitions from Sheffield and Birmingham in 1773, is confirmation of their continuing active presence in the cultural process. Williams goes on to say that organised religion and the monarchy are predominantly residual; accordingly, by pressing their relations with the Crown and the nation, and by the London trade casting their duties as 'sacred', in terms of their dominant position, the Worshipful Company of Goldsmiths relied a great deal on the residual.[38]

Boulton and Birmingham:
Emergence, the Market, Manufacturing and Metallurgy

During the 1773 campaign, if London was exemplary of dominance in the way in which Williams understands it, then Boulton and Birmingham represented an emergent force within the contemporary culture. Williams' notion of the emergent directly counters the traditional, the institutional and the residual since it is essentially related to the new. He writes, 'By emergent, I mean … that new meanings and values, new practices, new relationships and kinds of relationships are continually being created'.[39] Certainly, in existing scholarship, Enlightenment figures like Boulton are read as fundamentally new through their involvement in technological and scientific innovation and discovery, and even the creation of a new social class.[40] Certainly, Boulton and the Birmingham trade had no recourse to traditional, institutional or residual forms on which London's dominance relied,

---

[36]   Williams, p. 122.

[37]   BA&H, MS 3782/12/88/10. The term 'mistery' is widely accepted to refer to a trade or profession.

[38]   Williams, p. 122.

[39]   Ibid., p. 123.

[40]   On these aspects of Boulton's life and work in general, see Dick (ed.) and Mason (ed.). On Boulton as an example of a new social class see Mason, 'A New Species of

when arguing their case during the campaign. Peter Jones has noted the absence of regulation by guilds and the extent of free trade as a result in the West Midlands during the period; something that Boulton openly celebrated in mounting his challenge to London.[41] In his *Memorial relative to Assaying and marking wrought plate at Birmingham*, he stated that the lack of any guild or incorporation in the town should not present an obstacle to the establishment of an assay office there and that 'the Inhabitants are too sensible of the Disadvantages of such societies to wish for any'.[42] The Birmingham petition and the supporting documents circulated by Boulton, in addition to what survives of the correspondence between Boulton and Samuel Garbett during the campaign, reveal a great deal in terms of what might be considered new or emergent about Boulton and Birmingham.[43] In particular, Boulton's perception of the market, the techniques he was using at Soho in manufacturing silver and the understanding of what we would now call 'metallurgy' he shared with Garbett are indicative of Williams' notion of the emergent.

In the first counter-petition they presented, the London trade declared their protectionist stance towards the contemporary market for silver. They were able to supply the greatest demand that was likely to be made in terms of both the range and volume of products the market required, and, furthermore 'if the Goldsmiths Manufactory in the Towns of Sheffield and Birmingham should be increased that of London must suffer a proportionable decrease by which means the Condition of thousands of the working part of the trade here must grow daily more and more distressful'.[44] Claiming that the market would be sent into decline by the establishment of assay offices in Birmingham and Sheffield and the resulting expansion of the trade in the two towns, the London workers were fearful that their trade could not withstand competition from the provinces. Boulton's perception of the market could not have been more different.[45] He appears to have agreed with the London trade that the market was at risk, arguing that this was owing to '... the Inferiority of the Work, and the *exhorbitant Prices*, in spite of the *very Prevailing Taste for Plate* ...' that made London silver '... unsaleable to all but *a few rich* People'; the emphasis here is Boulton's.[46] Robinson has noted that Boulton's attitude to the market was largely determined by his current business

---

Gentleman' and David Brown, 'Matthew Boulton, Enclosure and Landed Society', in Dick (ed.), pp. 30–44 and 45–62.

[41]   Peter Jones, *Industrial Enlightenment: Science, Technology and Culture in Birmingham and the West Midlands 1760–1820* (Manchester, 2008), p. 141.

[42]   BA&H, MS 3782/12/89/23.

[43]   Samuel Garbett proved himself a highly dependable ally during the campaign. See Baggott, 'I Am Very Desirous of Becoming a Great Silversmith'.

[44]   BA&H, MS 3782/12/88/11.

[45]   BA&H, MS 3782/12/88/11.

[46]   BA&H, MS 3782/12/89/13.

interests, but in this instance, he was firmly in support of free trade.[47] In direct opposition to London's protectionism, he was confident that the growth of the trade in Birmingham and Sheffield would not have an inversely proportional effect on the market in London at all. He contended that should the three towns of London, Birmingham and Sheffield, '... make better and cheaper Plate than London now does, most infallibly every one of them would make and sell more'.[48] That Boulton's thinking was emergent is evident in the new values and practices he was advocating in terms of the market; competition not monopoly would benefit the trade by bringing prices down and improving quality, thus driving up demand and increasing employment in the trade.

When it came to manufacturing, the London trade were in no doubt that the manufacture of plate in the metropolis was carried out by '... the most exquisite and skilful Artists in all its variety of useful and curious Branches', and that their work possessed 'every Quality which can contribute to Elegance Utility and true intrinsic value'.[49] When Garbett was questioned by the Parliamentary committee about activities at Soho, he stressed that Boulton was producing silver that could equally meet the demands of the market in terms of design, use-value and range: he replied that Soho's output included '... many Ornamental Utensils – Tureens, Candlesticks, Vases, Coffee pots, all of solid silver'.[50] Quickenden casts doubt on the scale on which such articles were made at Soho, but he does concede that Garbett was giving a fair representation of the range Boulton was manufacturing.[51] There was nothing particularly new or emergent, then, in *what* Boulton was manufacturing, but there is evidence to indicate that *how* he was making it entailed new, emergent elements in terms of both design and process. William Bingley, who had been employed at Soho, stated before the Parliamentary Committee of 1773 that Boulton's patterns were more handsome than any he had seen in London.[52] Garbett was equally confident about Birmingham's capacity for design. When he was asked by the Committee 'Are there not Silver Buckles made in London of every Pattern that are made at Birmingham?' he replied, 'Not one in a hundred – [I] believe not one in five hundred'.[53] During the hearing, he also stated that 'There is not a day in Birmingham but a new Pattern of Buckle is made'.[54] There is documentary evidence that Garbett's claims about innovation in design was not

---

[47]   See Robinson, *Boulton*. Boulton may have been an advocate of free trade during this campaign, but he was most definitely in favour of protectionism when he lobbied Parliament in relation to the granting of patents for the Boulton and Watt engines.

[48]   BA&H, MS 3782/12/89/13.

[49]   BA&H, MS 3782/12/88/11.

[50]   BA&H, MS 3782/12/88/13.

[51]   Quickenden, 'Samuel Garbett' in this volume.

[52]   BA&H, MS 3782/12/88/18; Ken Quickenden, 'Boulton and Fothergill Silversmiths', *The Silver Society Journal*, no. 7 (Autumn 1995), pp. 342–356, p. 349.

[53]   BA&H, MS 3782/12/88/13/11.

[54]   BA&H, MS 3782/12/88/13/11.

just rhetoric, but if Boulton was capable of giving the Londoners a run for their money on the design front, the sources relating to the campaign in 1773 suggest they were not overly concerned about it.[55] They were much more caught up in the challenge presented by the new, emergent manufacturing processes Boulton was using in Birmingham.

Boulton sold silver in the London market, therefore the London contingent would have been aware that new modes of production were in operation at Soho.[56] Though Garbett, in his evidence relating to the Soho manufactory, made much of large articles that required the traditional, skilled hand-techniques such as hand-raising, chasing and engraving, by far the largest quota of domestic wares made at Soho comprised candlesticks manufactured by die-stamping.[57] The London faction seized on this aspect of Boulton's operation and, consequently, the Parliamentary Committee focused much attention on the question of Birmingham manufacturers using die-stamping. Demonstrating the tension between old and new, the Londoners took a dim view of die-stamping as it required less skill than the technique of casting which they still used widely. Silver made in Birmingham, they stated in their first petition, was 'wrought and done … in a specious but very slight and unserviceable manner'.[58] Furthermore, John Wakelin, a London silversmith, left the Committee in no doubt as to the difference between cast and die-stamped candlesticks when he told them, '[candlesticks] of 30 ounces [cast] would cost more than those stamped on account of Workmanship'.[59] Despite Wakelin's inference that this was a matter of 'workmanship', the actual terms used during the hearing uncovered the fundamental reason for London's objections. In the documents, reference is made again and again to whether the Birmingham trade were producing 'heavy' or 'slight' candlesticks, where 'heavy' was used to refer to candlesticks manufactured by casting, whilst the term 'slight' referred to candlesticks produced by die-stamping.[60] The real issue here was not a matter of workmanship but weight and, therefore, cost. Candlesticks made by stamping were indeed 'slight'; for example, the 'Lyon-faced' candlesticks Boulton manufactured at Soho were some 70 ounces lighter per pair than cast; they were also considerably cheaper at £27.9s. less to the customer.[61] This was about the challenge that new emergent and cheaper manufacturing processes presented to the London trade.

---

[55]   Quickenden, 'Lyon-faced Candlesticks and Candelabra', *The Silver Society Journal*, 11 (Autumn 1999), pp. 196–210.

[56]   Quickenden, 'Boulton and Fothergill: Business Plans and Miscalculations', *Art History*, vol. 3, 1983, pp. 274–294, 277.

[57]   Quickenden, 'Samuel Garbett' in this volume.

[58]   BA&H, MS 3782/12/88/11.

[59]   BA&H, MS 3782/12/88/15.

[60]   See accounts of the Committee's meetings BA&H, MS 3782/12/88.

[61]   Quickenden, 'Samuel Garbett' in this volume.

Boulton and the Birmingham faction arguably correspond most closely to Williams' notion of the emergent, in their understanding of what we would now call 'metallurgy'. The London trade, though, was quite certain that Birmingham's skills, abilities and knowledge were insufficiently developed to allow the trade in Birmingham to conduct the business of an assay office with sufficient rigour. In their successive counter-petitions and to bolster their case, the London trade made accusations relating to irregularities in the Birmingham trade. They claimed that Birmingham silver contained a higher proportion of iron than was legally permissible, and that Birmingham manufacturers were placing marks on their Sheffield Plate that resembled hallmarks in order to pass the work off as 'real plate marked at an Assay Office' and thus defraud the customer.[62] On the first charge, the Parliamentary Committee reported 'There hath been only One Fraud alledged [sic] against any Manufacturer at *Birmingham, viz.* That a Snuffer Maker there had made Silver Snuffers, in which were concealed Thirteen Pennyweight of Iron or Steel more than was necessary'.[63] Though the Parliamentary Committee's report made no mention of it, Boulton and Fothergill could arguably have been in the frame themselves on account of the second charge as the marks they applied to Sheffield Plate at Soho looked very much like the sponsor's mark they had registered for use on silver at the Assay Office at Chester.[64]

Despite the Worshipful Company of Goldsmiths' insistence that they were the best equipped to carry out the important duty of assaying and hallmarking with 'care and fidelity' they fared worse than Birmingham during the hearing.[65] It came to light that the Goldsmiths' Hall had been in the practice of passing and marking silver that was two pennyweights below the legally required standard. Needless to say, Boulton made a great deal of this in his *Observations Relative to the Standards of Wrought Plate*. Here, Boulton displayed his understanding of silver refining, knowledgeably picking apart the practice at Goldsmiths' Hall. Apparently incensed by London's arrogance in the face of proven ineptitude, he stated 'The *London* [sic] Goldsmiths Company assume the Character of Guardians for the Preservation of the Standards of Wrought Plate ... The Company, however daily authenticate by their Standard Mark Silver Plate worse by Two Pennyweight

---

[62]   Cited in Westwood, pp. 15–16, 17–18. The implication was that iron components were concealed in articles made from sterling silver, not that the actual silver alloy contained too high a proportion of iron.

[63]   BA&H, MS 3782/12/88/16, 36 and 36b. Benjamin May was the snuffer maker in question, and Joseph Wilkinson wrote first to Boulton about this on 27 February 1773. Wilkinson defended May on the grounds of ingenuity rather than damning him for any attempt at fraud. May himself wrote to Boulton in April, apologizing profusely for any embarrassment or trouble he had caused. In his own defence, he stated that silver buckles often contained more iron than the offending articles due to the chape being made of steel. The chape is the part of the buckle to which the strap or ribbon is attached.

[64]   See Baggott, 'Real Knowledge and Occult Misteries', pp. 206–207.

[65]   BA&H, MS 3782/12/88/10.

in the Pound than the legal Standards'.[66] Moreover, Boulton calculated that by this error '... The *Londoners* [sic] have an Advantage over the Country Silversmiths of nearly one *per cent*'.[67] In his letters to Boulton during the campaign, Garbett derided the Worshipful Company of Goldsmiths and the London refiners for their lack of knowledge, writing '... the London Refiners have lost the Art of making silver quite fine'.[68] Instead, he was secure in the understanding of the processes involved that he and Boulton possessed, and was certain that as a result they were better placed to conduct the business of an assay office. In a letter to Boulton dated 3 May 1773, Garbett wrote that he was about to dispatch 50 ounces of silver by fly to Boulton, which he said 'shall be as fine as we can make it and I don't doubt that 11 oz. 2 dwt. of it and 18 dwt. of copper will be better than the Standard of 11 oz. which is allowed to pass at Goldsmiths' Hall'.[69] In relation to the transfer of knowledge in the Enlightenment, Jones argues 'that the know-how component of useful knowledge was often constituted in such a way as to prevent or deter transfer from person to person and from place to place', and mindful of a need for secrecy, in the same letter, Garbett continued, '... I expect it will be better than the Standard Plate of the Kingdom for Reasons not prudent to put on Paper even to you'.[70] Reversing the traditional and institutional terms with which London had constructed their case and setting himself and Boulton in direct opposition, he concluded, '... when you have obtained the Law I suppose you and I shall be made acquainted with their most sacred Art and Mistery – but don't let Us give any of them our Real knowledge in exchange for their Shabby Occult Misteries'.[71] Boulton and Garbett's kind of knowledge, then, was the new emergent 'Real knowledge' that would challenge the 'sacred Mistery' of the Goldsmiths' Company. In turn, that was now denigrated as 'Shabby' and 'Occult'.

Conclusions

The Bill finally passed through the Commons on 18 May 1773 and gained Royal Assent in the Lords on 28 May.[72] The Birmingham Assay Office opened on 31

---

[66]   BA&H, MS 3782/12/89/6. Westwood clouds the issue with the following note, 'The London refiners then sold two qualities of silver. One known as "old sterling" was commonly used for making plate. The other, "upright sterling" or silver of the legal standard. Old sterling silver was usually 2 ½ dwts [pennyweights] worse than upright silver'. The point he should have stressed is that 'old sterling' at 11 Troy ounces in the pound had never, in 1773, been a legally recognised standard for silver. Westwood, p. 24.

[67]   BA&H, MS 3782/12/89/6.

[68]   BA&H, MS 3782/12/88/40.

[69]   BA&H, MS 3782/12/88/37.

[70]   Jones, p. 16 and BA&H, MS 3782/12/88/37.

[71]   BA&H, MS 3782/12/88/37.

[72]   Westwood, p. 25.

August, and Boulton and Fothergill were its first customers.[73] Acknowledging the
skill with which Boulton had countered the Londoner's objections, on 11 March,
Joseph Wilkinson wrote to Boulton '… [I] am happy to find your Manoueuvres
[sic] in Turning their own Batterys upon themselves – May that ever be the case
where such mean and dishonourable Arts are adopted!'[74] Certainly, Boulton's
direct challenge of the London case does seem to correspond with Williams'
idea of an emergent cultural form. However, Williams is clear that emergence
is difficult because the dominant works to incorporate, delimit and condition the
emergent, and I want to conclude by considering how far Boulton's approach was
really emergent in the sense that Williams intends. It is arguable that his attempt to
protect and expand the trade in Birmingham by establishing an assay office in the
town was not any different to the privileges the craft guilds, like the Worshipful
Company of Goldsmiths, sought to provide.[75] In the field of silver design, Boulton
was emergent to a degree, but, nevertheless, the silver business at Soho was a
failure economically. Boulton often underestimated the demand and miscalculated
the cost of the kind of silver he produced, therefore, financial success was not
possible when, in order to gain access to the level of skill required for his designs,
Boulton had to employ London silversmiths, rates of pay were often the same at
Soho as in London and he had to compete in the London market to achieve sales.[76]
In the end, London exerted its dominance in terms of the market. There is also the
question of how far Boulton's approach to manufacturing was emergent. Recent
practical approaches to the manufacturing processes used for silver at Soho have
shown that, in terms of larger, more complex pieces, Boulton made more use of the
traditional techniques of casting, chasing and engraving than the sources relating to
the establishment of The Birmingham Assay Office suggest.[77] Although, Boulton's
part with James Watt in the later development of the die-stamping process that
would eventually revolutionise production throughout the metals industries, and
the sophisticated understanding and knowledge of metals that he shared with

---

[73]   Register of Sponsors' Marks, 1773–1858 and Plate Register, 1773, The Birmingham
Assay Office.

[74]   BA&H, MS 3782/12/88/21.

[75]   My conclusions here have benefited from a discussion with Professor Chris
Evans, University of Glamorgan during the bi-centenary Conference at the University of
Birmingham in 2009 where I presented the paper from which this chapter derives.

[76]   See Quickenden, 'Business Plans and Miscalculations'.

[77]   On the processes involved in the manufacture of the tureens Boulton made for
Elizabeth Montagu, see Quickenden, 'Business Plans and Miscalculations', p. 283–284.
Quickenden has since confirmed his assertions; in 2009, a replica, based exactly on the
Montagu tureens in the collection of the Birmingham Assay Office, was made by a team at
Birmingham City University.

Garbett, should not be underestimated.[78] It is arguable that legally there was nothing really emergent about Boulton's victory in 1773. He did not succeed in creating a completely new system that replaced assaying and hallmarking, and, ultimately, the Birmingham trade was merely incorporated within the existing dominant institution of the law. However, Williams does write that, 'It would be wrong to overlook the importance of works and ideas which, while clearly affected by hegemonic limits and pressures, are at least in part significant breaks beyond them, which may again in part be neutralised, reduced or incorporated, but which nevertheless come through as independent and original'.[79] Notwithstanding the fact that the outcome of the 1773 campaign was ultimately limited by London's hegemonic power, Boulton was truly emergent in the very challenge he presented to London; in making a significant contribution to the processes that would, before long, see Birmingham itself established as a dominant cultural force during the age of industrialisation.

---

[78] On the development of automated steam-driven presses and advances in making and rolling metals at Soho, see Sue Tungate, introduction to Sue Tungate and Richard Clay (eds), *Matthew Boulton and the Art of Making Money* (Studley, 2009).

[79] Williams, p. 114.

Chapter 11

Dark Satanic Millwrights?
Forging Foremanship in the Industrial
Revolution: Matthew Boulton and the
Leading Hands of Boulton and Watt

Joseph Melling

Forging the Foreman: Industrial and Literary Models

There exist few detailed investigations of the evolution of workplace management during Britain's industrialisation. Remarkably little has been written since Pollard's *Genesis of Modern Management*, published almost 50 years ago (1965). In common with several other noted writers, Pollard suggested that British management science was poorly developed before the late nineteenth century.[1] Foremen often figured as dark 'tyrannous' overseers in accounts of early industrial life, though closer research often reveals that they figured in complex workplace battles.[2] Some of the most brutal supervisory practices were associated not with mechanised factories but with petty contracting, sub-contracting and self-employed workshops such as Sheffield's cutlery trades depicted by Pollard himself.[3] Less critical accounts of early industrial management included Eric Roll's (1930) celebrated study of the partnership of Boulton and Watt, in which he detects in the innovations of

---

[1] Sidney Pollard, *Genesis of Modern Management* (London, 1965), pp. 142–143, 147, 176–255, 189, 242, 269, 291; Karl Marx, *Capital,* Translated by Moore-Aveling (London, 1912), pp. 174–175, 382, 423; J.R. Harris, 'Skills, Coal and British Industry in the Eighteenth Century', *History*, LXI (1976), pp. 167–182; Joel Mokyr, 'Introduction: The new economic history and the industrial revolution' in J. Mokyr (ed.), *The British Industrial Revolution: an Economic Perspective* (Boulder, 1993), pp. 1–132, 35, 40–41, 45.

[2] Richard Price, *Masters, Unions and Men* (Cambridge, 1980), pp. 34–38, 175–178.

[3] *Second Report of the Royal Commission on Technical Instruction, Minutes of Evidence*, C.3981, Q7660, 'From the workmen we make the foremen. The cleverest workman becomes the foreman'. 'Rattening' *Foreman Engineer and Draughtsman* (January 1878), 18; C.R. Dobson, *Masters and Journeymen: A Prehistory of Industrial Relations 1717–1800* (London, 1980), pp. 39, 60; S. Pollard, 'The Ethics of the Sheffield Outrages', *Transactions of the Hunterian Archaeological Society*, VII (1953–54), pp. 118–139, 128–132, 138–139.

the second generation of partners the genesis of the scientific management of his own day.[4] Roll describes the breaking of an established journeyman system as Boulton and Watt Jr. struggled to school their hands in specialist skills and based key reforms on a rationale that grew directly from technological development, rather than from market pressures.[5]

The control and utilisation of technology has long been recognised as a reference point for the success of management as well as entrepreneurship in the industrial revolution. Recent research on early industrial technologies affirms the importance of incremental advances in technology.[6] James Watt himself acknowledged that the 'true inventor of the crank rotative motion was the man (who unfortunately has not been deified) that first contrived the common foot lathe'.[7] Historians of technology increasingly emphasise not only the institutional setting in which invention is recognised but also the cultural milieus where mechanical genius is celebrated. Christine MacLeod's recent account of technological change in manufacturing industry during the industrial revolution depicts the heroic qualities with which invention was endowed as well as the contribution made by older artisan skills and the impact of capital-labour conflict on the pace of invention and innovation.[8] Reputations were more readily constructed where entrepreneurs and inventors commanded, or could enlist to their aid literary and artistic talents in support of a tasteful narrative of success and benevolence to mankind. Not all of the luminaries of the new industrial age possessed Matthew Boulton's talent for suave publicity and promotion of his business connections. The history of entrepreneurial genius in Birmingham has been well-served by the monumental archive left by the Boulton and Watt firm as well as the prolific correspondence which Boulton composed. Boulton and Watt left little doubt of the scale of their achievement, though others relied on the energies of industrial evangelists such as Samuel Smiles to burnish their reputation. In responding to

---

[4]   Eric Roll, *An Early Experiment in Industrial Organisation, being a History of the Firm of Boulton & Watt, 1775–1805* (London, 1930), pp. 271–272.

[5]   Roll, p. 150: 'Custom, too, had established a certain system in engine building. Engineers acted as designers and supervisors only …'. Ibid., pp. 266–268, 270–272, for scientific management techniques 'apparent from the very beginning of machine industry'.

[6]   Mokyr, 'Introduction', pp. 11–15, 86–87, 91–94, 110; P. Scranton, 'Labour and Technology', *Technology and Culture*, 29/4 (1988), pp. 719–743; P. Scranton, *Endless Novelty: Speciality Production and American Industrialisation, 1880–1925* (Princeton, 1997), pp. 18–23. N.F.R. Crafts, *British Economic Growth during the Industrial Revolution* (Oxford, 1985), pp. 84–88; H. Belofsky, 'Engineering drawing – a universal language in two dialects', *Technology and Culture*, 32/1 (1991), pp. 30–34.

[7]   Birmingham Archives and Heritage (BA&H), MS 3219/6/1/319, James Watt (JW) to James Watt Junior (JWJ), 'Rotative Motions' attachment to letter of 10 November 1808, quoted in John Townley, 'Research Notes on Richard Cartwright' (unpublished, 2009), generously shared with me.

[8]   Christine MacLeod, *Heroes of Invention. Technology, Liberalism and British Identity, 1750–1914* (Cambridge, 2007), pp. 98–101, Watt's monument.

the plaudits of contemporary biographers and journalists, many inventors of the industrial revolution were disarmingly modest about their public careers, James Nasmyth replying in 1863 to Smiles' intention to include him in the pantheon of heroic engineers, agreed that 'destructive heroes' had been given excessive glory in contrast to the 'substantial benefactors of mankind'.[9]

Nasmyth also recalled the high reputation which Boulton's foreman, William Murdock, enjoyed during his earlier career and the Soho archive is more remarkable in acknowledging the original genius as well as practical contribution of Murdock.[10] The Scots artisan remains a prime and proud exemplar of a journeyman artisan-inventor inside and outside the Boulton enterprise (see Figure 11.1).[11] Murdock's career at the Birmingham enterprise also throws a keen light on the problems of developing labour management in this period and opens a window on the complicated and often diffuse progress of industrial technology. Relations with tractable and truculent labour were forged by Boulton and his Scots partner in the civic culture that informed the developing identity of British business. Employers as well as innovators contributed to a practical culture in which heroic invention was framed in terms of masculine qualities. Publicly subscribing to the civic virtues which he saw as embodied in his ventures, Boulton defended these virtues ruthlessly in striving with Watt to severely restrain, if not destroy the capacity of foremen, as well as the artisans, to challenge the interests, authority and reputation of the enterprise. Credit for invention was recognised by Boulton as part of the vital intellectual capital of an enterprise which was founded on a capacity for ingenuity and design rather than a mere competence in manufacturing. In this regard, Boulton's battles to isolate and accentuate the genius of his partner arose from the same strategic rationale which powered his negotiations with his foremen before and after the foundation of Soho.

Not only Boulton, but also those he employed, were able to call on an industrial and practical culture in celebrating inventive genius and individual character to fire their imagination and ambition. Long before Samuel Smiles eulogised the engineering genius of Boulton, Watt and other industrial pioneers, narratives of masculine success had become a decided feature of autobiographical as well as biographical genres composed to commemorate heroes of mechanical power. Edinburgh's scientific enlightenment raised literary monuments to engineers such as Telford, framing the hero's progress to fame and fortune from humble pastoral

---

[9]   British Library (hereafter BL), ADD 71075 635B, James Nasmyth (JN) to Samuel Smiles (SS), 29 April 1863 in Nasmyth Papers, within Samuel Smiles Papers (hereafter SP).

[10]   John Griffiths, *The Third Man. The Life and Times of William Murdoch 1754–1839, the Inventor of Gas Lighting* (London, 1992), pp. 299–300.

[11]   S. Smiles, *Industrial Biography: Iron Workers and Tool Makers* (London, 1886), pp. 241–243, 293–294; Paul Mantoux, *The Industrial Revolution in the Eighteenth Century* (London, 1964), pp. 321, 331–332; C.H. Wilson and W.J. Reader, *Men and Machines: A History of David Napier and Sons, 1808–1958* (London, 1958), pp. 3–4.

Figure 11.1   Print of William Murdock
Source: Courtesy of Birmingham Archives and Heritage.

origins to public renown and his peaceful conquest as a triumph of personal character
and ability.[12] It was this practical and poetic culture of mechanical capacity which
enabled foremen such as Murdock to resist Boulton's efforts to place Watt above
all other hands. By sheer force of ingenuity and masculine character, Murdock
forged his brilliant career as foreman and independent inventor. The purpose of
the present chapter is to locate Murdock among a larger group of able mechanics
who entered the service of Boulton and Watt at the end of the eighteenth century.

---

[12]   National Library of Scotland, Telford Papers (TP), MSS 19977/1, letter from
Colonel G.W. Pasley, 28 March 1835, concerning Telford's statue; TP, MSS 19977/55,
Tattenshall to Rickman, 1 November 1837; 19977/132; George May to Rickman, 23
March 1838, Manuscript 'Memoir of the early part of the life of the late Thomas Telford'.
Rickman noted that Telford 'erected for himself in the works which he has left behind him
a monument which shall last through distant ages'; SP, BL ADD 71075 635B, Letter to SS,
16 November 1860, with a covering letter from Cawley of 8 February 1861.

This enables us to see more clearly how the famous partners managed the talents and personality of their most able employees. It will be argued here that the second generation of management at Soho did not undertake a revolutionary transformation of workplace relations but rather extended and formalised social relationships which were made between masters, foremen and workmen in the early days of the partnership. The journeyman division of labour was resilient largely because the firm continued to depend on highly skilled labour and peripatetic artisans to build, erect and maintain machinery. This continued to be the case even when artisans proved as difficult as the maverick journeyman, Richard Cartwright, whose adventures included industrial espionage and culminated in a death sentence for theft (commuted to transportation on a prison ship).[13] Forceful personality was essential to survival in such dangerous rapids.

The following discussion is organised around an examination of the physical fabric and development of the Soho works, mainly through the eyes of Boulton, Watt and Murdock. The latter part of the chapter considers the question of intellectual property rights and the respective claims of the owners and their ingenious employees in regard to the development of inventions.

Titles and Deeds: Early Foremen at Boulton and Watt

It has been noted earlier that Pollard suggests the position of 'supervisor' was unknown to early industrialists, many managers being commonly identified as 'clerks', while Roll argued that journeyman production imposed few disciplines on the skilled workmen. Before accepting either point, it is worth recalling that relatively small numbers of workmen were employed from the 1770s in the construction of engines for the original partners, there being a shortage of gifted engineers able to complete orders to specification and on time. During these early days of design and experimentation, the most pressing need was for men who were able to make, erect (or repair) engines from delicate and unpredictable components. Since the firm's profits were tied by contract, not to the price of equipment but to the fuel economies achieved by the steam engines supplied, it was essential that the new engines were fitted well and performed efficiently. The bespoke orders and their assembly in remote mining localities and similar sites gave a formidable responsibility to 'erectors', whether local millwrights entrusted with complex installations or artisans despatched from Birmingham, often requiring negotiation

---

[13]   Roll, pp. 150–153; Jennifer Tann, *The Development of the Factory* (London, 1970), p. 81: 'Boulton and Watt did not have men to spare; they had enough problems getting skilled fitters for themselves ...'. BA&H, MS 3147/3/2/16, Matthew Boulton (MB) to Henderson, 10 February 1781, for the incident, quoted in Townley (2009). See H.W. Dickinson and R. Jenkins, *James Watt and the Steam Engine* (Cambridge, 1927), p. 281, for Cartwright's betrayal of Watt's crank invention.

with purchasers on allowances to be made for Birmingham or local craftsmen.[14] Customers in such localities frequently harboured bitter resentment of the patent royalties imposed by the Birmingham partnership and disaffected artisans might be induced to betray the secrets of the designers or even embark on industrial piracy on their own account.

To deliver and bind their journeymen to the firm, employers were forced to rely on legal contract, surveillance and wage incentives as well as trust and affection. The fastidious James Watt wearily complained persistently about the poor standards of workmanship he encountered in England as well as his native Scotland, particularly when erectors of 'fire engines' were expected to fit nozzles and other key components to his specifications. Boulton recognised the value of personal diplomacy in the handling of gifted hands as well as the need to preserve good reputation as part of the public capital of the emergent business. It was as a remarkably capable and flexible erector of engines that Murdock also came to the fore in the 1770s, Boulton again smoothing the ruffled sensibilities of his partner in urging the irascible inventor to be less grudging in showing generosity to such talented hands. Cautioning Watt at the end of 1778 on Murdock's completion of another order, Boulton commented: 'We shall stand in need of such a man & I wish you would send another [Murdock] and not part for trifles as the want of Proper men to work our Engines has brought in a loss of many reputations'.[15]

Workmen possessing such technical understanding were able to converse with engine makers such as Watt on questions of mechanical principle as well as practical application. Among key figures who emerge from the firm's correspondence were family groups such as the three Murdocks (William, James and John), and the Rennies who established their own business (John, George and John Jr) and John Walker, later foreman to George and John Jr. The first foreman working with Boulton and Watt is usually identified as John Hall in 1778, the same year that Boulton commended Murdock, though the formal supervision of foundry work appears to have waited on the appointment of Abraham Storey as head of the Soho Foundry. Joseph Harrison and other erectors remained independent craftsmen even after the number of retained engineers (as distinct from local men hired for particular jobs) increased with the creation of a network of regional 'offices' in the 1790s.

The career of Murdock has attracted considerable attention and throws a keen light on the problems of developing labour management, including problems of distinguishing the functions and aspirations of those organising different aspects of production. Murdock made his reputation installing and maintaining pumping engines in the metal mines of Cornwall where the peculiarities of capitalist venture investment, cost book accounting and workplace bargaining complicated labour relations as well as business dealings. Displaying indefatigable energy and

[14]   Roll, pp. 155–150. Even the basic supplies and parts were often made elsewhere before the establishment of a boring mill in 1796.

[15]   BA&H, MS 3147/3/2/16, MB to JW, 17 December 1778.

dedication to the completion of jobs as well as outstanding technical understanding, Murdock frequently applied to Watt for (and was almost as frequently refused) wage increases in recognition. Among his contributions were improvements of the sun-and-planets mechanical motion, the steam wheel, and numerous other Watt inventions.[16] The latter clearly shared Boulton's suspicions of Murdock's ambitions and was more determined that Murdock should not claim a partnership in the firm, even if it became necessary to share some of its profits with him, Watt insisting 'the man must always be our servant not [us] his'.[17] Boulton soothed the darker anxieties of Watt by appealing to his troubled partner and in language the Birmingham business elite of the Enlightenment would have approved, contrasting the reasoning genius of the scientist with the uncertain physical passions of Murdock. Appealing to his partner not to descend to the mental and physical impulses of the mere artisan, Boulton urged Watt to concentrate rather on protecting his intellectual property by patenting any improvement his foreman might later claim.[18]

The defence of intellectual rights remained important as the partnership moved from designing and adapting engines under patent to manufacturing products previously built on licence. Boulton and Watt found that Murdock became more rather than less essential to their progress. For their brilliant assistant possessed undoubted qualities of character as well as intellect that were invaluable to the enlargement and development of the Soho Foundry. The new Soho Foundry developed in the 1790s itself comprised different workshops or (in modern parlance) departments of manufacturing production. There was the 'Foundry' itself, as well as the 'Boiler Shop' and the 'Fitting Shop'. The making of boilers and the assembling or 'fitting' of engines together were quite distinct from the forging and boring of metal, though the boring and 'turning' of metal into machine parts was subsequently undertaken by a specific group of engineering workers. Whereas in the early days of Boulton and Watt, the engine millwright was expected to possess the skills of cutting, shaping and fitting together machinery components, fitters and turners were later divided into two sub-branches of the engineering workforce.[19] Although the Soho Foundry workforce was moderate rather than large, numbering

---

[16]   Griffiths, pp. 101–103, 107–109. Boulton made the celebrated comment in 1784 correspondence that 'We want more Murdocks for of all others he is the most active man and best engine erector I ever saw'. Ibid., p. 102.

[17]   Watt added 'as he has been a very useful servant to us, though his conduct lately has been very disagreeable'. Griffiths, p. 151. Griffiths claims Murdoch possessed 'superior mechanical flair' to Watt and played the key role in the development of cranking rods for the planetary system. Ibid., pp. 115–118, 125, 130, 134, 135–136.

[18]   'When a man is mad in any way it is in vain to reason with him about his disorder'. Quoted in Griffiths, p. 148. MB added 'He may be a useful tool in our hands if we make proper use of him'.

[19]   Keith Burgess, *The Origins of British Industrial Relations* (London, 1975), chapter 1.

scores rather than hundreds of hands, it could be argued that the works provided a significant template for the subsequent evolution of engineering labour in Britain.

The moving spirit in the organisation of the new boring mill at the Foundry was that exemplary millwright Murdock. He assumed the superintendence of the new works, though when James Watt Jr tightened control through more systematic management in the workshops there was less forbearance of craft customs.[20] From 1801, this 'general reformation' of labour superintendence at the Soho Foundry was contemplated and relations with workmen were strained for some years after, particularly when a strike of moulders was broken and ringleaders dismissed in 1814.[21] Murdock's exceptional qualities enabled him to flourish in these two distinctive periods of inventive design and of organised manufacturing production, and he was able to demonstrate his value inside the modernised workshop as well as in the field of practical installation. His capacity for commanding the respect of the journeyman workforce lay in his personal abilities as a leading mechanic, demonstrating to his fellow craftsmen by chalk and action how a job could be completed by practical understanding.[22]

Exceptional though Murdock's talents were, he was not a unique innovator or outstanding personality in either period of the enterprise. Correspondence reveals Boulton and the acidic Watt resorting to similar manoeuvres of control in their treatment of other ingenious artisans, particularly before the Soho regime was consolidated under the junior partners. In these later years, the responsibilities of supervision were clarified and a distinct group of industrial foremen emerged alongside a tighter system of piecework and penalties, designed to control costs and increase profits. Roll's argument that these reforms marked a radical departure in divisions of labour should be treated with care, for his own evidence reveals a complicated and uneven set of arrangements at Soho, including the persistence or even growth of internal contracts and piecework bargains that delivered substantial rewards to foremen leading gangs of skilled workers. The rationale for such contracts can be found in the continuing diversity of orders and demand for bespoke goods requiring a coordination of handicraft skills which remained difficult to standardise and monitor.[23] As noted earlier, the workforce at the Soho Foundry was not large and its members included temporary as well as permanent employees: in summer 1808 there were perhaps 50 hands employed in the Foundry Shop at the Soho Foundry, the largest group being moulders.[24] The Smithy Department and

---

[20]   Griffiths, pp. 288–292.

[21]   Griffiths, pp. 306, 308.

[22]   Murdock was described as 'the best turner that ever lived … he used to simply pick up a piece of chalk and draw [a job] on the door'. Griffiths, p. 299.

[23]   Roll, pp. 275–276.

[24]   BA&H, MS 3147/8/42/5 Lists of Employees: 'List of Workmen at Soho Foundry 19 July 1808' list. In July 1808 in the Foundry Department were 15 moulders (five of whom were assistant moulders), five apprentice moulders, six furnace workers and 10 smiths or similar, along with six patternmakers and three apprentices.

the Boiler Shop workforce were smaller, while the Fitting Department workforce of 40 was headed by William Harrison and largely employed skilled fitters (as they had come to be known).[25] Most of these skilled artisans were personally bound by a five-year contract to serve the partners. The management 'system' at Soho resembled, therefore, more of a patchwork of individual contractual arrangements and practices than a tightly engineered system of scientific central control. The workforce was a mix of many distinct personalities with a range of skills and they were directed by a group of foremen who resembled a ministry of many talents rather than a uniform band of sub-managers. As foremen, they necessarily enjoyed significant autonomy and were relied on by their employers as petty contractors as well as lieutenants who commanded the ranks of labour.

If internal contracts provided one instrument of discipline, they did not replace the basic articles of employment agreed between all significant employees from journeymen to superintendents developed in the age of Boulton's leadership. As noted earlier, the distinctive but complex rank of 'engine erector' was filled by a large number of individuals who included Robert Muir, David Watson and Thomas Wilson (John Southern being involved with the Drawing Office at Soho). Skilled artisans were personally bound by a five-year contract to serve the partners, metalworkers signing 'memoranda of agreement' setting out the standard terms of employment. The evidence indicates that Boulton and Watt were hardly generous in their wages: they responded to labour market conditions but gave opportunities for developing an expertise that may have enabled them to attract workmen at more moderate rates. Thomas Harrison accepted employment at the end of 1802 as a 'Workman or Journeyman', assisting his employers in their 'business of Fitting Filing and Turning of Iron and other Metals'. Harrison bound himself for five years not to act in an unreasonable manner or absent himself at unusual times, nor embezzle or spoil or neglect tools, in return for weekly wages rising from 14 shillings in the first year to 17 shillings by the fifth.[26]

Leading hands were selected from the best of such journeymen. Some foremen came to the firm from a distance but the demand for skilled labour was such that internal promotion became a key source. Boulton complained bitterly to Rennie in 1791 that he was 'reduced to making foremen of men scarcely fit to be hindsmen', though there is little evidence of poor quality among the supervisors.[27] One notable foreman was David Thomson, Foreman of the Soho engine yard in 1790, of whom

---

[25] BA&H, MS 3147/8/47 'List of Workmen at Soho Foundry 19 July 1808'. The Boiler Shop had at least two smiths, four apprentices, five strikers or similar, and 10 boiler makers (one riveter). BA&H, MS 3147/8/42/6 'Lists of Employees at Fitting Department' (July 1808) lists 24 workmen and 16 apprentices. A small 'Establishment' complement of a few sawyers and labourers amounted to 10 employees.

[26] BA&H, MS 3147/8/33/12, Memorandum of Agreement with Thomas Harrison, 3 December 1802. 'Deductions to be made for absence from sickness or other causes'.

[27] MB to Rennie, quoted in Griffiths, p. 292.

it was said: 'I doubt not his abilities proved adequate to his business'.[28] Personal integrity was evidently held to be as, possibly more, important than mechanical aptitude. Another gifted foreman, Peter Ewart (1767–1842), was born in Dumfries in Galloway, educated at Edinburgh University and had some connection with Josiah Wedgwood before serving as Albion Mill foreman, where it was reported as early as 1789 that he significantly improved the manufacturing process.[29] By the time John Haden and Hollins Hunt were directing labour at Soho, the scope for significant innovation on the workshop floor had diminished from the heroic period of experimental improvement and adaptation in which not only Murdock but leading hands such as Matthew Murray had shone by contributing ideas on the improvement of castings used in engine construction.[30] One of the most significant and controversial qualities which early 'foremen' brought to the enterprise was the kind of practical genius for which Murdock and others were commended.

The supervisor's rewards were more handsome than those reaped by the men he oversaw. John Haden's 'articles' (amended to 'memorandum') of agreement were signed with Robinson Boulton, James Watt Jr, and Southern on 23 July 1813 but the terms were similar to earlier contracts with foremen and with workmen.[31] Haden bound himself to serve the three principals as 'Foreman or Superintendent of Workmen for the space of five years', with the best of 'his Ability, Skill and Judgement' while obeying 'all lawful Commands'. In promising to be a true and faithful servant, Haden also undertook not to 'divulge, discover or disclose any Secret Art or Mystery whatsoever relative to the business'. Entering the position of foreman and superintendent, Haden received an annual salary of £120 in the first half of his service, rising to £130 for the remainder. On renewing his 'covenant' in 1818, Haden agreed with the partners to 'bind themselves firmly to each other by their presents', which included an increased salary of £150, though by 1823 Haden refused another contract and left the firm under a cloud in 1825, apparently in debt to journeymen in the Engine Yard.[32] Hollins Hunt was a 'manager' who developed a complex and colourful relationship with the firm, extending to the education of his children.[33] The equity of such remuneration has to be gauged against the undoubted technical as well as personal contribution which many of these early

---

[28]   BA&H, MS 3147/3/295/59, John Rennie to [?], 31 August 1790. For biographical notes on Thomson see BA&H, MS 3147/3/295/57–63.

[29]   BA&H, MS 3219/6/1/78, JW to JWJ, 16 October 1790; BA&H, MS 3782/12/66/2, J. Lawson to MB, 27 June 1789, regarding improvements suggested by Ewart.

[30]   Griffiths, pp. 291–295.

[31]   BA&H, MS3147/8/33/2, Articles of Agreement of John Haden, Handsworth, 23 July 1808. Another Haden document was originally dated 1808 and amended to 1813 again suggesting an original contract/article in 1808, BA&H MS 3147/8/33/3.

[32]   BA&H, MS 3147/8/33/5a, John Haden to JWJ. It was noted on 3 June 1824 that Haden's debts amounted to £49.7s.1d., with more than £13 owed to one man.

[33]   BA&H, MS 3147/8/33/27, Memorandum of Agreement with Thos. Hollins Hunt, 23 November 1818. Hunt was contracted as a Clerk and Superintendent of Workmen.

foremen and labour superintendents made to the organisation of work as well as the improvement of design.

The final part of this chapter addresses this question of intellectual property by returning to the most exciting period of innovation during the introduction of new mechanical technology and considers the struggles surrounding the battle for reward and reputation in the days when Matthew Boulton was at the height of his powers.

'A Peep at the Inward Man':
Leading Hands and Intellectual Intrigue at Boulton and Watt

It has been argued above that we can understand labour supervision at Soho not as a radical regime of scientific management but as a network of personal, legal and financial arrangements which evolved from the early days of the first partnership and were adapted after the building of the new works in 1795. The original partners were enmeshed in struggles not only with strong personalities who entered their service but with commercial partners and trusted suppliers such as the Wilkinsons, who undertook the essential tasks of moulding, forging and casting components. The personal inter-family conflict and crisis within the Wilkinson firm revealed piratical behaviour even among those with whom Boulton had close personal dealings. The subsequent decision to manufacture engine parts themselves presented the partners with unfamiliar and formidable challenges in maintaining high technical standards of workmanship while also dividing labour into specialist operations which promised greater control over effort, output and wage costs.[34]

The deeper issue involved in designing and making components was that of mastering the men, extracting ingenuity as well as effort without conceding any legal right to invention. The difficulties of dealing with an able worker were well illustrated by Robert Cameron, employed in developing ideas for a rotative engine in competition with Watt's. In February 1783, Boulton referred to the efforts of his lieutenant Buchanan to prevent Cameron from undertaking 'another rotative'. Boulton advised his partner to proceed with haste in getting,

> your designs cast, bored and forwarded as quick as possible, and it may be nearly finished there, before it is sent off without any body at Soho knowing anything of the matter. I think if it was pushed it might be got done almost as soon as R. C[ameron] has made the exp[erimen]t to try the heat power & consumption of Steam.[35]

---

[34] Roll, p. 273.

[35] BA&H, MS 3147/3/7/3a MB to JW, 15 February 1783. Boulton added that he was afraid 'of letting him [RC] loose in this great Town [London] until we have secured it by the erection of a few of our improved Engines'.

Boulton added that he would not object to paying Cameron 'his wages for a year to do nothing provided he could be locked up from doing mischief'. Boulton also composed a careful communication to Buchanan, explaining Watt was unwell and that as Boulton had given Cameron 'full scope in the experiment' in order that there should be no blame levelled at the partners for Cameron's failure, he was still disappointed that Buchanan had not halted 'proceedings that were obviously inferior to other matters of producing greater effects'. Insisting that it was time to bring Cameron's expensive folly to a close, Boulton sought to establish a key principle for the future in claiming that he had agreed that if 'upon trial Mr. Watt's construction was better, he would not wish to persevere'. His failure to match Watt in invention meant, claimed Boulton, that a new rule should be established in authority over design at the firm: 'I shall endeavour to prevail upon Mr. Watt to execute one after his drawings & can never consent to any person in our employ to say 'no – Mr Watt's plans shall *not be followed but mine shall take place*'.[36]

The early months of 1783 were absorbed in Boulton's parliamentary campaign to defend their patent rights against Cornish enemies, following legal advice to 'make as many contracts and create as many Engines as possible i.e. to make Hay while the Sun shines'. Their legal counsel, Palmer, also gave the opinion that Cameron could claim no property rights over his own designs since: 'he being our workman [whatsoever] machine he makes in our Service will be considered by all Courts as our act & deed but [Palmer] thinks we had better [ensure] & keep him on until we have firmly established our intentions'.[37]

It was against this background of commercial and parliamentary battles with Cornish and Welsh adversaries that Watt himself drafted two letters to Cameron, though Boulton was doubtful about some of Watt's 'midnight projects'.[38] In fact, Watt appears to have been unusually conciliatory towards 'Robert' while Boulton fed him news of further mischief introduced into the Engine House by Cameron in 1784.[39] Here again Boulton framed his own bitter reservations as the reflections of an enlightened man on the irrational impulses of the dark personality he had observed in a careful scrutiny of a wayward employee, suavely assuring Watt that:

---

[36]   BA&H, MS 3147/3/7/3b, MB to John Buchanan, 15 February 1783.

[37]   BA&H, MS 3147/3/7/4, MB to JW, 23 February 1783. Boulton complained, 'But in this hurrying City one's attention is so much divided that I think I had better return home & with you further digest this matter'.

[38]   BA&H, MS 3147/3/7/10, MB to JW, 19 May 1783.

[39]   BA&H, MS 3147/6/10/17, MB to JW, 24 April 1786: 'There is a young man that R. hath introduced into [ ] A.M. Engine House as to A. Mitchell but who could not get [the] engine going one night when Mr. Wyatt and several workmen were present, upon which Mr. Wyatt said he thought he wd not assure but he replyd that he already knew the principle much better than that foolish fellow who we had sent to erect N.R. Engine. The case is exactly the same with Malcolm & there you see for various reasons he will do us more mischief than good in the capacity of such as agent'.

Your reasonings about R[obert] are very just. I agree to every thing you say of him & have conducted myself toward him exactly as you would have done, having never had any Dispute, nor shown any want of confidence in him but it would have been stupidity in me not to have made many observations or taken a peep at the inward man, when ever he hath layd himself open. I have my doubts …[40]

Boulton was determined not only that Cameron should be deprived of further resources to develop his projects but that he should be kept under surveillance when visiting London in order that the good name of the firm be protected and that their enemies should not be given any further ammunition to use in the continuing battle over patents, telling Watt he was afraid of letting the artisan inventor 'loose in this great Town until we have secured it by the erection of a few of our improved Engines'.[41] Murdock was prized, admired and feared as a brilliant leading hand but he displayed remarkable loyalty to the partners and their successors. Cameron revealed more independent spirit, if less ingenuity, and was regarded with greater suspicion by Boulton, whether or not his designs led to success.

Boulton's commentaries on Cameron and other leading personalities in this young enterprise were composed in an unfolding industrial culture where social conflict brought impetus as well as impediments to industrial innovation. Artisans drew on an awareness of mutual traditions as well as contemporary market values to combine in resistance to employers' schemes. Boulton did not recognise a clash of legitimate values in such battles. His correspondence reveals a consciousness that such rude hands might seize control of production and rupture supplies of essential parts. This fear helped to convince him that the firm should consider manufacturing their own components rather than depending on other workshops swept by restive movements 'interfering in the management of them'.[42] Boulton's account of artisan smiths turning out at Mr Virgin's works was characteristic of this enlightened authoritarianism when he concluded, 'It is folly and madness and must be injurious to all parties'. The entrepreneur added that he could find no rational explanation for disputes over amounts of money that he considered trivial. He reached again to give a darker varnish to the movement in his sinister reflection, 'I fear they have some other more diabolical reason at the bottom'.[43] Where artisans chose to move away from the clear light of economic reason it appeared to Boulton a short step towards demonic calculation. His workers' capacity for an alternative, intelligent reading of the age of reason in which they lived was not seriously contemplated. The menace of such social explosions helps to explain an interest in the progress and

---

[40]   Ibid.

[41]   Ibid.

[42]   BA&H, MS 3147/3/6/23, MB to JW, 10 April 1782.

[43]   Ibid.

regress of both established and younger workmen who remained as keen as his observation of the race to develop newer engines.[44]

Darker Shades in the Birmingham Enlightenment: Conclusions

Pollard and Roll provide contrasting assessments of labour management during the industrial revolution. Pollard follows Marx in detecting systems of class control within the development of technology and workplace discipline while Roll views the Soho enterprise as a laboratory for management science as well as rational innovation. Both historians draw on management models of the early twentieth century to interrogate management practices at the dawn of mechanised production. The present chapter attempts to develop another perspective on both scientific rationality and industrial culture as they informed and shaped the management of labour in the world inhabited by Boulton and Watt. The material conditions for the progress of industrial technology and the growth of that practical culture, depicted in heroic terms by evangelists of the new engineering, were laid down by diverse and uneven customer demands which required bespoke production and piecemeal innovation by specialists who could adapt to shifting markets and changing order books. The institutional setting for the advance of workplace management came not from the technical genius of Watt but rather the commercial, legal and cultural acumen of Boulton as he fought to maximise income and extend intellectual property before the early patents were exhausted. Boulton recognised the political as well as commercial value of reputation and bolstered Watt's claim to scientific genius as an instrument to promote the partners' claim to intellectual property created within the walls of the firm.

In the early days, efficient construction of 'fire engines' was a necessary but secondary adjunct to the preservation of patent rights, hence the arduous political and legal actions in which Boulton engaged during the 1780s and 1790s. Fending off those who sought to pirate their designs and reap the fruits of their ingenuity, Boulton extended his lethal rationale to the handling of his employees. Presenting his concern for Watt's prestige and pre-eminent scientific reputation, Boulton displayed a ruthless determination to discipline foremen as well as journeymen when their technical aptitude and strong personality threatened the interests of the firm. In the dynamic and shifting conditions of the industrial revolution, a capacity for practical innovation was welcomed and feared, anticipated and suspected, by the Birmingham entrepreneur. Murdock's genius troubled Boulton as well as Watt, though Boulton worked by skilful stratagems to keep his talented foreman in loyal subordination without offering him the partnership that he was able to demand later. Cameron was less tractable and was effectively excluded from fraternity within the enterprise as Boulton worked to prevent his trips to

---

[44]   BA&H, MS 3147/3/10/17, MB to JW, 24 April 1786, regarding orders for a mine engine.

London and to undermine his claims to intellectual prowess. Such struggles were more than squabbles over scientific reputation or tokens of intellectual vanity in an age of competitive enlightenment. They were integral to the process of cultural production in which different narratives of industrialisation were being written and heroic endeavours by men of genius and character became the predominant motif of economic success among the middling classes.

There is little doubt of Boulton's superior skills in the Parliamentary lobby, the law courts and the drawing rooms of polite society. Similar concerns to segregate the men of genius from the talent of practical mechanics can be detected in the tensions and exclusions in the elaborate articles of service drafted by industrialists to bind their leading hands, the mutual promises separating while appearing to link the world of Boulton with that even of Murdock. If Boulton thought to crush the irrational ambition of the artisan innovator, he mistook the force of the romantic narrative which was cultured within the academy of reason. The heroic biography which was to become a staple of the age of Smiles was already being composed in the mechanical enlightenment of Watt's planetary motion. The practical culture of the artisan was also sustained in the work of the brilliant Murdock and though he never found his own voice in the Royal Society, his inventiveness was recognised even there.[45] The demonstration of the power of mechanics and the value of craft also inspired the kind of social movements among labour which Boulton attributed to the satanic forces of unreason. As Nasmyth wrote thanking Smiles for his life of Watt and Boulton in 1889, 'our modern Tools have given us some power to rapidly test the practical value of all grand Notions!'[46] Artisans and foremen in Victorian Britain continued to draw different conclusions from the heroic narratives of industrialisation than their masters.

[45] Quoted in Griffiths, p. 134. For the exclusion of Murdock and his paper to Royal Society see ibid., pp. 196, 238–239.

[46] SP, BL ADD 71075 635B, JN to SS, 17 October 1889.

Chapter 12

Workers at the Soho Mint (1788–1809)

Sue Tungate

As a pioneer in the developing metal industries in eighteenth-century Britain, Matthew Boulton (1728–1809) is well known for his Soho Manufactory, where he installed mechanical devices and automated machines to improve his workers' efficiency. In the late 1780s, he set up a new business, the Soho Mint, intending to mass produce a regal copper coinage sufficient for the growing industrial workforce in Britain. Towards the end of the eighteenth century there was an uneven distribution and considerable lack of small change and many counterfeit coins in circulation.[1] Boulton was concerned about workers who were being offered counterfeit coins in their wages. For example he wrote to his son:

> The Public has sustained a great loss by the illegal practise of counterfeiting halfpence, which has been lately carried out to a greater height than was ever before known and seems still to increase, in so much, that it is now too common a custom among many of the lower classes of Manufacturers and traders, to purchase these counterfeit halfpence at little more than half their nominal value, and to pay with this money their workmen & labourers, greatly to the injury of the honest part of the community, and to the detriment of Trade.[2]

Once an engine producing rotary motion had been developed, Boulton saw that steam power could be harnessed to solve this problem and he was also aware of the commercial opportunities offered in producing a copper coinage.[3] The Soho Mint became the first steam-powered coining operation in the world. Boulton was

---

[1]   This situation is fully discussed in C.E. Challis, *A New History of the Royal Mint* (Cambridge, 1992) and Sir John Craig (1953), *The Mint: A History of the London Mint from AD 287–1948* (Cambridge, 1953).

[2]   Birmingham Archives and Heritage (BA&H), MS 3782–13–36–37, 12 November 1789, MB (Soho) to MRB (Langensalz, Germany).

[3]   Many instances of technical improvements concerning the steam engine and the mint are found in Boulton's correspondence with Watt. Indeed, in 1783 Boulton discussed reports of him being credited with its invention alongside Watt. MS 3147–3–7–7, 16 April 1783, MB (London) to James Watt (Birmingham). Boulton was also heavily involved in the copper industry when he initiated the idea of a copper coinage. See Sue Tungate, 'Matthew Boulton and the Soho Mint: Copper to Customer' (unpublished University of Birmingham, PhD thesis, 2011).

able to introduce a completely new coining process without losing the support of his workers, whereas antagonism to innovation, both technical and organisational, was common elsewhere, especially at the Royal Mint.[4] He applied modern industrial methods to minting and struck 600 million coins for Britain and other countries around the world.[5]

This chapter explores the ways in which Boulton's employees helped to set up the Soho Mint and how they were used, recruited and retained. The work is based on research in the Mint record books and correspondence held in the Archives of Soho.[6] Previous work by Eric Roll and Jeremy Knowles includes studies of Boulton and Watt's steam-engine business, where innovative managerial techniques were developed. Roll also discusses the division of labour at the Soho Manufactory.[7] A fuller discussion of the management techniques and organisation of labour at Soho Manufactory is given by Sidney Pollard, who uses it as the best documented example of organisational planning in the developing industrial revolution. Regularity of work, division of labour and delegation of management were introduced, especially in the Soho Foundry.[8] The Soho Mint did not introduce many changes in labour practises that were different to those at the Soho Manufactory, but the introduction of the steam-powered press meant a greater need for more skilled labour, including engineers, engravers and technicians. The Soho Mint was under the sole control of Matthew Boulton and was not as regular in its work as other divisions at Soho, due to the difficulties of obtaining orders that could utilise the steam-powered press. The benefits of its largely automated processes could only be obtained when large orders were undertaken, as is discussed later in this chapter.

Kenneth Quickenden has discussed the conditions of employment and methods of organising the work force in Boulton's silver business and the biographical

---

[4]   D. Sellwood, 'The Trial of Nicholas Briot', *British Numismatic Journal*, 1986, 56, pp. 108–123.

[5]   Tungate, 'Matthew Boulton and the Soho Mint', p. 521.

[6]   Research into the correspondence of Matthew Robinson Boulton, James Lawson, John Southern, William Cheshire, William Brown, Zaccheus Walker Sr and Jr, William Brown, Richard Chippindall, and Boulton's bankers, Charlotte and William Matthews, was funded by an AHRC Collaborative Doctoral Award. The Archives of Soho have been described by F. Tait, 'How do we know what we know? The Archives of Soho', in S. Mason (ed.), *Matthew Boulton: Selling What All the World Desires* (London, 2009), pp. 108–116; F. Tait, 'Coinage, commerce and inspired ideas: The notebooks of Matthew Boulton', in Malcolm Dick (ed.), *Matthew Boulton: A Revolutionary Player* (Studley, 2009), pp. 77–91.

[7]   E. Roll, *An Early Experiment in Industrial Organisation: Being a History of the Firm of Boulton and Watt 1775–1805* (London, 1930; 1968), pp. 189–136; Jeremy Knowles, 'An Assessment of the Application of James Watt's Reciprocating Steam Engine in Cornwall on the Development of the Firm of Boulton and Watt, 1775–1795' (unpublished University of Birmingham, BA dissertation, 2008).

[8]   S. Pollard, *The Genesis of Modern Management* (London, 1965).

details of key workers, some of whom were later important at the Soho Mint.[9] Stebbing Shaw, as early as 1798, describes the working of machinery and employment at Soho Mint.[10] But little has been written about the workforce. Many of the workers worked in other areas of the Soho concerns and were drafted as needed. Boulton relied upon highly skilled Mint workers to make it a success, though apprentices, women, carriers and outside suppliers were important. Less skilled workers were employed, but as many unskilled operations such as loading blanks into the coining press were automated, none had the security of tenure that the more skilled workers enjoyed. Unskilled workers, such as girls and women were required for certain jobs such as packing coins when very large orders needed to be dispatched.

Boulton's business developed from the small manufacturing and mercantile firm set up by his father.[11] At the Soho Manufactory the employees were organised in a variety of workshops, each with a manager responsible for a particular area of production. Machines such as fly presses were used to make goods cheaply and standardised methods of manufacture were introduced wherever possible. Eventually, there were around a dozen businesses under the Soho umbrella, including steam engines, plate manufacture and a copper warehouse, run by Boulton, generally in association with others.[12] In 1788, he set up the Soho Mint without a partner, though sharing contacts and agents with other businesses at Soho reduced overheads and workers could be moved easily from one section to another. Boulton was heavily involved in the design of the Soho Mint and in the choice of personnel, including engineers, engravers, clerks, craftsmen and apprentices. There was little input from his partner James Watt (1836–1819), but his son, Matthew Robinson Boulton (1770–1842), and other relatives were involved.

Setting up the Soho Mint

Manufacturing coinage required specialist workers to make dies, roll metal, cut out blanks and stamp coins. The Royal Mint had carried out these processes for generations, but their supply of small change was very limited by the late 1780s and many of the copper coins in circulation were of poor quality. Boulton was determined to raise the standard of operations and make uniform coins, which had not previously been produced. To do this he automated processes, introduced

---

[9]   Kenneth Quickenden, 'Boulton and Fothergill's Silversmiths', *Silver Society Journal* (London, 1995), pp. 342–356, p. 351.

[10]   Stebbing Shaw *The History and Antiquities of Staffordshire* (London, 1798–1801), vol. II, p. 118.

[11]   C. Behagg, *Politics and Production in the Early Nineteenth Century* (London, 1990), p. 43.

[12]   H.W. Dickinson, *Matthew Boulton* (1936), p. 209.

steam-powered coining presses and organised the delivery of copper and the distribution of coin.[13]

Orders for coins were obtained via agents and salesmen; then metal, generally copper, had to be bought and transported to Soho, from suppliers in Cornwall or Anglesey, for example, where Boulton had contacts. The metal was annealed several times and then rolled to the correct thickness.[14] Accurate coining blanks were cut out using presses; and cleaned by being pickled in oil of vitriol or urine to remove the scale. The blanks were shaken in bags with an abrasive to remove burrs, loaded into tubes and fed automatically into the coining press. They were then struck with specially engraved dies to the required design, and the completed coins were wrapped in rouleau (cylindrical roll of coins), packed into barrels, and dispatched to a variety of customers.[15]

The Soho Mint was run initially by Boulton himself who organised the whole operation. His workers included competent clerks who co-ordinated the orders for copper, the building of the mint, payment of wages, dispatching coins and so on. He also required very skilled and experienced engineers and engravers for jobs, such as designing and building the coining presses, and making highly detailed dies. Other less skilled workers were needed to prepare blanks for the coining press or to pack the completed coins, but they had to be dependable and honest, so it was important to select suitable individuals. Boulton employed women and girls, as well as boys for the less skilled jobs.

Boulton's first experience of coining was to roll copper and cut blanks for an order from the East India Company (EIC) in 1786.[16] In 1785 he had erected a new water-powered rolling mill to roll metal for the Soho Manufactory, and the order for Sumatran coins was used to keep this new mill in full employment.[17] This led him, from 1787, to seek a regal coinage contract from the British government and to develop the Soho Mint with its steam-powered coining press. He already had a talented workforce with men experienced in rolling metal and operating presses, forging dies and making blanks for button manufacture.[18] His workers had the

---

[13]   See Sue Tungate, 'Matthew Boulton and the Soho Mint'.

[14]   Annealing is a heat treatment where metal is heated to above a critical temperature, and then cooled, in order to change its properties such as strength and hardness. It is used to make the metal more homogenous by softening it, and making it more ductile for further work such as rolling or stamping. It also relieves internal stresses and improves cold working properties. For copper, the metal may be cooled, either slowly in air, or quickly by quenching in water. Ibid.

[15]   Oil of vitriol is sulphuric acid, and scale is copper oxide. Rouleau are cylindrical piles of coins made up to a specific value, which were then wrapped in paper. Ibid.

[16]   MS 3792/12/63/43, JH (Soho) to MB (Chacewater), 23 September 1786.

[17]   G. Demidowicz, 'Power at the Soho Mint', in Dick (ed.), *Matthew Boulton*, pp. 121–123.

[18]   See Sue Tungate, 'Matthew Boulton and the Soho Mint'.

ability to respond quickly to change, and most of those employed to set up the Soho Mint were long-standing trusted employees.

Engineers and Craftsmen

Among the most important workers at the Soho Mint were the steam-engine erectors and engineers who were keys to its development, as without their expertise, it would not have been possible to install the steam-powered coining equipment. Boulton and Watt steam engines were at the cutting edge of technical innovation; they were erected on a commercial basis from 1777, and were used, for example, to pump water from the Cornish copper mines, which eventually provided most of the copper used at the Soho Mint. In 1784, James Watt was responsible for the development of the rotary engine, which meant that steam power could be applied to uses other than pumping water. William Murdock (1754–1839), one of Soho's most trusted and reliable employees had supervised the erection of steam engines since September 1779. Boulton continually praised his work, saying that the firm needed 'more Murdocks for of all others he is the most active man and best engine erector I ever saw'.[19] Around 1784, Murdock discussed plans with Boulton for applying steam power to coining, and he came to help at the Soho Mint in 1789.[20] John Rennie (1761–1821) was also involved early in his career as an engineer and provided drawings and estimates for the new Soho Mint building, which was under construction by January 1788.[21] He also rebuilt the rolling mill at Soho in the same year. Another engineer, Peter Ewart (1767–1842), helped to develop the mint equipment, and introduced a new method of cleaning the blanks.[22] He had worked intermittently for Boulton, starting as an apprentice to Rennie.[23] Important contributions were also made by James Lawson (1760–1818), who had been employed along with Murdock in Cornwall from 1779 as an engine erector and draughtsman.[24] By 1789 Lawson was supervising the installation of Soho Mint machinery, and in 1790 was involved in drawing up the steam-powered coining

---

[19]   BA&H, MS 3147/3/8/36, MB to JW, 8 November 1784.

[20]   BA&AH, MS 3782/13/120, Mint Inventions and Improvements Volume 1, Folder 8, Minutes of a conference held at Soho House, 7 January 1810 with William Murdock, James Watt, John Southern, Peter Ewart, James Lawson and Matthew Robinson Boulton present.

[21]   Demidowicz, 'Power at the Soho Mint', pp. 122–123.

[22]   BA&H, MS 3782/12/66/2, JL (Soho) to MB (London), 27 June 1789; Item 4, 6 July 1789.

[23]   BA&H, MS 3782/12/66/106, John Roberts (JR) (Soho) to MB (London), 27 January 1788.

[24]   Letters from James Lawson will be found in the General Correspondence files for the years 1782, 1783, 1785, and 1786 and from 1789–1799 in BA&H, MS 3782/12/66/ 1–57; BA&H, MS 3782/13/43/1–113.

press patent.[25] He was also an expert in die multiplication and responsible for important modifications to the coining process, developing an automatic layer-in which placed coining blanks automatically onto the coining press.[26] Later he directed the erection of Soho's steam-powered machinery at the Royal Mint.[27] John Southern (1761–1815), working as an assistant to Watt from at least 1781, became head of the Drawing Office, and was involved from the start with the Soho Mint.[28] In 1798, Southern was in charge of installing new coining presses with an improved design. He was also concerned with plans for the Russian Mint in 1799,[29] the Danish Mint in 1805,[30] and by March 1806 was drawing up plans to supply up-to-date equipment to the new Royal Mint.[31]

All these engineers, along with Boulton himself, contributed to the development of the steam-powered mint equipment. The engineers were flexible, working in other departments at Soho and elsewhere, including also acting as salesmen on occasion.[32] Southern, Ewart and Lawson described their experiences of setting up the new Soho Mint in a series of letters and minutes in 1810. All agreed that the technical development of the mint equipment was a team effort, with Boulton providing the drive and the inspiration.[33]

Other important workers were the experienced craftsmen who had worked for a long time at the Soho Manufactory.[34] Joseph Harrison, an engine erector who had helped erect the original Kinneill engine in 1775,[35] was able to apply his skills to install the screw to a coining press in 1786. This had been turned by blacksmith, Anthony Robinson.[36] Later, both men made vital contributions to the new steam-

---

[25] BA&H, MS 3147/3/14/11, MB (London) to JW (Soho), 16 July 1790.

[26] David Vice and Richard Doty, *The Conder Token Newspaper*, 1/3, 15 February 1997 and 1/4, 15 May 1997.

[27] See BA&H, MS 3782/3/116 to BA&H, MS 3782/3/124, on the setting up of the Royal Mint from 1805–1822.

[28] BA&H, MS 3782/12/66/58–104, 26 February 1788 to 3 December 1799; MS 3782/12/66/118–119; BA&H, MS 3782/13/43/114–135, November 1800 to May 1815.

[29] BA&H, MS 3782/12/66/102, JS (Soho) to MB (—), 27 May 1799.

[30] BA&H, MS 3782/13/43/120, JS (—) to MB (—), 6 January 1805.

[31] BA&H, MS 3782/13/43/122, JS (Soho) to MRB (—), 21 March 1806.

[32] BA&H, MS 3782/12/66/53, JL (Leeds) to MB (Soho), 28 November 1797.

[33] BA&H, MS 3782/13/120, Minutes of a meeting held at Soho, 7 January 1810.

[34] BA&H, MS 3782/12/108/ 53 and 54, Mint Notebook 1788, pp. 10, 52, 87 and Mint Notebook 1789, p. 21.

[35] John Harrison was also involved. B.M. Gould, 'Matthew Boulton's East India Mint in London 1786–88', *Coin and Medal Bulletin*, 612, August 1969, pp. 270–277.

[36] John Harrison was also involved. BA&H, MS 3147/3/12/5a, MB (London) to James Watt (JW) (Harper's Hill), 5 February 1788.

powered coining press.[37] John Peploe was a practical multi-talented employee, who made specialised nozzles and condensers for steam engines.[38] He could also make equipment such as edge-marking machines for the Soho Mint.[39] John Busch, though illiterate, played a significant role in forging dies which were then engraved by a variety of experts.[40] He also made important modifications to the automated coining press.[41] Other specialist parts were made by specific individuals such as John Middlehurst, who provided tubes to load the coining press, making at least four dozen monthly.[42] Another reliable employee, Thomas Kellet was in charge of the rolling mill,[43] which was used to roll metal for other businesses when the Soho Mint was not busy.[44] These individuals contributed in various ways to the success of Boulton's coining venture and were valuable members of his workforce.

Engravers

Other vital employees at the Soho Mint were the engravers, as Boulton could not make high quality coins without good dies. He placed great importance on obtaining the best designs, and spent time in finding the best engravers he could employ. He had great difficulty obtaining suitable workers. Dies were already used at the Soho Manufactory to manufacture a variety of items, and apprentices were taught to draw and engrave by John Eginton.[45] However Boulton thought that none of the resident Soho die makers were capable of making coining dies. Nor were any native engravers available, 'except bungling hands as you have near home who he [Wilson] says are a *stigma on the trade* [sic]'.[46] Before the Soho Mint was set up, freelance workers, including John Westwood Senior, were

---

[37]   Robinson was employed to turn the screws for the steam-powered coining presses which were cast by Dearman and Wilkinson. BA&H, MS 3782/13/36/58, MB (London) to Matthew Robinson Boulton (MRB) (Soho), 4 June 1791.

[38]   BA&H, MS 3782/12/72/3/20, John Scale (JSc) (Soho) to MB (London), 8 February 1778.

[39]   BA&H, MS 3782/12/66/11, James Lawson (JL) (Soho) to MB (London), 27 May 1791.

[40]   'I have sent for Busch to do me a job & as he can neither write or read, I beg of you to give him some instruction'. BA&H, MS 3782/13/36/151, MB (London) to MRB (Soho), 7 July 1800.

[41]   BA&H, MS 3782/12/66/21, JL (Soho) to MB (London), 20 January 1792.

[42]   BA&H, MS 3782/13/36/64, MB (London) to MRB (Soho), 21 March 1792; BA&H, MS 3782/3/13, Mint Day Book 1791–1795; BA&H, MS 3782/3/14, Mint Day Book 1795 to 1798.

[43]   Demidowicz, 'Power at the Soho Mint', pp. 122–123.

[44]   BA&H, MS 3782/13/36/84, MB (Truro) to MRB (Soho), 3 September 1792.

[45]   Quickenden, 'Boulton and Fothergill's Silversmiths', p. 344.

[46]   BA&H, MS 3782/12/59/22, RC (London) to MB (Soho), 29 March 1790.

employed to engrave medal dies, such as for Captain Cook's voyage in 1772,[47] and for Boulton's first coinage issue in 1786, only simple dies were involved, which were made by the London engraver William Castleton.[48] However, for the proposed regal coinage issue, Boulton needed an engraver who could produce a portrait die. He looked to foreign engravers as they had the best reputation, and in December 1786 Jean-Pierre Droz (1746–1823) from the Paris Mint, was engaged to engrave dies.[49] Droz, however, did not arrive at Soho to work until October 1788 and left again in June 1790.[50] His importance to the Soho Mint has probably been overemphasised as he cost Boulton over £1,700 and achieved very little, except for a series of pattern regal coins.[51]

When Boulton turned to token production in 1789, most of the initial dies were engraved by a local man, John Gregory Hancock (1750–1805) on a freelance basis. Hancock's importance as an engraver should not be underestimated. Not only did he make the dies for the 1787 Wilkinson and Anglesey tokens, but he was responsible in 1789 for the Cronebane issue, the first steam-struck tokens in the world.[52] However, he would not agree to be employed exclusively by the Soho Mint, although he later made further dies for Boulton, as a self-employed worker.[53] Other freelance specialists, such as Richard Phillips, a letter engraver from London, were employed.[54] A local man, Thomas Wyon (1767–1830) engraved various dies, including the 1791 George III pattern guinea and the 1795 pattern halfpenny.[55] Wyon was one of a family of talented engravers who received their initial training at Soho. Later in 1802, his twin Peter (1767–1822) engraved additional Soho dies and a memorial medal commemorating Boulton's death in 1809.[56]

When Droz left, Boulton again looked abroad for a suitable resident engraver. Rambert Dumarest (1760–1806) arrived at Soho in August 1790, but soon became

---

[47]   David Vice, 'The Resolution and Adventure Medal' *Format* March 1988, 36.

[48]   BA&H, MS 3782/12/59/2, William Chippindall (WC) (London) to MB (Kings Head, Truro), 12 September 1786.

[49]   L. Forrer, *Biographical Dictionary of Medallists* (6 Vols, London, 1904–1916), pp. 150–191.

[50]   J.G. Pollard, 'Matthew Boulton and J.P. Droz', *Numismatic Chronicle* 1968, 8 (Oxford), pp. 241–265.

[51]   BA&H, MS 3782/12/56/16, Sir Joseph Banks (London) to MB (London), 11 July 1790.

[52]   D.W. Dykes, 'John Gregory Hancock and the Westwood Brothers: an Eighteenth Century Token Consortium', *British Numismatics Journal*, 69, 1999, pp. 173–186.

[53]   BA&H, MS 3782/12/59/125, MB (Soho) to Richard Chippindall (RC) (London), 12 November 1802; BA&H, MS 3782/12/59/226, William Cheshire (Soho) to MB (—), 30 November 1802.

[54]   BA&H, MS 3782/12/66/9, JL (Soho) to MB (London), 23 May 1791.

[55]   BA&H, MS 3782/3/13, Mint Day Book 1791–1795, 9 July 1791.

[56]   Forrer, *Biographical Dictionary of Medallists*, p. 633.

homesick, and returned to France in June 1791.[57] He was responsible for making dies for Southampton, Cornwall and Glasgow tokens, issues for the Monneron brothers of Paris and 17 million EIC coins.[58] Another French engraver, Noel-Alexandre Ponthon (1770–1835), joined the team in summer 1791, and remained at the Soho Mint until July 1795. His employment was suggested by Dr Francis Swediaur, Boulton's agent in Paris.[59] In addition to Monneron issues, between 1792 and 1795 Ponthon engraved dies for at least seven token issues,[60] for coins for Sierra Leone, Bermuda and a further 22 million coins for the EIC.[61] Ponthon also made dies for a projected Bengal coinage in 1792. These had an unusual hexagonal shape, and were intended to cut down on the series of operations involved in making a coin.[62] This was a very different technological idea, and shows Boulton's ability to inspire innovative work from his employees.

Two further engravers joined the workforce in 1793. Conrad Heinrich Küchler (?–1810), escaping the dangers of the French revolution, sought a position initially to engrave dies on a speculative basis.[63] Küchler spent most of his working life at the Soho Mint, and was responsible for dies for the regal coinage contracts of 1797, 1799 and 1805–7, many pattern coins and for a series of beautiful medals.[64] John Phillp (1780–1815) came from Cornwall in 1793 to be apprenticed at Soho. He produced many images, including drawings of the coining apparatus in 1797,[65] and designs for medals and regal coinage.[66] He increasingly took responsibility for die engraving, especially during Küchler's absence in 1802.[67] He made dies for large coinage issues for the EIC and for

---

[57]   BA&H, MS 3782/13/36/60, MB (London) to MRB (Soho), 7 June 1791.

[58]   These coins, tokens and medals are discussed in Tungate, 'Matthew Boulton and the Soho Mint' and Tungate, 'Catalogue' *Matthew Boulton and the Art of Making Money*, pp. 56–86.

[59]   BA&H, MS 3782/12/91/117, Francis Swediaur to 'Andrew Smith' (pseudonym for MB), 27 February 1791.

[60]   The production of coins, medals and tokens at the Soho Mint has been detailed in Tungate, 'Matthew Boulton and the Soho Mint'.

[61]   BA&H, MS 3782/3/13, Mint Day Book 1791–1795, p. 121, 7 May 1793.

[62]   Sue Tungate, 'Technology, Art and Design in the Work of Matthew Boulton: Coins, Medals and Tokens Produced at the Soho Mint', in Dick (ed.), *Matthew Boulton*, pp. 189–190.

[63]   BA&H, MS 3782/12/59/45, RC (London) to MB (Soho), 12 February 1793.

[64]   Küchler's medals are listed in J.G. Pollard, 'Matthew Boulton and Conrad Heinrich Küchler', *Numismatic Chronicle*, Vol. X, 1970, pp. 259–318. His regal coinage is described by C.W. Peck, *Copper Coins of George III* (London, 1970).

[65]   Birmingham Museum and Art Gallery (BM&AG), 2003/0031/184a, Phillp Album,

[66]   Designs for halfpennies 1802, Phillp Album BM&AG 2003/0031/129.

[67]   Küchler left to work in London for a short time after a dispute with William Brown. BA&H, MS 3792/12/63/71, JH (Soho) to MB (London), 3 April 1800.

various Friendly Societies.[68] These engravers made excellent examples of numismatic art, which are still admired today.

Other Employees: Agents, Clerks and Managers

Boulton was often away from Soho and in addition to technical workers he needed individuals to market and run his businesses. He relied on trusted agents and salesmen to obtain the copper he needed for the coinage contracts and to sell his Mint products. One of his most important employees was his brother-in-law, Zaccheus Walker (?–1808), who worked as chief clerk to Boulton from at least 1760, keeping control of accounts from Soho's Birmingham Warehouse.[69] He reported on the progress of the Soho Mint machinery, problems with Jean-Pierre Droz and concerns about the profitability of the Soho Mint.[70] Walker was helped at times by his son Zaccheus (Zack) Walker Junior (1768–1822), who carried out various commissions for his uncle while travelling in Europe and America.[71] These foreign travels were essential in gaining orders; for example, Zack investigated the possibility of Soho Mint coining for the United States government in 1793.[72] Later Zack went to Russia between 1803 and 1809 to help erect the Russian Mint equipment, which was produced at Soho.[73]

Boulton also required a confidential secretary who could work at the Soho Mint. He described the sort of assistance he expected: 'I want an honest, faithfull [sic], and sensible man that can assist me in writing letters, that I can depend upon for the examination of our Cash, and all other accounts at Soho'.[74] A prospective employee needed to be 'A Person capable of writing his own language Correctly, in a plain hand, with a common knowledge of Book-keeping ... with satisfactory testimonials of his integrity & sobriety (two most essential requisites)'.[75] Boulton had suffered earlier with drunken employees, and did not want such individuals at the Soho Mint.[76] He also needed clerks who could translate in a variety of

---

[68]   BM&AG 2003–0031, Phillp Album.

[69]   BA&H, MS 3782/12/74/1–205, Zaccheus Walker (ZW), December 1765 – December 1790; MS 3782/12/75/1–178 ZW May 1790–January 1806.

[70]   BA&H, MS 3782/12/74/183, ZW (Birmingham) to MB (Buxton), 4 September 1789; BA&H, MS 3782/12/75/27, ZW (Birmingham) to MB (London), 26 March 1792.

[71]   BA&H, MS 3782/12/66/106, JR (Soho) to MB (London), 27 January 1788; BA&H, MS 3782/12/75/46, ZW (Birmingham) to MB (Truro), 25 September 1792.

[72]   BA&H, MS 3782/12/75/57, ZW (Birmingham) to MB (London), 18 April 1793.

[73]   Doty has detailed the establishments of the Russian, Danish and Royal Mint in Richard Doty, *The Soho Mint and the Industrialization of Money*, British Numismatic Society Special Publication, No. 2 (London, 1998).

[74]   BA&H, MS 3782/12/67/13, MB (—) to William Matthews (London), 26 May 1786.

[75]   BA&H, MS 3782/13/43/93, MRB (Soho) to JL (—), 28 March 1806.

[76]   BA&H MS 3147-3-2-13 MB to J. Watt, 1 September 1778.

languages and deal with agents in various parts of the world, who were sending information and orders. Accounts of his new steam-powered press were translated into French and German.[77] The Soho Mint eventually produced coins and tokens for around 20 countries in 14 languages.[78]

In London, Richard Chippindall worked as an agent for Boulton.[79] From 1793 he sold sets of coins and medals and dealt with token orders.[80] William Matthews, Boulton's London banker, was also involved. When he died in 1792, the business was continued by his widow, Charlotte, who helped distribute the regal coinage issue of 1797 from her house.[81] William Brown (1772–1820), who originally worked for Matthews, came to Soho in November 1791 as book-keeper and confidential clerk.[82] Later, in 1796, William Cheshire helped to run the Soho Mint and looked after Boulton's personal affairs.[83] These individuals promoted sales of Soho Mint products, as well as carrying out activities for Boulton's other businesses. Similarly, the engineers, Lawson and Southern travelled around the country fixing engines, but they also took orders for regal coinage until at least 1804.[84] It was an advantage to have flexible workers with a proven record of reliability.

One important contributor to the Soho Mint was Matthew (Matt) Robinson Boulton, Boulton's only son. Boulton educated his son to take over the Soho manufacturing empire and encouraged him to take an active interest in the business and scientific study from an early age.[85] Matt was expected to learn accounting, in particular double-entry book keeping, and from December 1786, he was in France to learn the language.[86] He undertook his first assignment for the Soho Mint at the age of 17 in October 1787, when he was asked to check on the state of dies for the

---

[77]   BA&H, MS 3782/12/75/3, ZW (Birmingham) to MB (London), 7 June 1790.

[78]   Tungate, 'Matthew Boulton and the Soho Mint'.

[79]   BA&H, MS 3782-12-59-1, RC (London) to JSc (Soho), 23 July 1782.

[80]   A notebook lists sales of silver, bronzed and tin medals, sales of sets and individual coins to dealers and individuals. BA&H, MS 3782/12/108/69, Medal ledger 1893–1816.

[81]   MS 3782-12-56 Item 92 MB (Soho) to Joseph Banks (London), 14 January 1798.

[82]   BA&H, MS 3782/12/67/70, Correspondence between MB and William and Charlotte Matthews, 1770–1802.

[83]   BA&H, MS 3782/13/43/94, MRB (Soho) to JL (—), 1 September 1806.

[84]   BA&H, MS 3782/12/66/83, to 87 John Southern (JS) to MB (Soho), 2–23 July 1797; BA&H, MS 3782/13/43/1 and 2, JL (Glasgow) to MB (Soho), 27 January 1800; 12 March 1800; BA&H, MS 3782/13/43/5–11, JL (Glasgow) to MB (Soho), 26 October 1801– 29 March 1802 and BA&H, MS 3782/13/43/ 21, JL (Manchester) to MB (Soho), 11 October 1804.

[85]   A.E. Musson and E. Robinson 'Training Captains of Industry: The Education of M.R. Boulton and James Watt Junior', in A.E. Musson and E. Robinson (eds), *Science and Technology and the Industrial Revolution* (Manchester, 1969), p. 204.

[86]   A series of letters detail what he was expected to learn. MS 3782/13/36, MB to MRB December 1787–October 1790.

proposed regal coinage being made by Droz in Paris.[87] Later Matt went to study in Germany, but on his return in October 1790 he played a major role in running the Soho Mint.[88]

Recruiting and Retaining Workers

Boulton relied heavily upon personal connections in recruiting his workers for the Soho Mint. He needed reliable employees as he was introducing new technology and using valuable resources such as copper and silver to make coins. Even workers loading blanks into the coining press and packing up the completed coins needed to be conscientious and trustworthy. Workers at the Soho Mint were often recruited via family ties, which were strong within the Soho workforce. For example Southern's father and three brothers worked for Boulton, as did Murdock's sons, and Chippindall's brother William assisted him in London.[89] Members of the Harrison family were employed by Boulton from at least 1766, including Mrs Harrison by 1773.[90] Joseph Harrison helped set up mint equipment in 1786, and other members of the family worked at the Soho Mint from the start: William coppered the roof, and John made the pistons.[91] In 1790, William Harrison Junior made coins and medals.[92] Similarly at least five members of the Jerome family were employed. John Jerome, apprenticed in 1790, joined his father John and other relatives, Joseph, Thomas and Timothy.[93] When work became scarce at the Mint this family was included in the list of key workers Boulton wanted to retain.[94] Many of Boulton's employees, such as the engineers, had previously worked for him. The major exceptions were the engravers who were frequently recruited from

---

[87]   BA&H, MS 3782/13/36/11, MB (Chacewater) to MRB (Versailles), 2 October 1787.

[88]   MS 3782/13/36/54, Resolutions on the Mint, MB to MRB, 26 May 1791.

[89]   Details of these relationships are seen in a variety of documents in the Archives of Soho.

[90]   Thomas Harrison to work at toy-making for five years, BA&H, MS 3782/12/95, 4 December 1766; Joseph Harrison to work as a whitesmith, for five years, BA&H, MS 3782/12/95, 23 September 1767; BA&H, MS 3782/12/72/3/ 8 and 9, JSc (Soho) to MB (London), 8 February 1773; BA&H, MS 3782/12/72/3/118, Memorandum by JSc February 1773.

[91]   BA&H, MS 3782/12/66/ 106 and 108, JR (Soho) to MB (London), 27 January 1788, 5 February 1788.

[92]   BA&H, MS 3782/12/95/32, 25 October 1790.

[93]   BA&H, MS 3782/12/95/33, Articles of Agreement, 6 December 1790.

[94]   'I will write to Mr Scale & consult him how I can employ some of the Mint hands'. BA&H, MS 3782/13/36/67, MB (London) to MRB (Soho) 4 April 1792; BA&H, MS 3782/13/36/84, MB (Truro) to MRB (Soho), 3 September 1792.

abroad. He also used a large number of apprentices at Soho, who might specialise in a particular product or process, but they were also expected to be adaptable.[95] The lack of trade guilds in late eighteenth-century Birmingham contributed to the flexibility of the workforce and enabled Boulton to train relatively unskilled workers in specific aspects of production, according to their ability.[96] He recruited 'uncorrupted boys from 14 to 20 years',[97] and employed both local orphan boys and poor children of the parish, but also some from further afield.[98] They were provided with food, lodging, clothing and education.[99] Girls were also apprenticed, but in general they were paid less and worked in less skilled jobs. They were valued for their nimble fingers and powers of concentration in tedious processes as well as their docility and cheapness.[100] Costs of apprentices were considerable: between 11 January and 5 May 1792, for example, £53.7s. was spent on board and lodging, and £16.17s. for clothing and shoes for 10 named Soho Mint apprentices, who were housed with respectable employees, such as Richard Bentley. He boarded up to five Soho Mint apprentices from November 1791.[101]

Well-educated workers meant that Boulton could produce items of the quality he required. Some at Soho gained an excellent training, which enabled them to progress in life. For example the engraver Hancock and his brother William were originally apprentices,[102] but by 1778, Hancock had set up work on his own, engraving dies.[103] Another apprentice, Edward Thomason (1769–1849), started work at Soho in 1786, but eventually took over his father's business, making coins and medals. In his *Memoirs* Thomason wrote of his experience as an apprentice: 'I was initiated in this scientific school at Soho which induced in me a versatility of taste for mechanics and to cultivate the arts and sciences'.[104] This shows the experience some received using new technology at Soho and several of the

[95]   Quickenden, 'Boulton and Fothergill's Silversmiths', p. 9.

[96]   Peter Jones, 'Birmingham and the West Midlands in the Eighteenth and Early Nineteenth Centuries', in Dick (ed.), *Matthew Boulton*, p. 23.

[97]   BA&H, MS 3782/13/36/60, MB (London) to MRB (Soho), 7 June 1791.

[98]   Apprentices came from Liverpool, Gloucester, Dublin and London for example. BA&H, MS 3782/12/23

[99]   Indenture with the overseers of the parish. BA&H, MS 3782/12/95/22, 8 January 1770.

[100]   Maxine Berg, *The Age of Manufacturers 1700–1820* (London, 1994), p. 156.

[101]   BA&H, MS 3782/3/13, Mint Day Book 1791–1795.

[102]   BA&H, MS 3782/12/95/3, 25 August 1794.

[103]   In 1778, John Gregory Hancock was listed as 'Modeller, Die Sinker and Chaser' in Bartholemew Row. In 1787 he was listed as an 'Artist' in Edmund Street, *Pye's Directory* (1778–1787). In 1803, he was listed as an 'Artist' in Hospital Street, *Chapman's Directory* (1803).

[104]   Edward Thomason, *Memoirs* (London, 1845), p. 3.

engineers associated with the Soho Mint had distinguished careers, for example Lawson and Rennie.[105]

Carriers and Suppliers

In addition to his regular workers Boulton engaged reliable suppliers to make items for the Soho Mint and trusted haulage firms to carry copper and coins. For example, Richard Bicknell provided casks, used for packing coin orders, from at least 1791 until 1799.[106] Other items, such as cutting files from James Bissell, and gloves from Robert Blood, were ordered on a routine basis. George Haden and Leonard Tyson supplied soap and candles, Richard Skeldon made shaking bags in which the blanks were abraded, and the Lucas brothers delivered sawdust for drying the coining blanks after cleaning.[107] Recurrent payments were made to Hugh Henshall, Charles Broadley and Edward Doughty, who moved items by canal.[108] For fast bulky deliveries, Boulton used Thomas Sherratt, who operated 'Sherratt's Flying Waggon', to deliver items to London in less than a day.[109] By using known suppliers, Boulton could ensure continuity of provision and contract for a good price.

Wages and Incomes

The Soho Mint was first created in expectation of a large regal coinage contract which did not materialise until 1797, but by 1789 the Soho Mint was functional, so Boulton started to produce coins and tokens for a variety of customers in order to recoup some of his investment. After an initial rush of orders from 1789–1790, a subsequent fall meant that the Soho Mint closed temporarily.[110] Then a large consignment was ordered by the EIC in February 1791 and work remained regular until September 1792, when events in France meant that the Monneron issues

[105]   James Lawson worked on the new Royal Mint from 1805–1822. See BA&H, MS 3782/3/116 to BA&H, MS 3782/3/124. John Rennie became a fellow of the Royal Society and designed many bridges, canals and docks. See Cyril T.G. Boucher *John Rennie: The Life and Work of A Great Engineer 1761–1821* (Manchester 1963).

[106]   BA&H, MS 3782/12/75/19, ZW (Birmingham) to MB (Soho), 19 February 1791; BA&H, MS 3782/3/15, Mint Day Book 1798–1799, 15 November 1799.

[107]   BA&H, MS 3782/3/13 to MS 3782/3/15, Mint Day Books 1791–1799.

[108]   Payments can be seen from at least 1792 for Henshall and from January 1793 for Broadley. Doughty moved cargoes south to Bristol from 1794. BA&H, MS 3782/3/13, Mint Day Book 1791–1795.

[109]   BA&H, MS 3782/12/59/169, William Cheshire to MB, 26 April 1798.

[110]   Boulton wrote in his diary on Sunday 12th September 1790: 'Turned off all the men at the Mint'. BA&H, MS 3782/12/107.

could no longer be made. Workers were again laid off until in 1794, a further two large orders for the EIC meant that the Soho Mint was busy again. This pattern of peaks and troughs of production can be seen in the Mint Record books, where the total wages bill went from an average of £28 per week in 1791 to a low of £9 in 1795 when there were few orders. When the Soho Mint was in full production in 1797 producing blanks for the United States, coining over 16 million pieces for Madras and nearly 90 million coins for the British government, wages rose as high as £90 per week.[111] By the end of July 1800, many Mint workers were again discharged, though in 1808 the total wage bill for nine months was over £2,322.[112] Fluctuating employment made it difficult to retain suitable workers for the Soho Mint.

One of the ways in which key workers were kept employed, was in making items for collectors including sets of tokens, and medals for special occasions. For example, Middlehurst applied his skills to bronzing coining blanks, in addition to his work making tubes.[113] Küchler engraved over 30 medal dies, including the famous Trafalgar medal of 1805, as well as engraving regal coinage dies.[114] These specialised mint items were sold by Chippendall and other agents. Boulton's clients included a range of collectors, and a special record book with medal sales was kept where requests were recorded till 1816.[115] Soho Mint items are still popular with collectors today, as can be seen by their prices in various sales and they provide a valuable historical record of eighteenth-century numismatic art.[116]

When work was scarce at the mint, some employees were moved to alternative jobs in other Soho businesses.[117] The list included familiar names: 'Youl [sic] see by my letter to Lawson that I intend to continue in my employ Messrs Bouch

---

[111] The information is taken from various Mint books. BA&H, MS 3782/3/13 to BA&H, MS 3782/3/15, Mint Day Books 1791–1799, and Mint Ledger BA&H, MS 3782/3/3/1, 1791–1809.

[112] BA&H, MS 3782/3/3, Mint Ledger 1808–1819, 1 October 1808 until 10 June 1809.

[113] BA&H, MS 3782/13/36/64, MB (London) to MRB (Soho), 21 March 1792; BA&H, MS 3782/3/13, Mint Day Book 1791–1795; BA&H, MS 3782/3/14, Mint Day Book 1795 to 1798.

[114] For a full list of medals produced by Küchler, see Sue Tungate. Ibid.

[115] The notebook lists sales of silver, bronzed and tin medals, sales of sets and individual coins, to dealers and individuals. BA&H, MS 3782/12/108/69 Ledger of Medals, coins etc. furnished to Sundry Persons from the Soho Mint 1793-1816.

[116] Sales of Soho Mint coins are detailed by Bill McKivor online via thecopperman@ thecoppercorner.com

[117] At least 30 mint workers were offered alternative employment in 1800. BA&H, MS 3782/12/58/95, 96 and 98, William D. Brown (Soho) to MB (London), 9 July, 20 July and 23 July 1800.

[Busch] Bill Harrison, Hodgetts ye forger & such of the Jeroms as are hired'.[118] Key engineers were employed on other projects: Lawson erected steam engines in a variety of locations and Murdock was employed on a range of schemes, including gas lighting. By this approach Boulton was able to ensure that these individuals would be available for future contracts. It also ensured that details of mint technology did not get passed to competitors, as workers with mechanical skills were scarce and much sought after. Other companies offered incentives to able operatives with technical proficiency developed at the Soho Mint.[119]

The wages at Soho were also relatively high and most were paid weekly, though Boulton preferred piece work payment, such as was introduced in the engine works by the 1790s.[120] In 1787 the *Birmingham Gazette* noted that a workman in Birmingham received, on average, 9s. or 10s. per week. The weekly agricultural wage was around 7s.6d. as late as 1793, though women were more poorly paid.[121] In comparison, the Soho workers were relatively well paid: by 1790, some experienced workers at the Soho Mint were earning up from 18s. to £1.15s. per week and wages of the clerical staff, such as confidential clerk William Brown were considerably more. He received 10 guineas per month from September 1792, which rose to £300 per annum by July 1800.[122] William Cheshire was paid 100 guineas per year from summer 1797 and 104 guineas from 1 January 1798. These figures contrast with an average lower-class wage of £40 per annum and middling-class incomes of over £100.[123] Some Soho Mint workers were given benefits in kind, such as housing or coal. For example, Busch was paid 10s.6d. per week initially, but his rent of 2s. was paid and he received extra payments amounting to around 10–13 guineas per year.[124] Boulton was, however, averse to the truck system, which could be abused by employers.

The more highly skilled employees, the engineers and engravers, were also well paid. In 1779, steam-engine erectors such as Lawson were only paid 10s.6d. per

---

[118]   'I will write to Mr Scale & consult him how I can employ some of the Mint hands'. BA&H, MS 3782/13/36/67, MB (London) to MRB (Soho) 4 April 1792; BA&H, MS 3782/13/36/84, MB (Truro) to MRB (Soho), 3 September 1792.

[119]   Pollard, p. 168–170.

[120]   Roll, p. 224.

[121]   Witt Bowden, *Industrial society in England towards the end of the eighteenth century* (York, 1925), p. 219; Selgin says that most male industrial workers earned between one shilling and two shillings per 13 hour day: George Selgin, *Good Money: Birmingham Button Makers, the Royal Mint, and the Beginnings of Modern Coinage 1775–1821* (Ann Arbor, 2008), p. 9.

[122]   Wages can be found in a variety of documents in the Archives of Soho including the Mint books.

[123]   F. Bedarida, *A Society History of England 1851–1990* (London, 1976, 1991), pp. 214, 216.

[124]   BA&H, MS 3782/3/13 to MS 3782/3/15, Mint Day Books 1791–1799.

week, a similar rate to Busch, and Murdock earned 15s per week.[125] But Lawson's wage rose rapidly, and by August 1791 he earned £63 per year 'over & above his Board' and was paid an extra three guineas for 'Experiments on Coinage'.[126] By 1800, Lawson's pay had risen to £300 per annum plus 1 per cent commission, with Murdock and Southern also receiving a similar rate of pay. Southern became a partner at Soho in 1810 after Boulton's death.[127] Murdock was also offered a partnership in 1810, but he did not accept.[128] Boulton appreciated loyal employees and they reaped the reward of their loyalty. The engravers at the Soho Mint were another well-paid group of workers, which showed the importance that Boulton placed on obtaining excellent dies. Droz had proved expensive, but Dumarest, and later Ponthon, were more valuable because they produced dies for many more products.[129] Dumarest was paid £137 for 11 months' work,[130] and Ponthon was employed from summer 1791 until 4 July 1795 on an annual salary of £80, but he often received extra payments.[131] Küchler preferred to be paid a piece rate for engraving dies, rather than receiving a monthly salary like Ponthon, Dumarest and Phillp. He had an agreement that Boulton would supply dies, which he would engrave with: 'such subjects as might be likely to sell', and the profits would be split.[132] In 1808–1809 Küchler was paid £106 for nine months' work, compared to Phillp's annual salary of £48.[133]

Health and Welfare

Boulton also provided health care for his employees. Medical men were engaged to care for workers, partly so that they could continue to be productive and ensure Boulton's investment, but also due to his paternalist relationship with his staff and genuine interest in their welfare.[134] For example Busch had an 'inflammatory

---

[125]   AD1583/1/3, Balance Sheet, JL in account with Boulton & Watt c1784; Cornish Archives available at http://www.cornish-mining.org.uk/story/boulton.htm

[126]   BA&H, MS 3782/3/13, Mint Day Book 1791–May 1795, p. 8, 13 August 1791.

[127]   Individual biographies listed in the index to the Archives of Soho at BA&H.

[128]   John Griffiths, *The Third Man: The Life and Times of William Murdoch 1754–1829* (London, 1992).

[129]   Dumarest engraved dies for 14 different issues and Ponthon for 31 issues for Boulton.

[130]   BA&H, MS 3782/3/13, Mint Day Book 1791–1795, p. 8, 13 August 1791.

[131]   B. Gould, *Noël-Alexandre Ponthon Medallist and Miniaturist (1769/70–1835)*, Coin and Medal Bulletin (London, 1972), pp. 312–319; BA&H, MS 3782/3/14, Mint Day Book 1795–1798.

[132]   BA&H, MS 3782/12/59/46, MB (Soho) to RC (London), 16 February 1793; BA&H, MS 3782/3/267/40, C.H. Küchler, Die Sinker, June 1799.

[133]   BA&H, MS 3782/3/3, Mint Ledger 1808–1819.

[134]   Roll, p. 224.

sore throat' in November 1798, and William Brown had a 'tumour on the back of his neck' a month later. Both were treated at Boulton's expense.[135] Boulton also ensured that the works were clean, well lit and well ventilated and where possible that processes were carried out safely. A special chimney to remove mercury vapour was installed in an area where medal blanks were gilded.[136] A further incentive for key employees was the works insurance scheme. This was funded by workers' contributions of one sixtieth of their wages, which paid benefits of up to 80 per cent to staff who were sick or injured. Boulton also contributed and any tips received from visitors to the Soho Manufactory were added to the fund. This helped to retain key workers, as no refund was available if an individual left Soho.

Boulton clearly trusted most of his work-force as during the Priestley riots in 1791, cudgels were issued to Soho workers, so that they could defend the Mint and Manufactory against rioters.[137] This was in contrast to Cornwall, where in October 1787, there was a serious threat of rioting and Boulton and Murdock had to flee to Truro to avoid a mob of 400 miners.[138] He was also tolerant of the political views of his employees when reformist ideas were suspect during Britain's war with France: for example, in 1798, John Southern, manager of the Soho Mint thought that 'a reform in parliamentary election was necessary'.[139] This caused Boulton no problems, and contributed to the good relations he had with his employees.

Conclusion

Boulton had a reputation as an enlightened employer, and his selection of reliable, honest and highly skilled staff to operate this innovative establishment, led to its eventual success and the spread of its minting technology around the world. In addition to engineers, who applied the newly introduced rotary engine to the coining process, there were also excellent die-makers, engravers and clerks. In general he recruited his initial Soho Mint workers from other areas of the Soho Manufactory, but he needed to import foreign engravers. He also employed apprentices whom he could train in his own methods. Most of his workers were long-standing employees, and his skilled mint operatives could carry out other activities within the Soho complex, which made them valuable members of staff. Family connections within his workforce were important, as can be seen by

---

[135]   BA&H, MS 3782/12/59/174 and 180 William Cheshire (Soho) to MB (London), 8 November 1798, 5 December 1798.

[136]   BA&H, MS 3782/12/46/109, MB (Soho) to Charles Hatchett (Hammersmith), 26 March 1801.

[137]   'This manufactory has always been distinguished for its order and good behaviour and particularly during the great riots at Birmingham', quoted in Shaw, vol. 2, p. 121.

[138]   BA&H, MS 3147/3/11/ 11–14, MB (Truro) to JW (Harper's Hill), 5–8 October 1787.

[139]   BA&H, MS 3782/12/66/89, JS (Soho) to MB (London), 24 April 1798.

generations of families remaining at Soho. He also employed women and girls to carry out less skilled tasks such as packing coins and outside contractors to carry coins to customers and to bring in supplies of copper. All these individuals were essential to the success of the Soho Mint.

Boulton was able to retain the loyalty of most of his key workers by recruiting through family connections and personal contacts. His patriarchal attitude was partly driven by self-interest, but he was also an enlightened employer who provided for his workers' health and welfare. It appears that his generous attitude was appreciated, as a reported 10,000 people turned out to follow his funeral procession.[140] Boulton's choice of workers proved to be essential for the success of the Soho Mint. His ability to organise collaborative effort, and to communicate his ideas to a team of individuals, enabled him to introduce new practices within the workplace. His employees were inspired by his enthusiasm, as shown by the glowing praise he received after his death.[141] His capacity to motivate his workers enabled them to produce hundreds of millions of coins during his lifetime, using new technology, to send to customers all over the world.

[140]   BA&H, MS 3219/4/49/88, J.T. Tuffen (Bristol) to JW (Glasgow), 29 August 1809.
[141]   BA&H, MS 3782/13/120, Minutes of a meeting held at Soho, 7 January 1810.

Chapter 13

Matthew Boulton's Jewish Partners between France and England: Innovative Networks and Merchant Enlightenment

Liliane Hilaire-Pérez, Bernard Vaisbrot

Whereas we benefit from a large historiography on minorities' networks like the Huguenots and Jacobites across France and England during the 'industrial Enlightenment', Jewish connections at the beginning of industrialisation have seldom been the focus of historical research. Studies of the activities of Jews have tended to focus on overseas trade in the seventeenth century, and on the role of Sephardic merchants as 'cross-cultural-brokers'.[1] But not much attention has been devoted to the role of Jews in the economic growth of the eighteenth century,[2] as the age of merchant diasporas was coming to an end.[3] In his book on the Prager family, Gedalia Yogev considered that 'in the eighteenth century, English Jews took almost no part in developments which were to revolutionise economic life in Britain'.[4] Thanks to recent studies belonging to the 'economic turn' as termed by Gideon Reuveni,[5] we are now more aware of the double-sided images of Jews 'as economically retrograde' and 'as economically avant-garde'.[6] But whereas

---

[1]   Philip D. Curtin, *Cross-Cultural Trade in World History* (Cambridge, 1984); Jonathan I. Israel, *Diasporas within a Diaspora: Jews, Crypto-Jews, and the World Maritime Empires (1540–1740)* (Leiden, 2002); David Cesarani (ed.), *Port Jews: Jewish Communities in Cosmopolitan Maritime Trading Centres, 1550–1950* (London, 2002); Richard L. Kagan and Philip D. Morgan (eds), *Atlantic Diasporas. Jews, Conversos, and Crypto-Jews in the Age of Mercantilism 1500–1800* (Baltimore, 2009); Francesca Trivellato, *The Familiarity of Strangers. The Sephardic Diaspora, Livorno, and Cross-Cultural Trade in the Early Modern Period* (New Haven and London, 2009).

[2]   One exception is Holly Snyder, 'English markets, Jewish merchants, and Atlantic endeavours. Jews and the making of British transatlantic commercial culture, 1650–1800' in Kagan and Morgan (eds), pp. 50–74.

[3]   Israel, *Diasporas within a Diaspora*.

[4]   Gedalia Yogev, *Corals and Diamonds. Anglo-Dutch Jews and Eighteenth-Century Trade* (Leicester, 1978), pp. 20, 21.

[5]   Gideon Reuveni and Sarah Wobick-Segev (eds), *The Economy in Jewish History* (New York, Oxford, 2011).

[6]   Jonathan Karp, *The Politics of Jewish Commerce. Economic Thought and Emancipation in Europe, 1638–1848* (Cambridge, 2008) and 'Can Economic History Date

authors question anew 'the economy in Jewish history', most of them deal with the nineteenth century, and with trade and consumption, rather than industry, echoing the revisions of the (long-lasting) Industrial Revolution, as Jonathan Karp has pointed out.

In this context, Matthew Boulton's networks provide an opportunity for reassessing the part of Jewish merchants in the growth of industrialisation: for connecting '(Jewish) commerce' and '(Gentile) industrialisation'.[7] Archives of the firm, known as the Archives of Soho, as well as the records of partners and commissioners, specifically allow the study of the involvement of Ashkenazi Jewish entrepreneurs in trading new and fashionable luxury goods across the Channel – goods that came from innovative manufacturers such as Matthew Boulton at Soho. These networks also help us to understand the cultural contributions of Jews to the Enlightenment in terms of 'the experiences' of 'the bulk of English Jews rather than small numbers of notables, intellectuals, and synagogue functionaries'.[8] Until now Eric Robinson was the only author who had focused on Matthew Boulton's Jewish partners, such as Moses Oppenheim, and Solomon Hyman.[9] Nevertheless, he did not pay attention to their identities as Jews in England.[10] Robinson focused on Oppenheim's bankruptcy in 1765, which prompted Boulton to travel to France, and which became a useful voyage for trade and obtaining secrets.[11] He would go on to develop links with Hyman in Paris, who became his most important customer in France. What chiefly interested Robinson was Hyman's role in transferring know-how, tools and workers from Paris. His findings were pioneering in their attempt to assess the two-way exchanges between France and England, at a time when historians tended to emphasise the superiority of England. Ignoring dogma, Robinson relied on the Letter Books of Soho which contained

---

the Inception of Jewish Modernity' in Reuveni and Wobick-Segev, pp. 23–42; Derek J. Penslar, *Shylock's Children. Economics and Jewish Identity in Modern Europe* (Berkeley, 2001).

[7]   Karp, p. 34.

[8]   Todd M. Endelman, 'Writing English Jewish History' in *Albion: A Quarterly Journal Concerned with British Studies*, 27/4 (Winter, 1995), pp. 623–636; Todd M. Endelman *The Jews of Georgian England, 1714–1830: Tradition and Change in a Liberal Society* (Michigan, 1999).

[9]   Eric Robinson, 'Boulton and Fothergill, 1762–1782 and the Birmingham export of hardware', *University of Birmingham Historical Journal*, VI/1 (1957), pp. 60–79; Eric Robinson, 'International exchange of men and machines, 1750–1800, as seen in the business records of Matthew Boulton', in A.E. Musson and Eric Robinson (eds), *Science and Technology in the Industrial Revolution* (Manchester, 1969), pp. 217–229.

[10]   This was coherent with the marginalisation of Jewish history in England, in the 1950s–1960s, as Endelman explained ('Writing English Jewish History').

[11]   Liliane Hilaire-Pérez, 'Les échanges techniques dans la métallurgie légère entre la France et l'Angleterre au XVIIIe siècle' in Jean-Philippe Genet and François-Joseph Ruggiu (eds), *Les idées passent-elles la Manche? Savoirs, représentations, pratiques (France-Angleterre, Xᵉ-XXᵉ siècles)* (Paris, 2007), pp. 161–183.

numerous letters written to Hyman. But Robinson only mentioned the Jewishness of Boulton's partners in a footnote: 'There seems to have been a large number of Jewish Merchants engaged in the "toy" trade', and he listed a small group of Anglo-Jewish merchants trading with Boulton, who had applied for naturalisation after the Jew Bill of 1753.[12] Interestingly, he mentioned Jews in the context of their integration and there was no reference to the repeal of the Jew Bill.[13] The status of a minority and the distinct identity of Jews were not an issue. This is also the impression given by a reading of the Soho Letter Books: Hyman was called our 'friend'[14] and, as all letters were written in English, identity went unquestioned.

There are two reasons for reconsidering this question. One is historiographical. Recent studies of minorities in the Enlightenment and on tolerance in the 'Industrial Enlightenment' invite us to re-investigate the Jewish identities of Boulton's partners.[15] They enable us to broaden the history of ordinary Jews,[16] and to analyse their connections with non-Jews and their role in the promotion of new products and processes in Europe.[17] In France, recent research on Jewish identities, especially the Sephardim involvement in long-distance trade and works devoted to migrants, communication and mobility, focuses on non-prominent merchants, and allows a better understanding of social and legal status in the Old Regime.[18] The second reason to revise Robinson's statements stems from the discovery of new archives in Paris, especially Hyman's papers, deposited because of his bankruptcy in 1776. These sources cast light on Jewish commercial networks in the Anglo-French trade, and on their merchant culture, especially their business language. The survival of a business ledger raises the question of Hyman's identity as an Ashkenazi merchant trading between Paris and London, and one whom Boulton

---

[12]   Robinson, 'Boulton and Fothergill, 1762–1782', note 42, p. 69.

[13]   For a reassessment see Dana Rabin, 'The Jew Bill of 1653: Masculinity, Virility, and the Nation' in *Eighteenth-Century Studies*, 39/2 (2006), pp. 157–171.

[14]   Birmingham Archives and Heritage (BA&H), MS 3782/1/37, Letter of 25 April 1765.

[15]   Adam Sutcliffe, *Judaism and Enlightenment* (Cambridge, 2003); Geoffrey Cantor, *Quakers, Jews, and Science. Religious Reponses to Modernity and the Sciences in Britain, 1650–1900* (Oxford, 2005); Peter M. Jones, *Industrial Enlightenment: Science, Technology and Culture in Birmingham and the West Midlands, 1760–1820* (Manchester, 2008); Peter M. Jones, 'Minorities and the culture of science and technology. A case study of Birmingham Dissenters 1760–1820', in *Documents pour l'histoire des techniques*, 15 (2008), pp. 58–73.

[16]   Todd M. Endelman, *Broadening Modern Jewish History* (Oxford, 2010).

[17]   Maxine Berg, *Luxury and Pleasure in Eighteenth-Century Britain* (Oxford, 2005).

[18]   *Évelyne Oliel Grausz, 'Relations et réseaux* intercommunautaires dans la diaspora séfarade d'Occident au XVIIIe siècle' (University of Paris-I-Sorbonne, unpublished PhD thesis, 2000); Jean-François Dubost, 'Les étrangers à Paris au siècle des Lumières', in Daniel Roche (ed.), *La ville promise. Mobilité et accueil à Paris (fin XVIIe-début XIXe siècle)* (Paris, 2000), pp. 221–288.

trusted so much that he dealt with him for 10 years. This chapter stresses that this identity mattered and Boulton had reasons to credit Hyman.

The Merchant of Paris: Hyman's Business

Hyman's archives in the Paris National Archives are schedules of debts and credits and a ledger of 176 pages, written on both sides and mixing orders and payments.[19] These documents are complementary to the Soho Archives which contain numerous letters sent to Hyman between 1764 and 1776.[20] It is therefore possible to trace links between the Soho letters and the orders and payments recorded in the ledger. The documents provide information on how Boulton set up his main Paris network, and they also help us to understand the person whom Boulton chose as a trustworthy partner.

An Anglo-French Business

The documents concerning bankruptcy reveal Hyman's extensive creditworthiness in Anglo-French trade, an attribute that he shared with a few other Parisian merchants, like the Blakeys who belonged to the Jacobite network and traded with London, Sheffield and Birmingham;[21] the Orsel brothers, who were well connected to Boulton; and Charles Raymond Granchez, who dealt with Boulton's factor in London, Baumgartner.[22] They were all established in business on the north side of the Seine and developed trade in the 1760s, thanks to the fashion for English wares following the Treaty of Paris of 1763.

Hyman shared common features with the Blakeys, including, not least, his Anglo-French identity. In the bankruptcy schedules, Hyman was described as an 'English merchant ... usually living in London and, when he is in Paris, residing in rue Beaubourg', in the centre of Paris. According to the bankruptcy papers in England, he had set up as a merchant in London in 1761. Mention is made that he died in London in September 1783,[23] and that on 11 August 1784, his widow,

---

[19]   Archives de Paris: D5B6 4376; D4B6 57 3493.

[20]   BA&H, MS 3782/1/9; MS 3782/12/108/4; MS 3782/1/10; MS 3782/1/37; MS 3782/1/38; MS 3782/1/39; MS 3782/12/2. But only one letter sent by Hyman is still recorded: MS 3782/1/22/21, from Paris, 19 May 1773.

[21]   Liliane Hilaire-Pérez, 'Steel and toy trade between England and France: The Huntsmans' correspondence with the Blakeys (Sheffield-Paris, 1765–1769)', *Historical Metallurgy*, 42, 2 (2008), pp. 127–147.

[22]   Carolyn Sargentson, *Merchants and Luxury Markets. The Marchands Merciers of Eighteenth-Century Paris* (London, 1996).

[23]   The National Archives, Kew (TNA), Admons Book 1, Death Intestate, PROB 6/160, 1783, Hyman Solomon, London, September. The author is grateful to Elisabeth Antoni for communicating the results of her research in TNA.

Elizabeth Hyman, was granted the administration of the 'chattels and credits' of the deceased, whose last parish had been Saint James, Duke's Place, London.[24] Letters from Boulton, in the Archives of Soho, alternatively mentioned Paris and London. The transactions recorded in the ledger were also headed either Paris or London, which confirms the dual location of the firm. In London, as trade directories indicate, Hyman was based close to the Minories, Goodman's Fields. The ledger indicates that in London, Hyman relied on his wife and on his brother-in-law, Hirsch, and in Paris he had a partner, Gabriel Lefevre, who is also quoted in the Letter Books; Hyman coordinated two merchant houses across the Channel. The Soho Letter Books mention his travels to London and visits to Soho,[25] and in the bankruptcy records, considerable expenses: £5,300 within 13 years, are listed for travels to the 'house in London'. The intensity of Hyman's connections with England made him stand out in the small group of Anglo-French merchants in Paris.

This was confirmed by his two bankruptcies on either side of the Channel. The first one occurred in London in 1775.[26] In response to creditors, a Commission of Bankruptcy was convened to decide whether Hyman was eligible for bankrupt proceedings. On 6 May 1777, the Order Book registered the Court's judgment that the requirements were met, and submitted the name of Hyman's main petitioning creditor, Thomas Livesey, a hosier, in Bridge Street, Westminster.[27] An advertisement in the *London Gazette* alerted all other potential creditors, and notified the debtor to surrender himself and his possessions to the Guildhall, on 24 May, 3 June and 24 June 1777.[28] During the meetings, the creditors' claims were examined and the debtor's assets valued and distributed as dividends. Having satisfied all the legal requirements, Hyman was issued a Certificate of Conformity which effectively discharged him, although dividends might continue to be paid after this date, 31 December 1777.[29]

---

[24]   TNA: PCC Admons, 1750–1800, Solomon Hyman, 11 August 1784.

[25]   BA&H, MS 3782/1/37.

[26]   TNA: B4/21, 1771, Docket Book, no. 244, Hyman Solomon, Joulmin, solicitor, 6 October 1775.

[27]   TNA: B1/68, 1777, Order Book, p. 136, 20 June 1777; B4/21, 1771-Docket Book, no. 321, Hyman Solomon, Joulmin solicitor, 10 May 1777. A second entry in the Docket Books was registered on 10 May 1777. Another application followed as some dockets were initially struck only as threats against debtors in order to urge repayment rather than initiate a genuine attempt to get a declaration of bankruptcy.

[28]   *London Gazette*, issue 11769, 10 May 1777, invited creditors to prove their debts and summoned Hyman to appear at the Guildhall in order 'to make full discovery and disclosure of his estate and effects'.

[29]   TNA, B6/5, 1774–1781, Certificate of Conformity, p. 96, 31 December 1777; 'Bankruptcy Directory, Certificate of Bankruptcy, Society of Genealogists, Drawer 1279: 'Hyman, Solomon, late of Little Somerset Street, Whitechapel, Middlesex, merchant, May 13, 1777. C (Certificate) 10 December 1777'. The granting of the certificate suggests that

The importance of Hyman's trade across the Channel is also revealed in his bankruptcy accounts in Paris: 71 per cent of his creditors were English and they accounted for nearly 80 per cent of his debts: £12,156 (Table 13.1). In comparison, Blakey's figures were lower: £3,870, with only three partners mentioned in the bankruptcy and Granchez only dealt with four English merchants, to whom he owed £1,208. Among Hyman's creditors, the biggest Birmingham entrepreneur and Boulton's rival, John Taylor, was the most prominent. Hyman also had debtors elsewhere in England, suggesting that he was involved in extensive networks of credit.

Table 13.1   Hyman's Bankruptcy, 1776

	Creditors					Debtors			
Country	*Nb.*	*%*	*£*	*%*	*Country*	*Nb.*	*%*	*£*	*%*
England	35	71	12156	79	England	10	15	665	10
France	14	29	3319,7	21	France	57	85	6064,6	90
	49	100	15475,7	100		67	100	6729,6	100

The 35 English creditors		The 10 English debtors	
Names	*£*	*Names*	*£*
John Taylor	1214	Abraham Salomon	200
Beaston & Colson	831	Abraham Harris	92
Tho. Leversy	830	J. Cracknel	90
James Bond	682	Michel Salomon	83
James Touchet	668	Wilp Killon	63
Evans & Butches	629	Messelay	45
Boulton & Fothergill	591	J. Kentish junior	36
Samuel Barber	554	Dandelay	33
Henry Jones & Co.	533	Michel Le Grand	15
James Brees	475	Elias Abraham	8
Chs. Chadewick	438		665
Wm. Seddon Jr	434		
Tho. & Rich. Hild	424		
Tho. Bouwen Bank	406		

Hyman had been forthcoming in the disclosure of his effects and that the approval of four-fifths of the creditors had been secured. As a result, he was released from the liability of any debts contracted up to the act of bankruptcy, as a means to give him a second chance to restart his business. The certificate was introduced in Ann, c 17 and completed by the granting of an allowance to the bankrupt.

The 35 English creditors

Names	£
Wm Wenther	381
Gerson Isachs	340
Joseph Jenkinson	289
Thomas Webb	248
Joseph Berks	236
John Simions	236
Tho. Blinkhorn	236
Robert Farrand	232
Wm. Holden	226
Tho. Hanfield	177
Benjamin Hamfrey	150
Wm. Phillips	130
Abraham Iretans	110
H.H. Deacon	110
John Hallows	75
More & Thomson	73
March Homeguy	55
Joseph Barry	42
Joseph Green	39
J. Herdson	33
Isaac Harrops	29
	12156

The same was true on the French side of his business. The bankruptcy listed 57 French debtors, such as retailers, factors and bankers, like Moulin La Bourdonnais, his major debtor, who was mentioned in the Soho Letter Books. Moreover, Hyman bought snuff boxes from Parisian retailers, and sold them to England through Calais. Hyman, who was to become Boulton's major intermediary in France, was a well-connected Jew in a highly specialised cross-Channel business.

Hyman as a Parisian Jew

Parisian Jews were in a difficult position in the mid eighteenth century. Unlike London, where they accounted for roughly 12,000 individuals, there were no more than 800 in Paris, so the opportunities for contacts with non-Jews were very few.[30] Moreover, Jews in France faced economic exclusion, being deprived of property,

---

[30]   Daniel Roche, 'Juifs et gentils à la veille de la Révolution' in Bernard Blumenkranz (ed.), *Juifs en France au XVIIIe siècle* (Paris, 1994), pp. 177–190.

of shops, and excluded from guilds. These conditions were worse for Ashkenazim, who were recent migrants to Paris, less integrated into merchant networks and less assimilated than the Sephardim. The main area of Ashkenazi settlement was a particularly poor district around Rue Beaubourg.[31] The situation for Jews in London was probably easier. Although the naturalisation Bill of 1753 was rapidly repealed, contacts between Jews and non-Jews helped the process of integration. Toleration facilitated assimilation in a way that hardly existed in most places in Europe. Geoffrey Cantor states that many Jews 'saw themselves as English men ... [jewishness] just happened to be a quirk of birth'.[32] Our hypothesis is that Boulton, who was manifestly tolerant with respect to religious minorities, as Peter Jones has shown,[33] chose English Jews to break into French markets, because they were involved in an active process of integration. Their different identities meant that they were skilled culture brokers, which, associated with sound business abilities, helped them to secure esteem, reputation and credit.

The ledger reveals the way Boulton set up trade with Hyman. Robinson has shown that Boulton's first network in Paris relied on Moses Oppenheim,[34] who also lived in Paris and in London, and was a link between Boulton and Hyman, as stated in the Soho Letter Books in 1764.[35] In 1769, after Oppenheim's bankruptcy, Hyman was still trading with him. Boulton established close contacts with both men. Robinson thought that the connection with Oppenheim was direct, whereas the relationship with Hyman was mediated by the London factor John Cantrell until Cantrell's death in 1769. But the ledger shows that Boulton had a long-established connection with Hyman. Boulton's name appeared in Hebraic letters in an order on 25 September 1765, and subsequently, for goods and payments until 1775. It is possible that Boulton bypassed Cantrell. The comparison between the Soho Letter Books and Hyman's ledger suggests that Hyman was dealing with Boulton even without mentioning his name, by using the numbers on the pattern cards of Soho

---

[31]   François Lutsman, 'Les juifs de Paris sous la Terreur d'après leurs cartes de sûreté', in *Archives juives*, 35 (2002/2), pp. 115–123; André Burguière, 'Groupe d'immigrants ou minorité religieuse? Les juifs à Paris au XVIII° siècle', in *Revue de la Haute-Auvergne*, 50 (1985), pp. 183–200; Christine Piette, *Les juifs de Paris (1808–1840). La marche vers l'assimilation* (Laval, 1983).

[32]   Cantor, p. 41.

[33]   Jones, *Industrial Enlightenment*, p. 185.

[34]   The Oppenheim family, who were glassmakers from Pressburg and settled in England are better known than Hyman: Léon Khan, *Histoire de la communauté israélite de Paris. Les Juifs de Paris au XVIIIe siècle d'après les archives de la lieutenance générale de police à la Bastille* (Paris, 1894); Anita Engle, 'Mayer Oppenheim "de Birmingham"', in Anita Engle (dir.), *Readings in Glass History* (Jerusalem, 1974), pp. 61–71; Hilaire-Pérez, 'Steel and toy trade between England and France'.

[35]   BA&H, MS 3782/1/35, Cashbook 1763–1765; MS 3782/1/14/5, MS 3782/1/14/7, Letters received; MS 3782/1/9, MS 3782/1/37, Letter Books.

goods, which became a sort of code.[36] Boulton was trading in a direct way with Hyman since the beginning and that is consistent with the fact that he met Hyman in Paris during his trip in 1765, in the Café Maillard, Rue St-Martin, near Rue Beaubourg.[37] Hyman may have provided contacts, as is suggested by the names and addresses he wrote on some pages of his ledger, in French and in the Latin alphabet. Whilst the names for debtors and creditors were in Yiddish, like Boulton, the use of French for Parisian customers or furnishers might indicate a degree of integration and assimilation.

The ability to make contacts in this fashion was essential for technological transfer. As Robinson tells us on the basis of the Soho Letter Books, Hyman had organised for Boulton a meeting in a garret with a Parisian artisan to learn how to 'boyl brass work in color'. What underpinned the exchanges of recipes, tools and workers between Hyman and Boulton, were these direct connections. Boulton used personal contact and familiarity as a means of managing technical exchanges, when, for example, he urged Hyman again in 1769, to find workers to provide new recipes for gilt work.[38] For Boulton, Hyman was a transmitter of technological knowledge, because he was a Jewish merchant actively involved in a process of integration and in providing contacts.

Dealing with Hyman

Boulton seems to have highly praised Hyman, who was regularly invited to Soho.[39] It is possible that his familial network, with his brother-in-law and wife based in London and one brother in Calais, enhanced his reputation. Like the Sephardic Jews and the Quakers studied by Nuala Zahedieh,[40] Hyman was a member of a rejected community of Parisian Jews and would have tried to avoid any difficulties for himself and for the local community. Although it is impossible to know how religious Hyman was, community mattered. Jewish connections appeared clearly in the ledger, both in Paris and beyond: in Strasbourg, for example, where he sent goods to a rabbi. Hyman's business skills were what mattered to Boulton and other partners.

---

[36]   In several letters to Cantrell in the Letter Books, Boulton dealt with orders of steel chains numbered 1206, 1207, 1233 and 1250, BA&H, MS 3782/1/37. In the ledger, these numbers become a sort of code, sometimes with the name Boulton (1765) but also without his name (1766).

[37]   Hilaire-Pérez, 'Les échanges techniques'.

[38]   BA&H, MS 3782/1/10.

[39]   BA&H, MS 3782/1/37.

[40]   Nuala Zahedieh, 'Making mercantilism work: London merchants and Atlantic trade in the seventeenth century', *Transactions of the Royal Historical Society*, 9 (1999), pp. 143–158.

Hyman's abilities in making contacts derived from his specialisation in Anglo-French trade and from his familiarity with a limited range of products. Hyman's mastery of products, prices and languages came from his 'intense use of the system'.[41] Although 'toys' could encompass a huge diversity of products, Hyman traded specific goods which he selected with care. The main articles he bought from England were steel chains: *shtal keytn*, notably from Soho, for men and for women; small boxes: *kazilets*, in steel and gilt; and also lots of pencils, buttons, face-bow mirrors and trinkets. From Parisian retailers, he bought snuff boxes exclusively. There was a third traffic flow from Paris to London, for earrings and necklaces, but we do not know their exact provenance. Hyman provided detailed lists of the different prices for these goods, and sometimes their qualities as well. A regular dialogue developed with Boulton. In September 1765, Hyman precisely recorded the prices of chains he was supplying to a French merchant, which had been bought from three English sources: Webb, Boulton and Cracknel. Boulton's goods, the most numerous, were ranged by price from 21 French pounds per dozen to 30 pounds, roughly from 13 to 19 shillings per dozen, in six batches of four dozen. Webb's chains were mainly around 13 and 15 shillings. Soho Letter Books recorded at the same time, that Boulton had an exchange with Hyman about the prices of chains.[42] What was at stake was Webb's competition, for Boulton could not offer the same prices as Webb, especially as Hyman wanted one more link put into Boulton's chains. One link would cost a few extra shillings; the range of prices was narrow and the negotiation tight. The same issues underpinned other goods, like Parisian snuff boxes. Hyman listed them as per pattern, price and size and in couples of small and large items. He would always respect this classification in orders (Table 13.2). There was a rationality in Hyman's accounts, which was based on his sound knowledge of the peculiarities of the products. Specialisation allowed him to organise sets of products to fit customers' requirements. He could then make quality assortments of the boxes according to colour, decoration and appearance.

These organisational skills and technical understanding of the 'toy' trade were part of a wider cultural context. His sound practical knowledge and use of the system fostered skills for transposing and combining. Although Hyman was not part of the same culture as Boulton, they shared similar business languages.

---

[41]   Jean-Charles Depaule, 'Contacts et circulations: espaces et figures' in Jocelyne Dakhlia (ed.), *Trames de langues. Usages et métissages linguistiques dans l'histoire du Maghreb* (Genève, 2004), pp. 71–81, p. 75.

[42]   BA&H, MS 3782/1/37, Letter to John Cantrell, May 1765.

Table 13.2   Managing qualities in orders in Hyman's ledger

N°	Size	Orn. 1	Orn. 2	Colour	Material	Details 1	Details 2	Dozs.	Price/ doz. l.t.	Total price
1765 – Snuff boxes bought out from Mr Prévôt										
1	Small	Painted			Grey gold			24	5	120
2	Large	Painted			Grey gold			12	6	72
3	Small	Painted						24	6	144
4	Large	Painted						12	8	96
5	Small	Engraved						24	10	240
6	Large	Engraved						12	12	144
7	Small	Painted		2 colours		Double bottom		4	12	48
8	Large	Painted		2 colours		Double bottom		4	14	56
9	Small	Painted			Mat cast iron	Double bottom	With cartel	12	18	216
10	Large	Painted			Mat cast iron	Double bottom	With cartel	12	21	252
11	Small	Painted	Striped	White		Double bottom		7	30	216
12	Large	Painted	Striped	White		Double bottom		6	42	252
										1856

Language-shifting and Business Practices on both Sides of the Channel

This aspect of our work on Hyman's ledger relates to research on the languages of commerce and account keeping in early modern Europe, and more specifically to the use of Yiddish in economic practice.[43] Not only have very few eighteenth-century business ledgers been studied,[44] but where they have been analysed, this has been in terms of their transactions, and not as documents for the history of merchant culture. The challenge is to understand those skills which were developing via mercantile practice and what part these skills played in the modernisation of economic life.[45] One most interesting feature in Hyman's account keeping was his language-shifting, based on frequent transliterations from French and English into Yiddish. Embodied within his business language, Hyman was seeking to

---

[43]   Evelyne Oliel Grausz (dir.), 'Jewish Business Records, Early Modern Trade and Merchant Culture', The Rothschild Foundation Europe, and Gilbert Buti, Michèle Janin-Thivos, Olivier Raveux (eds), *Les langues du commerce* (Aix, forthcoming).

[44]   Helen Clifford, *Silver in London. The Parker and Wakelin Partnership, 1760–1776* (New Haven, 2004), pp. 12–13.

[45]   Franco Angiolini, Daniel Roche (dirs.), *Cultures et formations négociantes dans l'Europe moderne* (Paris, 1995).

express a rationality based on transposing and combining operational skills. This also reflected his ability to connect with non-Jewish culture, and assimilate within French society.

Tracing Hyman's Language

This section provides a methodology for coding Hyman's multilingualism,[46] and understanding the genesis of his Yiddish idiom, beyond the reduction of Yiddish language merely to a derivation of German.[47] There are four stages (Table 13.3). Firstly, Yiddish hand-writing is reproduced into Yiddish symbols, secondly, into a literal Latin alphabet inscription and thirdly, it is translated phonetically, for instance by adding vowels, so that the distinctiveness of Hyman's language is retained, like transliterated words from French and English. In the final stage, a French translation is provided. To assist, codes (colours) for distinguishing words were written in a form of internal bilingualism for Yiddish and Hebrew, like letters used for numbers or commercial vocabulary, for instance *shur* for goods, and external bilingualism for Yiddish and non-Hebraic languages, like French, English, Latin, such as ditto. If the ledger was simply translated, we would lose sight of his linguistic skills. For instance, Hyman's way of writing to London was a transliteration from English into the Hebraic alphabet. The same was true for words such as vessels, earrings, necklaces or 'Boulton'. French words were also transposed, phonetically, like the shop sign, 'à la ville de Versailles'. The aim is to trace the transliterations, which were often based on his phonetic understanding, especially for French words, as evidence of the process of acculturation.

This transcription methodology enables us to recover Hyman's linguistic world. For the Ashkenazim in Paris, linguistic seclusion provoked disdain and hostility from non-Jews. Acculturation to French culture was not as commonplace as among the Sephardim. Hyman's ledger shows that he was composing a language of his own which seemed to echo the linguistic strata of his cultural experience.

The methodology focuses on his use of enumeration, the calendar and transliterations. Hebrew, the traditional and religious language of the community, coexisted with Yiddish, French, English and Latin. It was used in specific cases, and ways, firstly, for example, for commercial transactions, such as I have received: *kibalti*, or debtor of: *khayev*. Secondly, for indicating the end of the month for receiving a payment, he would put *sof:* end, and then, the name of the month. For other transactions, he would use Yiddish, like *geshikt* for sent. He would never use French or English for business operations. Apart from transactions, Hebrew was reserved for relatives, like wife: *ishti*, brother: *akhi* and brother-in-law: *gise*.

[46] Few studies on practical business language skills exist. See Keram Kevonian, 'Numération, calcul, comptabilité et commerce' in Sushil Chaudhuri and Keram Kevonian (eds), *Armenians in Asian trade in the Early Modern Era* (Paris, 2008), pp. 284–367.

[47] Jean Baumgarten, 'Histoire de la grammaire yiddish (XVIᵉ–XXᵉ siècles)' in *Histoire Epistémologie Langage*, 18/1 (1996), pp. 127–149.

Table 13.3   Four steps methodology for translating Hyman's ledger

---

1. Hebraic block letters

1765   הבי גיקויפֿט מן מוזיע פראָאופֿאָוה

120   נומר 1   ד'ד דוצנט גלײנה גימאָלטה אורגרע ה' ליוור

2. Transcription in Latin alphabet

1765   hby giquyft mN mûzyE proufouh

numr 1   X'd ducnt glyynh gim'ûlth 'ûrgrE h' liwr 120

3. Phonetic transcription

1765   hobe gekoyft min mossye proufou

numèr 1   24 dutsent kleynè gemoltè orgrè 5 livr   120

4. Translation (here in English)

1765   have bought from Monsieur Prouvot

number 1   24 dozens small painted grey gold 5 pounds 120

Hebrew was most of all used for enumeration. There was a clear pattern between Hebraic numbers and Arabic ones, hence between traditional practice and non-Jewish dominant practice. This was the case for prices and quantities of goods, with Hebraic writing, such as letters having a numerical value: the alphanumeric system. For instance, *yud,* the tenth letter of the alphabet, is 10; *yud beth*, the tenth and second is 12. Hyman wrote 15 as 9+6: *teth + vov,* not *yud + he,* which was usual in Hebraic accounts. Within this traditional numeration, he nevertheless showed himself capable of adaptation: instead of using the ordinary form of the letter *kâf*: K, for 20, he used the final form of the letter, ך instead of כ, probably to avoid confusion with ב: *beth,* meaning two. For 100, he would use the Hebraic word: *mea,* whereas he could have used ק: *quf.* As well, instead of writing seven with the seventh letter of the alphabet: ז, he wrote it phonetically as *zayine*, with a *zayine*, a *yud* and a *nun*, which would make 7+10+60. He did the same with 30, letter L, ל: *lamed,* written in full. Hyman knew Hebrew well, and understood and respected the use of conventions transmitted by the community, such as transactions, family life and enumeration, but he adapted them in order to make his transcriptions clearer and to avoid errors. Far from being a rushed and approximate transcription of day-to-day practice, Hyman's language was highly rational.

This did not preclude flexibility. For instance, his use of the calendar was apt to change. Years were in Arabic numbers, but not always – numbers of days were either given in the Hebrew alphanumeric system or in Arabic numbers. In the very much codified area of the spelling of dates, community and religious language was used by Hyman, but it did not predominate to the exclusion of all else. This is confirmed by his extensive use of Arabic numbers for summing up prices and for the enumeration of patterns, like Boulton's, although quantities and prices were also sometimes in the alphanumeric system. His linguistic borrowings corresponded to types of activities. As in the case of Armenians in the seventeenth and eighteenth centuries, Arabic numbers were particularly used for accounting operations. For money units, Hyman transliterated sterling and *livre*: *shterling* and *livre* and then used both units for accounting. Yet Hyman also used Hebrew in a dynamic way, for transcribing his pronunciation of Yiddish: instead of writing an *ayin* at the end of words, in non-tonic syllables, he used the Hebraic 'H' ה: *hé* as in *kleyné*, written *glyynh*, meaning small, a spelling that was actually used by some Yiddish publishers as well. Phonetics played a major role in his transcriptions, which we can interpret as a sign of his effort to assimilate. This was reflected in his vocabulary for the months of the year. Whereas months were either in Yiddish or in English transliterations: December, January, February, and March, some changes occurred in the ledger, with the intrusion of French transliterations, like *april* becoming *avril*, *zebdember* becoming *zebdamber* and the same for *nofamber*. Hyman's pronunciation was evolving.

His use of French seems to have responded to the logic of day-to-day contact. This was the case for some French names and addresses written in the Latin alphabet. French customers would always be termed as *mossie*, that is *Monsieur* transliterated into Yiddish, and addresses would also be in phonetic French: *a la vil de Versail*. We can imagine the difficulties he experienced in pronouncing French names, like *Prévôt*, that he wrote *Prouvouh*, but English names were less altered.[48] Day-to-day practices, such as contacts and trading relations, like saluting and addressing with Monsieur, shaped Hyman's business language.

All this had an impact on his labelling of products, a field where Hebrew, the language of tradition, was not used. In the eighteenth century, products, especially 'toys', were quickly changing. Their shapes, colours and materials could be matched with no preconceived system of classification and their labelling was largely based on promotional strategies. 'Toys', emblematic of Boulton's trade at that time, was an arena for the elaboration of new linguistic skills and rationalities.

Merchant Taxonomies and Jewish Languages

Hyman was alternately using Yiddish, French and English for denominating products. He would use Yiddish for casks: *fas*, for steel chains: *schtal keytns* and *shtahl* for steel. But he also used English and French for a wider range of products, such as earrings, necklaces, including marcassite, pencils in English, trinkets: *berloques*, boxes: *kazilets* or cassettes, and cork-screws: *tirebouchon* in French. Different appellations could be used within the same lists; orders for goods were multilingual, however the vocabulary employed was fixed. They were not ambiguous or uncertain, but Hyman's own commercial idiom. Hyman was accustomed, in every day practice, to combining borrowings from different languages, and in the process would attribute a national or community identity to the products he traded in. His vocabulary was a function of the different markets

---

[48] On names, see Sonia Branca-Rosoff, 'Le métissage linguistique au Maghreb: trois chantiers pour une collaboration entre historiens et linguistes', in Dahklia (ed.), pp. 469–483.

in which he operated . It was as though familiarity with local practices and markets led to the forging of new words.[49] Hyman's Yiddish was a dynamic language, integrating and transmitting words of different origins, linked to new products, and to their particular qualities. Therefore it is likely that merchants' Yiddish was one language through which knowledge of new products was spread on the Continent.

These skills echoed his ability to make up assortments and price lists. Hyman was constructing a language for quality. In one order for snuff boxes of September 1765, he used Yiddish for the small and large sizes and for the external and surface decorations, like *gemolte*: painted, *gravirte*: engraved, *geshtrayft*: striped, and *vays*: white. French, however, was used for some external characteristics; for example: decorations such as festoon: *feston*, sun: *soleil*, and star: *etoil.* Hyman's language was immersed in an economy of quality and variety, a product-based economy in which 'toys' played a prominent part. As he was reporting to Boulton on the skilled workforce in Paris, and enticing artisans, his French vocabulary was used to evoke skills in finishing and modelling. In effect he was attributing to Parisian goods an identity based upon high skills, and thus contributing to their reputation abroad.

As a merchant involved in a process of acculturation and integration, Hyman was immersed in practices which were shaping his language; his language constituted a veritable technology, an adaptative and functional tool which was amenable to invention. In this way, Hyman's linguistic skills were expressing the vitality and richness of the practices that he was developing in his trade. Hyman's ledger reveals his abilities in adapting, combining and translating languages and modes of accounting, at a higher level than in the codification supplied by trade dictionaries that were gaining in reputation at this time. Practitioners like Hyman were looking for speed, utility and the added value of interpretation. In the economic sphere, therefore, languages were not only shaped by learned *encyclopédistes,* they were fashioned by business practices, by vernacular idioms, by oral communication and by deeds; they were responding to the increasing diversity of goods, vocabularies and identities in the Enlightenment.

Conclusion: Anglo-French Jewishness in the 'Industrial Enlightenment'

Boulton's network of relations with Anglo-French Jews tend to support the secularised Weber thesis proposed by Zahedieh for explaining the prominent place of Quakers and Jews in colonial trade in the seventeenth century. Religion or 'awe of God and conscience' counted less than the part played by the community, in securing their creditworthiness and reputation, in 'enforce[ing] good conduit and information flows'. Although this is probably true in the case of Boulton's networks with Jews also, the pattern seems to have been slightly different in the case study proposed here. What mattered was the mechanism of integration, that

---

[49]   Depaule, 'Contacts et circulations: espaces et figures'.

is, contact with non-Jews rather than the Jewish community itself. As Endelmann has stressed, the modernisation of Jewish society in Georgian England was driven by day-to-day practices which fostered contacts with non-Jews. For Boulton, who was a man with a tolerant outlook, Jewish identity may have mattered, chiefly as a measure of an ability to connect with the whole of commercial society, but there was also another advantage. Ashkenazim experienced difficult conditions in Paris, where the Jewish community as a whole were less numerous and had fewer occasions for contacts than in London. For Anglo-French Jews, who had another additional experience in England, the challenge was to develop connections with the French. That was probably Hyman's greatest advantage in the eyes of Boulton. It was an advantage consciously built up, by dint of focusing on a specialised trade: Anglo-French 'toys'. He probably inherited a Jewish economic involvement in 'toys', a trade associated with commercial mobility and the buying and selling of second-hand goods. But 'toys' was also the leading sector of eighteenth-century consumer society. By mastering this segment of the market, Hyman gained trust and acquired the skills that were essential to his commercial integration: making connections, composing assortments and prices and forging a language. Within Boulton's network, 'systems of significations and 'systems of relations' were brought together.[50] Hyman was probably not well-educated, but he had sound practical and relational knowledge that led him to develop operational skills. As such, he belonged to Boulton's world, a world in which merchant culture, hybrid identities and cross-Channel practices were blended to form a component of the Enlightenment.

---

[50]   Daniel Roche, 'Avant-propos. De l'histoire sociale à l'histoire des cultures: le métier que je fais' in *Les Républicains des lettres* (Paris, 1988), p. 22.

Chapter 14

Enlightened Entrepreneurs versus 'Philosophical Pirate' (1788–1809): Two Faces of the Enlightenment[1]

Irina Gouzévitch

Introduction

From its creation in 1762 and especially from the moment when steam power was introduced, the Soho Manufactory received hundreds of foreign visitors. This chapter will deal with one of them, Augustin Betancourt[2] (1758–1824), a Spanish engineer, inventor and traveller whose short visit to Matthew Boulton's undertakings in Birmingham (Soho) and London (Albion Mills) in November 1788 has a particular resonance in the history of the Industrial Revolution.[3] The classic studies of James Watt's double-acting steam engine usually cite Betancourt as the man who disclosed the fundamental principle of this invention. This 'exploit' which led to the rapid introduction of similar engines in France left its mark on Boulton and Watt. They harboured a grievance for many years, describing Betancourt as a 'thief', a 'rascal' and a 'philosophical pirate'. In the historiographical literature, his action is often considered as a perfect example of industrial espionage.

---

[1] Several materials used in this paper were collected thanks to the support of the research project: 'Enginyeria i cultura científica a Catalunya i Espanya (1720–2000)', HUM2007-62222/HIST.

[2] The name of the Spanish engineer is spelled here according to the French tradition. It was the way he designated himself and signed his works during his stays in France. However, many other versions of this name also exist due to various geo-linguistic traditions (Agustín de Betancourt y Molina in Spain, Avgustin Avgustinovič Betankur in Russia) or to the different orthographical practices (Bethencourt), not to mention erroneous transliterations (Bettancourt, Bethancourt, Bettencourt, Betincourt, etc.). In the quotations which follow, the authentic orthography used in the original is maintained.

[3] For an extensive bibliography and chronology relative to Betancourt's life and works, see: *Betancourt: Los inicios de la ingenieria moderna en Europa* (Madrid, 1996), pp. 111–114; Julio Muñoz Bravo, *Biografía cronológica de don Agustín de Betancourt y Molina en el 250 aniversario de su nacimiento* (Murcia, 2008).

In view of its sensitive nature, this episode has prompted many debates as to whether Betancourt should be praised or blamed. However, whatever the position of the different authors, their narratives are inspired by three sources: Betancourt's academic memoir of April 1789;[4] his correspondence with his family;[5] and the correspondence of Boulton and Watt. However, these invaluable testimonies are all those of interested parties. In addition, the story sounds quite different when expounded in a learned text as opposed to a personal letter. The further the account is from the event, moreover, the greater the risk of error. The discrepancies which come to light when these sources are confronted arise from these circumstances. The only way of breaking through this vicious circle is to revisit the facts without preconceptions and to adjust them in the light of other concomitant or parallel events. This is the task I would like to try here, using a considerable body of documentation which puts the story into context.[6]

In its spirit and its approach, this study builds on the research stimulated by two commemorations held in 2008 and 2009: the 250th anniversary of Augustin Betancourt's birth[7] and the bicentenary of Matthew Boulton's death.[8] It is inspired by the wish to deepen our knowledge of the 'Industrial Enlightenment',[9] and of its

---

[4]   Archives de l'Académie des Sciences (AAS Paris), Augustin Bétancourt, *Mémoire sur une Machine à vapeur, à double éffet*, 1789. Séance 16.12.1789, pochette. Copy: Bibliothèque de l'Ecole des ponts (BEP), MS 1258. Resumed: Gaspard Riche de Prony, *Nouvelle Architecture Hydraulique* (2 vols, Paris, 1790–1796). Reproduced: Jacques Payen, 'Bétancourt et l'introduction en France de la machine à vapeur à double effet: 1789', *Documents pour l'Histoire des techniques*, 6 (1967), pp. 190–191.

[5]   Juan Cullen Salazar, *La família de Agustín de Betancourt y Molina: Correspondencia íntima* (Islas Canarias, 2008).

[6]   The research has taken place in four countries: Great Britain: Birmingham Archives and Heritage (BA&H), Science Museum and Imperial College Library, London; France: Bibliothèque de l'Ecole des Ponts (BEP), AAS Paris, and private collections; Spain: Archivo Historico Nacional (AHN) Madrid; Archivo Historico Betancourt y Castro (AHBC); family archives located at Tenerife, and Biblioteca del Fons Històric de Ciència i Tecnologia de l'ETSEIB, Barcelona; Russia: National Library and various State archives, St Petersburg. The author thanks the staff of the BA&H, and in particular Fiona Tait and also Peter Jones, Antoni Roca-Rosell, Juan Cullen Salazar and Emmanuel Bréguet, whose help made this research possible.

[7]   Konstantinos Chatzis, Dmitri Gouzévitch and Irina Gouzévitch, 'Betancourt et l'Europe des ingénieurs des "Ponts et Chaussées": Des histoires connectées', *Augustín de Betancourt y Molina: 1758–1824* (*Quaderns d'Història de l'Enginyeria*, 10, 2009), pp. 3–18.

[8]   See: http://www.theassayoffice.co.uk/boulton2009celebrations.html; Shena Mason (ed.), *Matthew Boulton: Selling what all the world desires* (Birmingham, 2008).

[9]   Peter Jones, *Industrial Enlightenment: Science, technology and culture in Birmingham and the West Midlands, 1760–1820* (Manchester, 2008).

principal actors without subscribing to any heroic myths which, for a long time, have dominated the national historiographies of those involved.[10]

To begin, an important point needs to be underlined: Betancourt was neither the first nor the only person to try to penetrate the principle of 'double action'.[11] Among the foreign visitors who preceded him, two, at least, obtained similar results – the Russian mechanic, Lev Sabakin and the Spanish naval officer, Fernando Casado de Torres.[12] The first, having observed the new machine in London (1786), published a description of it in Saint-Petersburg (1787) and in Moscow (1788). The second published nothing, but in June 1788 he suggested to the Spanish Navy a proposal for a 'fire machine with double injection' intended to work a sawmill in the arsenal of La Carraca.[13] Nevertheless, it is Betancourt who is considered to be responsible for the introduction of the masterful innovation to continental Europe. Why?

Setting aside any concerns about priority, this question raises an issue of intense interest to scholars: the issue of technical mediation and its determining conditions. What relations did the actors maintain with the State agencies and the industrial world? What factors ensured or, on the contrary, hindered local effectiveness? We will return to these issues. In the meantime, it is worth mentioning that unlike Sabakin and Casado de Torres whose initiatives led nowhere, the mediation of Betancourt did not prove unproductive: it gave impetus to the rapid construction of the double-action steam engines in France and thus accelerated the rise of mechanisation on the continent.

---

[10]   The nature of the relationship between Betancourt, Boulton and Watt is still a lacuna. An attempt to reconstruct it was made by Peter Jones, in his unpublished paper 'Agustín de Betancourt and the Quest for Technological Knowledge in Great Britain, 1788–1804' which he kindly allowed me to use.

[11]   Dmitri Gouzévitch, Irina Gouzévitch, 'El *Grand tour* de los ingenieros y la aventura internacional de la máquina de vapor de Watt: un ensayo de comparación entre España y Rusia', in Antonio Lafuente, Ana Cardoso de Matos, Tiago Saraiva (eds), *Maquinismo ibérico* (Madrid, 2007), pp. 147–190.

[12]   Fedor Zagorskij, 'L.F.Sabakin – vydaûŝijsâ mašinostroitel' konca XVIII – naĉala XIX veka', *Trudy IIET*, 21 (1959), pp. 328–341; Juan Helguera Quijada, 'Transferencias de tecnología británica a comienzos de la revolución industrial', in Juan García Hourcade et al. (eds), *Estudios de Historia de las Técnicas, la Arqueología industrial y las Ciencias* (Junta de Castilla y León, 1998), pp. 89–106.

[13]   Juan Torrejón Chaves, 'Innovación tecnológica y reducción de costes: las máquinas de vapor en los arsenales de la Marina Española del siglo XVIII', in *Cambio tecnológico y desarrollo económico* (Barcelona, 1994), pp. 179–190.

The Threads of History According to Betancourt[14]

First of all, the engineer explains how, commissioned by the Spanish government to build a collection of hydraulic models, he went to London to see 'a steam engine which combines all the discoveries made until this day'. However, his discussions with 'different mechanics and Physicists' having yielded nothing, he decided to go to Birmingham where 'M. M. Wats [sic], and Bolton [sic] had made recent discoveries on the steam engine'.[15] These latter gentlemen received him 'with the greatest honesty' and showed him their factory of buttons and silver plate. As for the steam engine, they only told him that 'those which they currently made were Superior to all the others' because they were more economic and rapid, but they did not refer to the source of these advantages. Back in London, a friend obtained permission for Betancourt to see Albion Mills where he was finally able to observe one of these machines in motion. Three details seem to have struck his expert eye: the parallelogram replacing the chain which suspended the piston in the cylinder; two pipes and four valves which moved with every oscillation of the beam, whereas 'in the machines of Wats [sic], or that of Chaillot' there was only one pipe; and finally, the small size of the cylinder. Considering the extraordinary effect of the machine, all these factors made him suspect 'that there could be a double effect'.[16] As for the air pump, the condenser and the governor (the regulator of speed), Betancourt recounts that he had not seen them 'because all these parts were hidden'.[17] Once back in Paris, he endeavoured to reproduce all the parts accurately, by forming their 'different plans and profiles'. Thus, he managed 'to assemble a machine with double effect'[18] and to make a model of it which succeeded beyond all expectations.

To test the degree of authenticity of this heroic narrative, let us question its various points in the light of the other sources. The voyage to England was planned by the engineer to take place in February 1788. However, he was not able to carry out this commitment. In April, his Parisian workshop had been visited by the ambassador of Spain whose enthusiastic report resulted in a decision to create, on the basis of a collection of models, the Cabinet of Machines in Madrid and to entrust its direction to Betancourt.[19] Unfortunately, during the following months, Betancourt lost three of the eight collaborators who were in charge of this task in Paris, and this seriously hindered progress. The information concerning the dispatch of two new Spanish apprentices to

---

[14]   For the source of this account, see: BEP, MS 1258, Bétancourt A., *Mémoire sur une Machine à vapeur à double éffet*, ff. 2v–4r. The quotations are translated from French into English by the author.

[15]   Ibid., f. 3r.

[16]   Ibid., f. 3v.

[17]   Ibid., f. 3v.

[18]   Ibid., f. 3v.

[19]   AHN, Estado, leg. 4088, lib. 2, doc. 153.

be sent soon to France was communicated to Paris on 26 October 1788.[20] Two weeks later Betancourt left for England.

As regards the date of departure, one has to choose between the 8, 10 and 11 November 1788, dates specified in three different documents.[21] They also raise some interesting issues concerning the motive for the voyage. On one hand, they provide evidence that Betancourt was already acting as Director of the Cabinet of Machines; he wanted to produce a paper drawing of any device likely to enrich the collection. On the other hand, they confirm that he was fascinated by steam engine technology in particular, probably because of the model of a fire pump on which he was then working. Betancourt's statement which follows, however, suggests that the reasons for this interest may have gone beyond the simple desire to improve an object for the collection:

> My model of the fire pump underwent enormous changes with my voyage to London. From parts which were made, there is hardly a quarter which was useful and although I kept the same steam cylinder, it will have a double force. Messrs Périer have already seen the plans and are so content with them that they will execute a full-scale one with all the innovations I have introduced.[22]

The involvement of the Périer brothers in the introduction in France of the double action steam engine reinvented by Betancourt is scarcely a surprise.[23] Less well-studied has been the link between their swift visit to Betancourt's workshop and the 'privilege' (i.e. special licence) for the construction of the steam mill which Jacques-Constantin Périer was going to obtain hardly one month later – on 18 April 1789. All of this suggests that the French entrepreneurs were particularly interested in the results of Betancourt's inquiry. Arguably, they can be considered as the 'shadow' commissioners and perhaps even sponsors of his voyage to England.

---

[20]   Ibid., lib. 7, doc. 320.

[21]   AHBC, leg. 9332, José de Betancourt y Castro to Agustín de Betancourt y Molina, 26.11.1788; Cullen Salazar, *La família*, p. 100; Antonio Rumeu de Armas, *Ciencia y Tecnologia en la España illustrada* (Madrid, 1980), pp. 83–84; AHBC, leg. 9325, 10.1.1789; Cullen Salazar, *La família*, pp. 112–114.

[22]   AHBC, leg. 9534, Agustín de Betancourt y Molina to José de Betancourt-Castro, 17.3.1789; Cullen Salazar, *La família*, pp. 127–129.

[23]   These French mechanics became the first industrial builders of the steam engines of simple action in France (Chaillot pump, 1778) ignoring the French exclusive privilege obtained the same year by Watt. See Payen, 'Bétancourt et l'introduction en France de la machine à vapeur'; Jacques Payen, 'Documents relatifs à l'introduction en France de la machine à vapeur de Watt', *Revue d'histoire des Sciences et de leurs applications*, 18/3 (1965), pp. 309–314; Jacques Payen, *Capital et machine à vapeur au XVIII siècle* (Paris, 1969); H.W. Dickinson, *James Watt: Craftsman and Engineer* (Cambridge, 1935); James Patrick Muirhead, *The Life of James Watt* (London, 1858).

The 'Civilities' of Soho

We do not know how long Betancourt stayed in London and whom he contacted there. We do not know, either, whether he visited the Albion Mills, where at least one of three rotative machines was already in action,[24] on first arriving in London. It seems, however, that he went first to Birmingham – in a hurry, as he wanted 'to get acquainted with these famous artists',[25] Boulton and Watt. Such a decision seems the more plausible since the Soho Manufactory was all too well-known to the Périer brothers and for evident reasons, closed to them. The calculation made sense: Betancourt, a perfect unknown in the eyes of the partners of Soho, had no compromising background, and he was endowed with letters of recommendation as was the custom. Boulton and Watt thus did not have any reason to refuse him the civilities which they usually granted to any respectable traveller, including a guided visit of their famous 'toy' workshops.

Nevertheless, the aim of Betancourt's visit was different. At least one of the engines he longed to see was then functioning at Soho, the so-called 'Lap' engine which moved 43 polishing discs used in the workshops.[26] However, most of this machine, hidden by a brick wall, remained invisible. Its remarkable efficiency could only be measured by the variable speed transmitted to the discs, thanks to the system of transmission which transformed the alternating movement into rotary motion. We do know, however, that Betancourt's hosts left him hungry to know more about the sources of such advantages. Disappointed, but all the more determined to satisfy his curiosity, he went back to London, and this time he directed his steps to the Albion Mills.

The Albion Mills: 'Meeting-place for the Curious and Serious'[27]

Since its opening in 1786, this steam corn-mill which combined a number of innovations in the fields of architecture, mechanics and energy had experienced a spectacular influx of visitors.[28] Watt detected a dangerous frivolity in this fact and wrote to Boulton, on 17 April 1786: 'It has given me the utmost pain to hear of the

[24]   This hypothesis, suggested by Peter Jones is in agreement with logic but as the tangible proofs are missing, the question remains open. See: Jones, 'Agustín de Betancourt', p. 7.

[25]   BEP, MS 1258, f. 3r.

[26]   This engine is displayed today in the Science Museum, London, see H.W. Dickinson, *Matthew Boulton* (Cambridge, 1966), p. 125.

[27]   John Moss, 'The Albion Mills: 1784–1791', *Transactions of the Newcomen Society*, 40 (1967–1968), p. 51.

[28]   Samuel Smiles, *Lives of Boulton and Watt* (London, 1865), pp. 353–359; Wallace Reyburn, *Bridge across the Atlantic: The Story of John Rennie* (London, 1972); Moss, 'The Albion Mills', pp. 47–60.

many persons who have been admitted into the Albion Mills merely as an Object of Curiosity. ... R. [Rennie] no doubt has vanity to indulge as well as us, but ... Dukes and Lords, and noble peers will not be his best customers'.[29]

John Rennie had all sorts of reasons for not sharing this opinion. Besides being the builder of the largest installation of this kind ever attempted, he could be proud of having perfected a series of innovations which promised a great future, including the famous 'governor' applied to the steam engine, allegedly one of the glories of Watt's technology. The biographers of Rennie firmly assert that the idea was that of the young Scot, taken up and re-arranged by Watt from a sketch made by Boulton during a visit of inspection to Albion Mills.[30] Be that as it may, the objections of Watt seem to have had an effect and access to the mill was restricted. On this point, we have the testimony of Lev Sabakin whose attempt to penetrate there in 1787 had failed.[31] If Betancourt nevertheless got in, it is because he was introduced by a mysterious 'friend'. Who could this have been?

If we consider someone external to the mill, the choice is vast: a public personality, a dignitary, a diplomat, a scholar or a mechanic who was one of Betancourt's acquaintances? The likelihood of identifying the right person among all these people is negligible. It might have been someone on the mill's staff. To bribe an employee using a modest sum was a common practice among the hunters of secrets: even the going rates were known.[32] And why not Rennie himself? Watt's injunction to be less welcoming conflicted with his need to attract customers. Maybe the young mechanic decided to take revenge on his employer. However, it could equally have been, very prosaically, one of the partners of Soho, as the following letter from Watt suggests:

> Mr De Betancourt came recommended to us as an ingenious man & ... among other things he was shown the Albion Mill, which he was told acted by the power of steam both in the ascent & descent of the piston. How far the particular mechanism was explained to him I don't now remember; but he had the modesty to ask to have the engine stopt, on which 50 or 100 men were attendant, that he might see the inside of it which was refused. But the outside he was at liberty to examine & the principle of action was explained to him.[33]

---

[29]   Quoted by Cyril T.G. Boucher, *John Rennie: 1761–1821* (Manchester, 1963), p. 11.

[30]   Reyburn, pp. 37–38.

[31]   Lev Sabakin, *Pribavlenie k Fergûsovym lekciâm, soderžašee v sebe o ognennyh mašinah, sočinennoe tverskim gubernskim mehanikom L'vom Sabakinym* (Moskva, 1788), pp. 36–37.

[32]   According to the Russian ambassador Simolin, in 1780/82, this rate was half a crown. See: Pëtr Zabarinskij, *Pervye 'ognevye' mašiny v Kronštadtskom portu* (Moskva-Leningrad, 1936), p. 165.

[33]   BA&H, MS 3219/6/1/ 319/32, James Watt (JW) to James Watt Jr (JWJ), 10 (11?) November 1808.

On 7 January 1809, James Watt Jr. wrote on the same subject to John Playfair: 'My father does not recollect whether it was himself or Mr Boulton who showed the Albion Mills to M. Bettancourt [sic], but is certain it was one of them, and that they afterwards saw him together at Soho'.[34] These two interesting testimonies were both written at least 20 years after the events, and this fact raises questions. Why, indeed, recall this small episode after so many years? Why insist therein, unlike previously, that Betancourt *saw the two manufactories* on the *agreement* of the partners, that everything was *shown* and *explained* to him, including the secret principle of the new machine? And supposing that things had occurred this way, why complain 20 years later of his craftiness? It is tempting to read in these lines the desire of Boulton and Watt, not so much to discredit the 'honesty' of Betancourt as to undermine his remarkable mechanical ingenuity.

The formal aspect of the affair seems unambiguous, however. Betancourt committed an act of espionage, but the partners of Soho hardly scorned resorting to such practices. We have just seen an example: Watt did not refrain from 'borrowing' the young Rennie's idea of the governor. Is it likely that he would have hesitated to perpetrate a similar action with regard to Betancourt if his professional interests required it? Two documents drawn from the Archives of Soho offer a clear answer to this question. The first, an un-signed and un-dated manuscript (but most probably dating prior to 1790), seems to be a report addressed to Watt which contains a translation into English of some pages of the *Nouvelle Architecture Hydraulique* dealing with Betancourt's experiments on the expansive force of the steam. The second is a handwritten sheet entitled 'Betancourt's experiments'; it contains just three columns of figures, the first two representing the temperatures in degrees of Réaumur and Fahrenheit and the third one, the values of the expansive force of the steam in inches of mercury.[35] According to Maxime Gouzévitch, these last series of figures would be the non-definitive estimated measures of Betancourt copied out from a draft dating, most probably from 1788–1789. In other words, this was a draft dating from the period when Betancourt and Gaspard Prony were still working on the steam experiments. These two documents provide evidence of the intense interest that Watt brought to bear on the results of Betancourt's research. They also confirm, paradoxically, that he spied on Betancourt for his experimental data at the very moment when the latter spied on Watt for his machine.[36]

Any moral judgement would be anachronistic, however, inasmuch as we are dealing with a period when the legal status of the inventor and of intellectual property was insufficiently protected, and when the pirating of a foreign industrial secret was considered as a patriotic act. From this point of view, the action of

---

[34]   BA&H, MS 3219/4/113, p. 6, JWJ to John Playfair, 7 January 1809.

[35]   BA&H, MS3219/4/167.

[36]   See: Maxime Gouzévitch, 'Aux sources de la thermodynamique ou la loi de Prony/ Betancourt', in Chatzis, Gouzévitch and Gouzévitch (eds), *Agustín de Betancourt*, pp. 141–145.

Betancourt, like the actions of Watt and Boulton, was pretty commonplace. This fact being recognised, it is the intellectual aspects of the affair that we must now examine.

Some Thoughts about the Skills

Judgements on Betancourt's activities tend to converge upon three points of view. The first one which we may call the 'contesting' one, insists on the fact that Betancourt invented nothing but simply reproduced what had been shown and explained to him.[37] The second, the 'admiring' point of view, insists upon the total autonomy of his approach.[38] The third point of view, which might be described as the 'moderating' perspective, postulates that to recognise the perspicacity of Betancourt hardly requires us to be amazed at it. Why? Because, as the French historian Jacques Payen puts it, 'whereas somebody incompetent would have noticed nothing, a man like Betancourt could scarcely avoid deciphering what he saw, almost as easily as if it had been a clear-cut technical drawing'.[39] Our point of view would be a 'conciliating' one, however. It should be asked, indeed, whether there is really a contradiction between what Betancourt and the Soho partners said about what would have been *shown* and *explained* to the Spaniard during his visit to Boulton undertakings. Actually, everything depends on the sense which each party attributed to these two verbs. For Boulton and Watt, *to show* meant a guided visit of the factory with a permissive glance at the installations and the provision of some polite, if evasive comments. Insofar as Betancourt was entitled to these 'civilities', they were convinced that they had *shown* him the machine and *explained* its principle. For Betancourt, the opposite sense can be inferred. Since he was not authorised to look inside the machine, he was '*shown*' nothing. As for the explanations provided, we know that he was less than satisfied with them.

Those who had seen these machines in action could easily imagine what kind of opportunity had been offered to Betancourt. Be it in Soho or in the Albion Mills, the engines were of the same type: a rotative machine activating either polishing discs or millstones. Their size, when compared to the Newcomen or the Watt single-acting machine, remained considerable, however. Both were concealed in zones with restricted access. The request to 'inspect' the machine was equivalent

---

[37]   This point of view is widely shared by the English authors starting with the contemporaries of Watt. See, e.g. Robert Stuart, *A descriptive history of the steam engine* (London, 1824). The French version: Robert Stuart, *Histoire descriptive de la machine à vapeur* (Paris, 1827), p. 236, is used here. See also: Muirhead, pp. 267–268.

[38]   This point of view is typical of Spanish historians, e.g. José Antonio García-Diego, 'Agustín de Betancourt como espía industrial', in *Estudios sobre Historia de la Ciencia y de la Técnica* (Valladolid, 1988), 1, pp. 105–125; Rumeu de Armas, *Ciencia y Tecnologia*, p. 85.

[39]   Payen, *Capital*, p. 159.

to stopping the factory. One can understand that no works foreman would have accepted such a constraint.

Besides the document-based narratives, we also have the familiar descriptions and drawings illustrating the machinery at Albion Mills (1787),[40] plus two others, made by foreigners who visited the factories of Boulton several months apart. The first one comes from the hand of the Russian Sabakin, and appeared in his brochure published in 1788 in Moscow. The second, from the hand of Betancourt, appeared first in the *Mémoire* (1789), then in the *Nouvelle Architecture Hydraulique* (1796).

Sabakin versus Betancourt

Like Betancourt, Sabakin went to England at the expense of the Crown to improve himself in the mechanical arts (1784–86). Like Betancourt, he was interested in steam technology which led him to contact Boulton. Like the Spaniard, the Russian came equipped with solid recommendations. Both understood at once the importance of this invention for industrial development. By contrast, the drawings resulting from these visits are very different. That of Sabakin is very primitive, some important elements are missing such as the parallelogram and the air pump; moreover, it contains stop valves with hammers which were typical elements of the older machines. For a better comparison of the two engines, atmospheric and double-acting, Sabakin arranged their main elements in a symmetrical way.[41] However, we do not know if such a presentation resulted from a deliberate choice, or from a simple lack of information. As for the drawing by Betancourt, one finds there, worked out to the smallest detail, the construction of the double-action steam engine.[42] It has disturbing resemblances to the drawing of Watt's machine which Farey published in 1827. Betancourt's drawing shows only a different system of valves and it is not equipped with the governor. In the light of this resemblance, the question naturally arises: what did Betancourt actually see? Did he wish to hide some of his observations so as not to compromise his reputation for 'ingenuity'?

Recall the elements which he says he did not see: the air pump, the condenser and the governor. The first two were pre-existent to the double-action engine, and Betancourt reproduced them on his drawing. However, if he had seen the original machine close up, he would not have had to reinvent the valves which were less ingenious than those employed by Watt. His condenser and the air pump are also shaped and located differently. All this suggests that Betancourt did not observe these two elements, but rather added them by analogy. As for the governor, it was a true innovation and since Betancourt was already aware of its existence, one can

---

[40]   See W.K.V. Gale, *Boulton, Watt and the Soho Undertaking* (Birmingham, 1952), p. 9, which contains a facsimile of the original drawing published in J. Farey, *A Treatise on the Steam Engine* (London, 1827).

[41]   Sabakin, *Pribavlenie*, pp. 38–51, tab. XII–XIII.

[42]   The original Betancourt's drawing is held in: BEP, MS.Fol.104/2.

suppose that the information had filtered through. Not enough, however, to allow him to learn its construction, and he gave up trying to reproduce this device in his drawing. We have, therefore, no reason to believe that Betancourt had seen more than he admitted. Far from being a banal act of piracy, the visit to Albion Mills stimulated an intense intellectual effort which made it possible to merge together the visible and the invisible and thus to reconstitute, starting from the disparate elements, the device which principle and structural elements had already planted in his mind.

The Missing Link

This hypothesis arises out of a small sketch scribbled in pencil and India ink on the back of a drawing representing the elements of bridge design. This drawing, in the hand of Augustin, is preserved in the family archives.[43] The context indicates clearly its origins: it was a typical school drawing similar to those done by the students of the École des Ponts in Paris at the time of Jean-Rodolph Perronet who was director between 1747 and 1794. The sketch is barely legible.[44] However, its cognitive value is out of all proportion to the most perfect of school drawings. It is, indeed, the missing link which allows us to follow the progress of Betancourt's inventive thought process.

One sees on this drawing made twice in all probability, a sketch of a fire pump. The part made with pencil, and the part outlined with India ink represent two versions of one and the same device. The whole looks like an attempt to solve certain technical problems proper to this type of device. The original version contains two steam cylinders connected to a single boiler, and a valve alternating the direction of the injection. Betancourt seems ultimately to have opted for a single cylinder, and this is the element which he highlights by using India ink. In this last device, the steam was injected into the lower part of the cylinder, below the piston, while its upper part was connected to the pipe of the water pump by an opening made in the wall of the cylinder. In addition, the lower part was equipped with a valve to eject the used vapour and the upper one with a tap to evacuate the pumped water. A vertical rod fixed to the piston was completed by two bolts. One can conclude from the sketch that Betancourt tried to design a compact device free of a heavy beam and with two distinct cylinders, for vapour and water as was the case with the well-known machines of Newcomen and Watt. It was while ruminating in this fashion that Betancourt got the idea of using both surfaces of the piston, in other words the double effect. However, Betancourt inevitably had to meet the problem of the joint connecting the piston to the transmission mechanism. This joint, which had to ensure both the extension and the compression, must necessarily be rigid, and thus it needed to be a rod rather than the usual chain. At the same time, Betancourt

---

43   AHBC; Cullen Salazar, *La familia*, p. 126.
44   It was identified by D. Gouzévitch in 2006.

had to find how to bring the piston back: he first tried to resolve this problem by means of two cylinders operating in half-phase. Nevertheless, none of the known experiments of this kind had really succeeded, and Betancourt eventually gave the idea up in his turn: he outlined instead a single cylinder with a piston rod and two bolts. An element was still missing, however: some kind of device to guide the piston rod to the pump. It could be an 'extension piston rod' bolted to the top of the piston rod which would have connected with some form of 'parallel motion'.[45] It seems, however, that Betancourt did not know what he really needed. It was at this stage in his ruminations that an echo of Watt's achievements in this field reached his ears and prompted his decision to travel to England.

Let us now recall what Betancourt admits to have seen of Watt's machine: the parallelogram replacing the chain intended to raise the piston, two pipes where normally there should have been only one, four valves which moved in keeping with the beam and a cylinder of small size. Given the background described above, did he need something more for the 'click' to occur in his mind and for the principle of double action to dawn with the same brightness as though he had observed it, or had deduced it from the explanations given?

What Benefit for the Inventor?

Once in Paris, the Spanish engineer set out to reconstitute the machine and to build a working model. The minutes of the Academy of Sciences enable us to follow this process. The entry for 16 December 1789, notes that Betancourt had presented his memorandum and that the academicians Borda and Monge had been appointed to report on it. The record for 10 February 1790, shows that they presented their report, stating that 'the Academy must applaud the zeal and the intelligence of Mr. de Bétancourt who brought back to France the power of a discovery, knowledge of which would naturally only have reached it later'.[46] In the same year, Prony reported Betancourt's discovery in the first volume of his *Nouvelle Architecture Hydraulique* and promised to provide its description in the next volume:

> Bettancourt ... had a model of a machine executed ... The internal mechanism, by means of which the double injection takes place, is completely of the invention of ... Béttancourt [sic]; and although he is unaware whether the process is the same as that of MM. Wats [sic] and Bolton [sic], has seen the secret that these

---

[45]   I sincerely thank Jim Andrew, Collections Adviser, Thinktank, Birmingham Science Museum, and Ben Russell, Curator, Mechanical Engineering, National Museum of Science and Industry, London, who helped me to interpret this sketch.

[46]   AAS Procès verbaux. T.108, séance du 16.12.1789; T.109, séance du 10.2.1790; Dossier 'Bétancourt'.

latter made of it, there is reason to believe that the English artists have not achieved a greater degree of precision and simplicity.[47]

The first reaction of Watt dates to 23 July 1790:

> I have a letter from M. Levêque of July 2, he has seen P^ers [Périer's] Engine which he does not like, says Mr De Betancourt instructed him how to make Double Engines, & has sent model of them to Spain as he does of every thing he sees, & has written a Memoir upon the effects of them, which will be published in Prony's Hydraulogie, we must be more & more cautious in respect to foreigners, the greatest rascals among them are the Gentlemen.[48]

Prony kept his promise and integrated into volume 2 the description of two of Betancourt's works: his experiments on the expansive force of steam and the double-acting machine, highlighting 'those executed in Paris, on the isle des Cygnes, to activate a corn-mill. ... These machines, whose execution is so well conceived ... are an undeniable proof of the excellence of the sophisticated mechanism that has been substituted for that of the machines at Chaillot'.[49]

In 1802, Watt obtained an explanation from Prony which left him satisfied: 'the exceedingly friendly manner in which he received me & the character of the man make me perfectly [willing to] aquid [sic] him of any evil intention towards me he merely published what he had been told by Perier Betancourt and other thieves'.[50]

We think, however, that this condescending account may be a rather inadequate explanation: the letter in which Watt states that he had explained everything to Betancourt was written five months after his election as a 'corresponding member' of the Mechanics Section (first class) of the Institut National. This election took place on 20 June 1808, and the choice was made between three candidates, Watt, Betancourt and the Bavarian engineer Baader. Betancourt would be elected the following year. If we compare the date of this election (5 December 1809) with the dates of the quoted letters of Watt and his

---

[47]   Prony, 1, p. 572.

[48]   BA&H, MS 3782/12/79, JW to Matthew Boulton (MB), 23 July 1790.

[49]   Prony, *Nouvelle Architecture*, 2, p. 35.

[50]   BA&H, MS 3219/6/1, p. 319, JW to JWJ, 10 November 1808. Boulton seems much more intransigent, as the following letter suggests: 'in my intended answer to the Report of certain members of the Institute such as prony & perrier [sic] the former [illegible] a book upon Steam Engines in which he gives all the Merit of that inventing to the latter & to Monsr Bathincour [sic] who is by profession a Thief as appears by his publication after he had been [civilly?] treated at Soho & shown the double engine, which in a subsequent publication was claimed by him, as were the two engines we erected at Paris: by Monsr Perrier. It seems to me a waste or a misapplication of Time to assert our right to any of these trifles whilst the French [nation?] claim the very Root & branches of all our Country'. See BA&H, MS3782/12/56, MB to Joseph Banks, 23 January 1804.

son in which the skills of Betancourt were questioned (10 November 1808, and 7 January 1809), we obtain an explanation for the altered attitude of the British inventor with regard to his Spanish colleague. He had tried to undermine Betancourt on particularly sensitive ground, that of academic recognition! It seems, however, that the arrow did not reach the target: the disavowals of Watt did not prevent Betancourt from collecting the majority of votes (35 out of 43). Would Betancourt have remained untouched by these attacks? There is nothing to suggest that he felt afflicted. Moreover, between his first voyage in England and his election to the Academy of Sciences (the Institut National as it became), his relations with the partners of Soho went through some interesting developments. Evidence is found in the letter from Boulton to Watt of 4 September 1794: 'I have just rec[d] a letter from Mr James mentioning one you have received from le Chevalier de Betincourt [sic]. I suppose it is the same Gent[n] that came to see Droz at Soho[51] & whom Monsieur Proney [sic] mentions in his Book. If so I think he should have General answers but not particular prices of either Cast or wrought Iron'.[52]

On the back of this letter, an inscription of the hand of J. Watt Jr specifies: 'Chev[r] of Betancourt. West Indian Engines'. This comment sheds important light on the work that occupied Betancourt during this time. Resident in England for three years from 1793 to 1796, he was then trying out the application of steam power to the sugar mills on behalf of the rich planters of Cuba, and according to his own testimony, in December 1794, two at least of these steam engines were already under construction.[53]

Thanks to the document, lately discovered by Olga Egorova at the archives of Cuba, we know who was commissioned to build these engines: it was the iron master from Coalbrookdale, William Reynolds.[54] But evidence is also given that Betancourt tried at first to pass on an order to Boulton. However, memories at Soho were long. And when, by his letter of 3 September 1794, James Watt Jr, acting on behalf of his father, made enquiries in this connection through the Spanish merchant in London, Fermin de Tastet about the credentials of Chevalier de Betancourt, he added the following postscript: 'we formerly

---

[51]   Jean-Pierre Droz (1746–1823), Swiss medallist, inventor of the die replication devices; on the invitation of Boulton he worked at the Soho Mint between 1788 and 1791: see Richard Doty, *The Soho Mint and the Industrialization of Money* (London, 1998), pp. 25–45.

[52]   BA&H, MS 3219/6/2/132, MB to JW, 4 September 1794.

[53]   See José Antonio García-Diego, 'Huellas de Agustin de Betancourt en los archivos Breguet', *Anuario de Estudios atlanticos*, 21 (1975), p. 203.

[54]   Olga Egorova. *Agustín de Betancourt. Secretos cubanos de un ingeniero hispano-ruso* (La Habana, 2010), pp. 52–53.

saw him here [at Soho] & have some reason to complain of his proceeding as a philosophical pirate'.[55]

Enlightened Entrepreneurs versus 'Philosophical Pirate'

This expression deserves consideration because it raises the problem of technical mediation both in the broadest sense, and also from the more concrete angle of the actors in this process. The three significant players in this story belong incontestably, even if to varying degrees, to this category of mediators who, by their curiosity, their mobility and especially their enlightened sociability, contributed to the development of the 'trade in knowledge' which opened the way to modernity. By treating Betancourt as a 'philosophical pirate' Watt Jr affirmed, in actual fact, the enlightened intellectualism of this man of action. It was in some way a measure of respect, a kind of implicit granting of 'civility', which could only be applied to an equal. At the same time, it was also an indication of a certain embarrassment felt by the partners of Soho when wearing two 'caps', that of enlightened and cultivated gentlemen, members of the 'republic of the sciences', and that of successful entrepreneurs. A good illustration of this duality is the ceaseless invocation of 'thieves' and 'rascals' in their correspondence. There is nothing similar in the case of Betancourt, who does not even think of responding in kind. He seems to slide along on the surface of the events and, apart from his famous memoir he does not mention them in the documentation that has survived.

This difference of attitude finds its explanation in the conception that each of the three actors had of his professional mission, his aims and his social position.

What can be said in this respect of Watt? A mechanic endowed with an exceptional creativity, whose mind was well-anchored in the British empirical tradition, he combined vocation and mission by marketing his talent. However, the only way to succeed in the competitive conditions of industrial England was to proceed rapidly and efficiently to industrial application. In return, the inventor drew an income sufficiently large enough to guarantee a comfortable existence and a respectable social status. He remained, however, a man subject to a gnawing anxiety lest his achievements were overturned and his income jeopardised. Despite his position in the marketplace, he remained vulnerable to the misconduct of collaborators and to the intrigues of his competitors.

The same comment might be applied to Boulton, but it should be qualified. More sensitive than Watt to the 'lunar' spirit, he was for a long time reluctant to give up the civilities that an enlightened gentleman owed towards his peers, enlightened scientists, irrespective of their geographic origins. The Soho Manufactory seems to be one of the few to have kept its doors open to the numerous visitors until

---

[55]   BA&H, Timmins Collection, JWJ to Fermin de Tastet (FdT), 3 September 1794. See BA&H, MS 3147/3/524/120, Spanish Engineer, FdT to JWJ, 5 September 1794 for de Tastet's answer which probably contributed to the agreement not being concluded.

the end of the eighteenth century. Yet at its very climax, this domestic culture of Enlightenment in England was harshly tested as it came into conflict with the interests of the business. As contractors who wished to continue to thrive, Boulton and Watt had to face this difficult challenge. The actions perpetrated by Betancourt, if they did not deal a deathblow to the lunar dreams of Boulton, surely shook him strongly enough to rouse him against his peers, the 'gentlemen-philosophers', and thus, to accelerate his transformation from savant to business man.

As for Betancourt, the specific conditions of Spain gave his talent as a mechanic a vocational orientation very different from that of Watt and Boulton. Free enterprise, so little developed in this highly centralised country, was not a sphere of activity appropriate to a young nobleman. Thanks to his origins, his training and his influential friends in the Canary Islands, he could aspire to a high-ranking post in the State administration. A substantial salary and a solid social position being thus guaranteed, the vagaries of the market had no impact on his actions. The voyage to England was but a stage in the fulfilment of a larger State mission which he hoped would lead to a greater career as a technical expert in the service of the enlightened monarchy. The creation of a collection of models of general utility appeared at the top of this list. The steam engine was not a major concern of Betancourt, either before or after his discovery, but only one of several areas of interest.

Does this mean that the vocational choices which the Spanish 'philosophical pirate' Augustin Betancourt and the British 'enlightened entrepreneurs' Matthew Boulton and James Watt ultimately resulted in them belonging to two contrasting universes? Our answer may appear paradoxical but at a certain level of abstraction, we perceive between these two mentalities many more affinities than differences. Enough, in any case, to see in them two faces of Enlightenment.

Chapter 15

Creating an Image:
Portrait Prints of Matthew Boulton[1]

Val Loggie

During Matthew Boulton's lifetime, having a portrait painted became accessible to a much broader range of people.[2] Through the use of gesture, pose, dress, props, background and labelling, an artist could convey signals about the sitter for the viewer to interpret.[3] Boulton was aware of this and had his portrait painted by different artists for different audiences, each signifying different messages about his role and status. Copies were made of some of those portraits in the form of paintings, miniatures and prints so that friends and family could own an image of Boulton. Prints, including portraits, also became more accessible to wider audiences as they were produced in large quantities by publishers like John Boydell.[4] Print collecting was extremely popular in the middle and later eighteenth century; portfolio collections were kept by the wealthy connoisseur while framed prints were used as decorations in the homes of the middle classes.[5] Having a printed portrait was the only way of making an image of a person available to large numbers of people and became more important as fascination with the famous grew.[6] Portraits were reproduced as single sheet prints and as illustrations in magazines.[7]

---

[1]   The research for this chapter was undertaken as part of an AHRC-funded PhD, Valerie Loggie, 'Soho Depicted: Prints, drawings and watercolours of Matthew Boulton, his manufactory and estate, 1760–1809' (University of Birmingham, unpublished PhD thesis, 2011). I am grateful to Richard Clay and Antony Griffiths for their help.

[2]   Desmond Shawe-Taylor, *The Georgians: Eighteenth-Century Portraiture and Society* (London, 1990); John Ingamells, *National Portrait Gallery Mid-Georgian Portraits 1760–1790* (London, 2004), pp. xiii–xvi.

[3]   Shearer West, *Portraiture* (Oxford, 2004), p. 71.

[4]   Stana Nenadic, 'Print Collecting and Popular Culture in Eighteenth-Century Scotland', *History*, 82/26, 1997, pp. 203–222.

[5]   Nenadic, *Print Collecting*, pp. 209–210; Marcia Pointon, *Hanging the Head: Portraiture and Social Formation in Eighteenth-century England* (New Haven and London, 1993), p. 55.

[6]   William Vaughan, *British Painting: The Golden Age* (London, 1999), p. 47.

[7]   Images of Boulton were produced for inclusion in magazines but are not considered here.

Boulton worked hard to build his reputation locally, nationally and internationally; he cultivated a high profile by lobbying Parliament and seeking patronage from important and famous figures.[8] Visitors to Soho expected to be able to see the man as well as the factory; John Hodges told him, 'the generality of people of distinction and fortune that visit Soho, as well as foreigners' were disappointed when they could not see him.[9] Boulton's understanding of his own role in the promotion of the businesses is visible in the way he chose to have himself represented in portraits. His portrayal became increasingly sophisticated, and the later images were disseminated to much wider audiences. Boulton was painted by various artists; his friends, Birmingham artists, James Millar and Joseph Barber painted or sketched him, as did John Phillp, one of his own protégés. But he also had his likeness taken by artists of national renown and significance such as Tilly Kettle, J.S.C. Schaak, Zoffany, Carl Frederick von Breda, Sir William Beechey and L.F. Abbott.[10]

These portraits were produced to be displayed in different settings and portrayed different aspects of Boulton's life and character. The Tilly Kettle portrait of the early 1760s was one of a pair with a portrait showing his second wife Ann, and would have hung in a domestic situation.[11] The Schaak portrait was also one of a pair; the bill is for two paintings but does not identify the sitter in the second.[12] It is possible that it was of his business partner, John Fothergill and that they were painted to be hung at the Manufactory.[13] There, they would have been seen by potential customers so would have needed to portray successful businessmen with an understanding of taste and fashion. Boulton's portrait shows a fashionably dressed, confident young man, apparently disturbed in the act of reading. He is dressed in a court suit with elaborate buttons and frogging, his hand tucked into his waistcoat, a convention of portraiture rather than something people actually did, showing that Boulton understood and could make use of such conventions.[14]

[8] Eric Robinson, 'Matthew Boulton and the Art of Parliamentary Lobbying', *Historical Journal*, 7/2 (1964), pp. 209–229.

[9] Birmingham Archives and Heritage (BA&H), MS 3782/12/63/16, John Hodges to Matthew Boulton (MB), 12 September 1780.

[10] Painted portraits of Boulton are considered at greater length in Val Loggie, 'Portraits of Matthew Boulton' in Malcolm Dick (ed.), *Matthew Boulton: A Revolutionary Player* (Studley, 2009), pp. 63–76.

[11] The portraits are reproduced in Shena Mason, *The Hardware man's Daughter: Matthew Boulton and his 'Dear Girl'* (Chichester, 2005), plates 1 and 2.

[12] BA&H, MS 3782/6/191/56, J.S.C. Schaak to MB, receipted 13 October 1770.

[13] See Loggie, 'Portraits of Matthew Boulton', pp. 64–65 and plate 2 for further consideration of this possibility.

[14] West, p. 25.

Von Breda's Portrait of Boulton

As Boulton's fame grew, people outside the family circle expressed an interest in owning images of him. John Rennie, an engineer who had worked for Boulton and Watt in the early stages of his career, asked them to sit for the American artist Mather Brown in London in June 1792 but Boulton did not have the time.[15] However, the Swedish artist, von Breda was on a tour of Britain, painting portraits to fund his travels and in July he painted Boulton, Watt and William Withering, another of the Lunar men. The portraits of Boulton and Watt were shown at the Royal Academy exhibition in 1793.[16] Boulton was portrayed three-quarter-length, seated and dressed in black (Figure 15.1). This is the first portrait where he was shown with attributes specific to him rather than generic items such as an unspecified book, but Boulton's diverse range of businesses made it difficult to select a single symbolic object. He is shown holding a medal which is indistinct but presumably one of his own manufacture and represents the whole of his medallic output, the area of his greatest interest at the time. Highlighting the Mint was important, as Boulton was seeking the contract to produce the national coinage. He has a magnifying glass in his other hand which signifies the quality of the medal, suggesting it would withstand close examination, but also portrays Boulton as a connoisseur, a man of learning and taste.[17] The principal building, the well-known frontage of his manufactory, is shown in the background and a number of mineral specimens are placed on the table beside him.[18] These cannot be positively identified, but one may be intended to signify copper ore, the raw material for many of Boulton's products.[19] Thus the portrait can be read as representing the transition from raw material to finished product, the building in which this took place and the sorcerer himself at the centre. It signifies and unites learning, arts, science, and manufacturing.

Full-size copies of this portrait were made by von Breda for Watt, Withering and John Rennie.[20] A miniature, a more intimate and portable format for family or close friends, was made for Boulton's daughter, Anne.[21] In 1796, Boulton agreed to further duplication and circulation to a wider audience when von Breda asked

---

[15] BA&H, MS 3147/3/296/12, John Rennie to MB, 7 June 1792.

[16] Algernon Graves, *The Royal Academy of Arts A Complete Dictionary of Contributors and their work from its foundation in 1769 to 1904* (London, 1906).

[17] Harry Mount, 'The Monkey with the Magnifying Glass: Constructions of the Connoisseur in Eighteenth-Century Britain', *Oxford Art Journal*, 29, 2, 2006, pp. 167–183.

[18] For discussion on the circulation and promotion of views of the principal building as a symbol of Soho and its use in creating a brand identity see Loggie, 'Soho Depicted'.

[19] I am grateful to Dr R.A. Ixer for guidance on the mineral specimens and to Dr Richard Clay for discussion on this.

[20] BA&H, MS 3782/12/38/55, C.F. von Breda (CFvB) to MB, 27 March 1793; BA&H, MS 3782/12/38/169, CFvB to MB, 5 October 1793.

[21] BA&H, MS 3782/12/39/314, CFvB to MB, 18 November 1794.

Figure 15.1   *Matthew Boulton* by C.F. von Breda, oil on canvas, 1792
Source: Courtesy of Birmingham Museums and Art Gallery.

to produce a print after the portrait. The painter asked how best to publicise the print in Birmingham and suggested that Boulton should ask friends to begin a list of subscribers.[22] Von Breda hoped that the main purchasers would be Boulton's friends, acquaintances and admirers and recognised that much of the demand would be in Birmingham. The question of the audience for a print of someone who was not a famous actor, politician or royalty is one which later concerned Boulton's friend, Sir Joseph Banks. He suggested the sales of a print depended on three things, 'the Excellence of the Painter; the Talents of the Engraver, & the Notoriety of the Person it Represents'. He continued 'A man like me, who has never medled [sic] in Politics, & who Cannot, of Course, possess a Squadron of Enthusiastic Friends, is not likely to Sell a dear Print. A Cheap one will answer better among the men of Science'. Banks felt that soliciting a subscription for the publication of his portrait would be interpreted as vanity, but admitted he felt inclined to do so as an act of gratitude to the artist. He worried that people would

---

[22]   BA&H, MS 3782/12/41/50 CFvB to MB, 10 February 1796.

not understand his motives so dare not subscribe himself, 'nor should I venture, on account of the high Price put upon proofs, & Very little real superiority they have over Prints, to purchase privately more than a Few of them'. Proofs are the earliest prints taken from a plate which are of higher quality as the plate has not yet worn. They were sought after by collectors and were more expensive than the later, common prints. Banks also highlighted one of the important roles of portrait prints for their sitters, as presents, tokens of respect or affection. He planned to lay some prints aside for his family 'in hopes that they may become usefull [sic] to some one sometime hence as Presents, when difficult to obtain in the Shops'.[23]

The print of Boulton was to be a mezzotint, a form which was quick and cheap to produce and could reproduce the tonal effects of painting. The technique dominated portrait prints until the 1780s when other techniques such as line engraving began to become more popular.[24] Mezzotint plates wore quickly so earlier prints were of higher quality, and the difference between proofs and prints was more obvious than in other forms. Ten days after submitting his proposal, von Breda wrote that the plate was ready and 'it does in my opinion honor [sic] to the artist both as to the likeness and execution and will I hope meet with your approbation'.[25] Von Breda initially stipulated that the price of the proofs should not exceed 15s. and the prints 10s.6d. On publication they were fixed at 12s. for proofs and 7s.6d. for prints.[26] The number of copies printed or sold is not known.

Von Breda had included the principal building to represent Soho in the portrait. This made the assumption that viewers would recognise it or accepted a possible reading of it as a house.[27] The original portrait and its copies were owned by friends and family, they would have been hung in a domestic setting where, if necessary, they could be explained to visitors.[28] However, interpretation of the image passed beyond the control of Boulton's immediate circle when it was exhibited and the print was produced. How many viewers understood what the building represented cannot be known, but by the twentieth century it was sometimes misinterpreted as showing Boulton, 'a frequent host at his Soho home (represented in the background of the print)'.[29] This gives a different message, not necessarily one of which Boulton would have disapproved, he did have a

[23]   Joseph Banks to Thomas Phillips RA, 12 September 1808 in Neil Chambers (ed.), *The Letters of Joseph Banks: A Selection 1768–1820* (New Jersey, 2000), p. 287.

[24]   David Alexander, 'The Portrait Engraving in Georgian England' in *The British Face: A View of Portraiture 1625–1850* (London, 1986), p. 27.

[25]   BA&H, MS 3782/12/41/72, CFvB to MB, 20 February 1796.

[26]   BA&H, MS 3782/12/41/50, CFvB to MB, 10 February 1796; BA&H, MS 3782/12/41/72, CFvB to MB, 20 February 1796.

[27]   The principal building drew on country house architecture, see Demidowicz in this volume.

[28]   Loggie, 'Portraits of Matthew Boulton'.

[29]   Stephen Deuchar, *Painting, Politics & Porter: Samuel Whitbread II (1764–1815) and British Art* (London, 1984), pp. 43–4. Others have also interpreted it in this way.

reputation as a generous host. It also suggests someone who could afford a house much grander than the one he actually had, again probably not something to which he would have objected.

Production and Distribution of the Print after Beechey's Portrait

Boulton's portrait was painted in 1798 by Sir William Beechey, a favourite of the Royal family, and exhibited at the Royal Academy in 1799 (Figure 15.2).[30] This marked a further advance in Boulton's status, that he could command a portrait by an artist of the calibre of Beechey. Once again Boulton was portrayed seated, holding a medal in his left hand and a magnifying glass in his right. A mineral specimen is placed under a dome, more overtly a specimen collected for study, than the example in the von Breda portrait. This strengthens the reading of Boulton

Figure 15.2   *Matthew Boulton* by Sir William Beechey, oil on canvas, 1798
Source: Courtesy of Birmingham Museums and Art Gallery.

---

[30]   Graves, *Dictionary*.

as collector and intellectual; it alludes to a scientific rather than a manufacturer's interest. Here, there is no direct reference to the industrial building. Boulton appears more confident, holding the medal up for the viewer to see and gazing directly at that viewer rather than into the distance. By now, he was well known for the status of his Mint and was producing the national coinage, so he could expect the viewer to be more aware of who he was. Beechey's portrait of Boulton was again reproduced, in full-size copies after Boulton's death by Beechey's studio and in miniature.[31] Such copies and exhibition at the Royal Academy widened the audience for portraits of Boulton, it allowed them to be seen by those other than visitors to the house or the manufactory.

Figure 15.3   *Matthew Boulton F.R.S. & F.S.A.* by William Sharp after
Sir William Beechey, line engraving, 1801
Source: Courtesy of Birmingham Museums and Art Gallery.

---

[31]   Loggie, 'Portraits of Matthew Boulton'.

By 1799 Boulton felt able to encourage the creation of a print after Beechey's portrait (Figure 15.3). Producing a print ensured wider circulation and meant that the image was available for longer than the duration of an exhibition. This new print would be distributed far more carefully than Reynold's mezzotint after the von Breda. This paper will demonstrate that it would be used as an affectionate gift to family and friends, a present for those Boulton wished to impress and as a mark of status. The production and distribution was nominally organised by Matthew Robinson Boulton as a present for his father but as will be seen was very much encouraged by Boulton senior. It was to be a line engraving, a slow and expensive form to undertake, but one which was considered more prestigious than mezzotint. The most sought after practitioner of the line engraving, William Sharp, was asked to undertake the commission.[32] A line engraved plate could withstand the printing of around 2,000 copies but needed very high sales to make it commercially viable.[33] This print was never intended as a commercial venture, the costs of production and distribution were borne by M.R. Boulton, a method of funding a print which was relatively common, particularly for portraits, in the eighteenth and nineteenth centuries, and which was known as a 'private plate'.[34] The private nature of the commission is not made apparent on the print, it reads as if published by Sharp as a commercial enterprise. This raises the question of how many other portrait prints were produced as private plates and have not yet been recognised as such because detailed archival evidence is not available.[35]

Boulton enlisted his friends to encourage the project: Charles Dumergue, dentist to the Royal family told M.R. Boulton that Boulton senior 'gave me to understand that he believed you had desired Mr Sharp to Engrave his portrait & that you would make him a present of it'.[36] Sharp had told Dumergue he would engrave Boulton senior for 300 Guineas, half to be paid in advance, the remainder on completion.[37] M.R. Boulton agreed terms and made an initial payment of 100 guineas in February. 'As artists are not in general men of business or accurate accomptants' he suggested the money should be handed over in person and a receipt obtained immediately.[38] This proved justified when Sharp wrote in November that he had 'immediate occasion for money' and asked for a further hundred guineas

---

[32]   Richard Sharp, 'William Sharp' in *Oxford Dictionary of National Biography online* [accessed 29 August 2008].

[33]   Tim Clayton, 'Figures of Fame: Reynolds and the Printed Image' in Martin Postle (ed.), *Joshua Reynolds: The Creation of Celebrity* (London, 2005), pp. 49–59, p. 50.

[34]   Antony Griffiths, *Prints and Printmaking: An Introduction to the history and techniques*, revised edition (London, 2004), p. 149.

[35]   I am grateful to Antony Griffiths, Keeper, Department of Prints and Drawings, British Museum, for discussion on this.

[36]   BA&H, MS 3782/13/9/103, Charles Dumergue (CD) to Matthew Robinson Boulton (MRB), 17 June 1799.

[37]   BA&H, MS 3782/13/9/103, CD to MRB, 17 June 1799.

[38]   BA&H, MS 3782/13/13/116, MRB to Charlotte Matthews, 23 February 1800.

although he had not yet completed a print.[39] Sharp worked from Beechey's portrait and his sketch, gridded for transfer to the plate, survives.[40] He suggested that the printing should take place in stages, 100 proofs printed, the plate examined, and a few days later another hundred ordered. This was 'to check the carelessness of Printers, who, often like to make quick work to earn their money the more easy and many impressions are often spoil'd'. He asked for the wording of the title and concluded with a postscript, 'The prints ought not to be sold for less than One Guineas, proofs two Guineas'. He was conscious of the limited market for the print, proposing to print only 500 copies although the plate was capable of far more. He was anxious to ensure that the status of the print (and the art of line engraving) was not compromised by setting a low price. Although he had no financial interest in its sale since it was a private plate, the status of the print was directly linked to Sharp's own status and vice versa. He aimed to sell to 'Philosophers, connoisseurs and others *who have money*', rather than in large quantities.[41]

Sharp wrote the letter quoted above in January 1801; M.R. Boulton finally replied in September with a letter brought to town by Matthew Boulton who was to consult with Sharp regarding the inscription and distribution; Boulton senior was still very much involved in the project.[42] M.R. Boulton agreed to the proposals for printing and asked for copies for distribution among his friends in the Midlands. He suggested providing a list of those in London who were to receive copies and asking them to call on Sharp as this would make distribution easier and allow his friends the opportunity to view and purchase Sharp's other works. M.R. Boulton agreed that the price of the print should not be any lower than Sharp had suggested and asked that Sharp exercise his judgement to select two of the best proofs for M.R. Boulton's own collection. He wished to avoid a verbose inscription, believing that

> A man must not effect to live in the memory of posterity by his titles but by his deeds & if the print is not sufficiently recommended by the names of the subject & the artists, with its own intrinsic merit as a specimen [...] I should despair of adding to its value by a string of titles however long.[43]

He felt F.R.S. (Fellow of the Royal Society) could be added and that Esqr. was a 'valuable appendage', although he was prepared to defer to Sharp's 'better judgement and taste upon this point'. He recognised the importance of the title, the way in which the sitter was named could immediately impart information

---

[39]   BA&H, MS 3782/13/15/32, William Sharp (WS) to MRB, 25 November 1800.

[40]   BM 1853,1210.492 British Museum Collection Database, www.britishmuseum.org/collection, British Museum [accessed 24 April 2010].

[41]   BA&H, MS 3782/13/15/33, WS to MRB, 7 or 8 January 1801.

[42]   BA&H, MS 3782/13/15/148, MRB to WS, 19 September 1801. The date on the print is 1 May 1801, but such dates are not always accurate, pers. comm. Antony Griffiths.

[43]   BA&H, MS 3782/13/15/148, MRB to WS, 19 September 1801.

which could anchor the impression given by the image. Reynold's mezzotint had been titled *Matthew Boulton Esquire*, signifying that he was a gentleman but leaving further information, such as manufacturer or connoisseur, to be interpreted graphically or through prior knowledge of the sitter. The Sharp print, which more clearly depicted Boulton as a connoisseur, would be titled *MATTHEW BOULTON/ F.R.S. & F.S.A.* The inclusion of reference to his fellowship of both the Royal Society and the Society of Antiquaries also signified acceptance by the London establishment.

The print was considered one of Sharp's best with Sir Joseph Banks among its admirers.[44] Although the print was nominally arranged by M.R. Boulton, Boulton senior was clearly involved in the choice of engraver and plans for distribution, and asked his friends to encourage the project. He had succeeded in arranging a print which was highly regarded and discussed by connoisseurs, in linking his own name with a high quality product by the finest line engraver of his time, which, although not of Boulton's own making, would help associate his products with taste and excellence. However, he did this as a privately funded venture, not one which made a profit. The print was shown to Boulton's friends at an early stage. Richard Chippindall, a Soho agent in London, saw it and wrote to Boulton, asking who the artist was, 'whether or not the plate was *his property*' and how to acquire a copy.[45] Boulton and his son began drawing up lists of people to receive copies.[46] This process of categorising friends and acquaintances was clearly a difficult one as the lists were altered. Decisions had to be made, not just who was to receive prints and who would not, but also who were to receive the more prestigious proof prints and who the 'common prints'. Boulton had long been in the habit of categorising potential customers in order to determine how much attention they should receive. At the time of a sale at Christie's James Keir wrote of sending one letter only to 'such lords, &c., as have, or pretend to have, taste' and a shorter letter to others.[47] Time and effort should be spent on people who were most likely to generate a profitable return or exert their influence in Boulton's favour.

Those to be given copies of the print included close friends like Charlotte Matthews and Charles Dumergue who had helped with its production, Lunar men like Erasmus Darwin, Samuel Galton, James Keir and Joseph Priestley for whom the prints would be a token of affection and friendship, and family members George Mynd and Ann Holbrook. The print was considered so important that it was specifically mentioned in the wills of family friend Amelia Alston and the

---

[44]   Chambers, *Banks Letters*, pp. 287–288.

[45]   BA&H, MS 3782/12/59/101, Richard Chippindall (RC) to MB, 13 March 1801.

[46]   BA&H, MS 3782/12/107/29 MB diary 1801 pp. 6, 10–11; BA&H, MS 3782/13/41/114, MRB to RC, 3 February 1802. A summary of these lists along with biographical information on recipients can be found in Loggie, 'Soho Depicted', table 1.

[47]   BA&H, MS 3782/12/65/2, James Keir to MB, 1 April 1771.

aforementioned George Mynd.[48] Long-term business associates or staff like Watt, Murdoch, John Southern, John Hodges and some of the Soho clerks were listed, for many of them it was a token of the man who had started them on their careers. Birmingham men like Dr. Carmichael, Samuel Garbett and Richard Lawrence were also to be given copies as a mark of respect, friendship or a business relationship. Established artists like Beechey, Flaxman, Longastre, Peter Rouw, and James and Samuel Wyatt were given copies, through friendship but also to enhance Boulton's status as a man who recognised and commissioned high quality art. Members of the London establishment like Lord Glenbervie and Sir Joseph Banks were given copies, again through friendship, but also to reinforce that view of Boulton as someone who mattered to important men. Copies were sent abroad, to Russia, the United States and Europe, associating Boulton's name with a high quality product across the world. The Russian Ambassador, Count Simon Woronzow was given copies when he travelled to Russia to meet the new Emperor Alexander I; Boulton told Woronzow:

> I am not so vain or presumptuous as to offer one to the Emperor; but if his Imperial Majesty should have any collection of prints in his library I should feel myself highly honoured by having that print which is marked on the back side (a proof) placed among them merely as a Specimen of good engraving.[49]

This distribution was not just about the recipients; Boulton would have been conscious that the print could have been seen in the collections of important men like the Emperor of Russia and that this would provide status and encourage business.

In spite of M.R. Boulton's suggestion that allowing people to collect copies of the print direct from Sharp would generate additional sales, Sharp argued that this would interfere with his professional engagements so Richard Chippindall, the Soho agent who had asked where to obtain a 'good' print distribution. M.R. Boulton explained that 300 proofs and 139 prints had been taken, some of which had already been distributed. Chippendall sent the plate to Soho on 5 February along with copies of the print.[50] For a commercial print the plate would have stayed with the artist or publisher who could organise the production of further copies as they were required.[51] Sending the plate to Soho meant that the Boultons could control the number of prints made and be certain that no others were taken without their knowledge. Sharp had received his fee and had no commercial interest in

---

[48] Public Record Office, PROB 11/1866 1833; PROB 11/1548 1813. I am grateful to Shena Mason for these references.

[49] BA&H, MS 3782/12/47/112, MB to Count Woronzow, 6 April 1802.

[50] BA&H, MS 3782/13/41/114, MRB, to RC 3 February 1802; BA&H, MS 3782/13/41/52, RC to MRB, 8 February 1802; BA&H, MS 3782/13/37/31, RC to MRB, 22 February 1802.

[51] Griffiths, *Prints and Printmaking*, p. 140.

the print. However, as indicated above, he did want to ensure that it was not sold at a price likely to damage his reputation or make purchasers query the prices of his other prints. Chippendall told M.R. Boulton 'You will observe the *modesty of Engravers* by his [Sharp] informing me of its being the *custom* of the Trade to leave with the Engraver *6 proofs & 6 prints* as a present for himself & Friends'.[52] Sharp retained 12 proofs and prints for sale as well as his 'presents', suggesting he did not expect it to sell in any quantity. There was confusion in the lists between prints and the proofs; recordkeeping with regard to this print was unreliable from the outset, another indication that the primary motive was not to make a profit.

People who do not appear on the distribution lists were also sent copies; the artist, Amos Green wrote that he would 'never look at it but with pleasure, or without a pleasant recollection'. As has been demonstrated, the print had multiple authors in Beechey, Sharp, Boulton and M.R. Boulton, all of whom shaped its production in some way. Green was a further author of one particular copy; he returned Boulton's daughter's copy of the print to which he had evidently been making some additions, 'I think you will approve of the lines round her Portrait of you; I have not wrote your name underneath thinking she might add some lines appropriate to her own feelings'.[53] Green had taken an exclusive copy of the print and added to it to make a unique and sentimental version, arranged by Boulton for his daughter. He also added some lines of text and left the option for Anne Boulton to add a title or inscription of her own about her father, personalising it even further. An object that had been used to raise Boulton's profile nationally and internationally was also used to create something highly personal.

These lists and evidence from letters provide an idea of how the print was distributed but cannot tell the full story; others were undoubtedly given away. Prints were to be made available for purchase through Sharp or the printsellers John Boydell or Ryland. Boydell expected to take them on M.R. Boulton's account and to return any copies not sold, suggesting that, like Sharp, he did not believe the print would sell well.[54] No evidence has been found for Birmingham printsellers carrying the print, perhaps they were expected to be purchased direct from Soho. In order to publicise the print more widely, Chippendall advertised in newspapers and magazines.[55] A copy of the print was provided to Richard Phillips, publisher of the *Monthly Magazine* for 'advertising'.[56] Phillips and the *Monthly Magazine* were known to Boulton as they had previously printed a description and illustration of the Manufactory and Phillips produced the *Public Characters* volumes which had

[52]   BA&H, MS 3782/13/37/31, RC to MRB, 22 February 1802.

[53]   BA&H, MS 3782/12/47/163, Amos Green to MB, 6 June 1802.

[54]   BA&H, MS 3782/13/37/32, RC to MRB, 25 February 1802.

[55]   BA&H, MS 3782/13/37/31, RC to MRB, 22 February 1802; BA&H, MS 3782/13/41/74, RC to MRB, 12 March 1810.

[56]   BA&H, MS 3782/13/41/74, RC to MRB, 12 March 1810.

included Boulton.[57] John Lowe, a Birmingham bookseller wrote that a customer wished a print of Boulton and asked the cost.[58] People outside Boulton's immediate circle were being made aware of the print and actively seeking to obtain a copy. Lowe's customer could have known of the print through the press advertising or through word of mouth. The print was probably being discussed, particularly in the Midlands, by those who knew Boulton personally and those who knew of him, perhaps through owning something made at Soho.

Boulton was ill and confined to bed for much of 1802, so it is likely that less work was undertaken to promote and distribute the print than had originally been intended.[59] Distribution did continue, but in a haphazard manner. In 1808, it was offered to Lord Muncaster who was collecting portraits of distinguished men of modern times for his library. He asked Boulton to sit for a painted portrait but as Boulton was too ill Sharp's print was offered as an alternative.[60] There was a long tradition of portraying a number of admired figures together, often as busts in libraries or gardens. More often this was philosophers, writers, composers, thinkers or religious figures and could include contemporary figures as in the Temple of Worthies at Stowe.[61] Muncaster's Library was more unusual in that it included a number of manufacturers. There was a growing realisation of the importance of manufacture to the country and a sense of pride in the accomplishments of British manufacturers like Boulton and Wedgwood. By the time of Boulton's obituary in 1809, it was expressed even more strongly, 'we ought not to omit paying a just tribute of applause to those who have promoted arts, industry, and commerce and diffused plenty and comfort through the realm, by cultivating science, and applying it to the useful arts of peace'.[62]

This was a print with limited sales potential, particularly at the prices Sharp suggested. Boulton's closest friends and acquaintances would have expected to be given a copy. While those of the middle rank often collected and displayed prints of those they admired, but did not necessarily know, this was a print for connoisseurs, by the best line engraver and priced accordingly.[63] It was a considered purchase; a customer would have been wealthy or particularly interested in Boulton, Beechey or Sharp. In January 1809, Chippindall produced an account which

[57]   *The Monthly Magazine and British Register*, No. XVII, Vol. III, May 1797; (Richard Phillips), *Public Characters of 1800–1801* (London, 1801).

[58]   BA&H, MS 3782/12/47/171, John Lowe to Z. Walker, 14 June 1802.

[59]   H.W. Dickinson, *Matthew Boulton* (Cambridge, 1936), p. 191.

[60]   BA&H, MS 3782/12/53/23, Lord Muncaster to MB, 23 September 1808; BA&H, MS 3782/6/137/. Copy letter, William Cheshire to Lord Muncaster, 27 September 1808.

[61]   West, *Portraiture*, p. 87; Alison Yarrington, 'Popular and imaginary pantheons in early nineteenth-century England' in Richard Wrigley and Matthew Craske (eds), *Pantheons: Transformations of a Monumental Idea* (Aldershot, 2004), pp. 107–122; 107–108.

[62]   *Monthly Magazine*, 1 October 1809, p. 330.

[63]   Nenadic, 'Print Collecting', p. 214.

showed that from 1802 to 1808, three proofs and 43 prints had been sold.[64] If Sharp and Chippindall's original figures were correct, 81 percent of the proofs and 15 percent of the prints remained undistributed.[65] Boydell's reticence in carrying them suggested that he felt they were unlikely to sell well and he had no wish to incur any risk if he stocked them, even if it meant offending Boulton. By this time his firm was in serious debt so he would have been cautious.[66] There would have been a market for the print in Birmingham and the Midlands through local pride. Chippindall's accounts show he sold the proofs at 32s. each and prints at 16s, less than Sharp had suggested and M.R. Boulton had agreed, with an income of £39.4s. The full print run of 300 proofs and 139 prints would have had a retail value of over 500 guineas at these prices which would have covered costs, but the number of copies given away meant that this print could never recover the expenditure on its production.

The cost of producing the print cannot be accurately determined; Dumergue outlined a fee of 300 guineas for Sharp with half in advance but Sharp was paid 100 guineas in November 1800 which may suggest a fee of 200 guineas was agreed. Payments were made to cover expenses but there was some dispute about these. It is not clear if any payment was made to Beechey to allow the use of his painting, no evidence has been found of one, and as Beechey's records do not survive for the time of the original painting, it is not clear how much he was paid for that and if the right to produce an engraving was included. Further expense was incurred in advertising so the venture undoubtedly made a large loss. However, it was never intended to be a commercial venture but a mark of respect, an indication of Boulton's importance and a way of highlighting that importance to a wider audience. The exhibition of Beechey's portrait at the Royal Academy in 1799 had brought it to notice. The production, circulation, sale and advertising of the print made a wider audience aware that Boulton had had his portrait painted by a significant artist and engraved in high quality. Offering the print for sale was more to do with extending awareness of its existence and making it accessible than generating income. Selling copies provided a reason for advertising that such a print had been produced and offered the possibility of display in printshops. Boulton would have been highly conscious of the potential the print offered and assisted its production by mentioning it to friends so that they encouraged M.R. Boulton.

M.R. Boulton and his son, Matthew Piers Watt Boulton continued sending out these prints for years. Sir George Chetwynd of Grendon Hall in Staffordshire wrote in 1838 to acknowledge receipt of a copy adding 'It is not only an important addition to my Collection of "Staffordshire Worthies", but most interesting

---

[64]   BA&H, MS 3782/13/41/74, RC to MRB, 12 March 1810.

[65]   The figures suggest 53 proofs and 75 prints were given away or sold through suppliers other than Chippindall.

[66]   Timothy Clayton, 'John Boydell' in *Oxford DNB online* [accessed 24 June 2007].

with reference to my Cabinet of Gems from Soho Mint [...]'.[67] M.R. Boulton's grandson offered copies to the National Portrait Gallery and Birmingham Museum and Art Gallery, stating that they had been taken from a private plate which had subsequently been destroyed by fire. He believed it was a rare print as he had not seen it in any dealer's catalogues and the copy offered to the NPG had been damaged by damp.[68] This suggests he did not have access to the large remaining stock of prints and raises the possibility that they too were destroyed in the fire.

Most manufacturers did not need to prove that they could understand, display and influence taste and fashion. It was men like Boulton and Wedgwood who needed to be able to do this in order to suggest that their products were also tasteful and fashionable. The print of Boulton by Sharp can be read as Boulton looking to show that he understood, could identify, appreciate and commission high quality work. Prints of manufacturers would not be expected to have high sales to collectors and connoisseurs but more often would have been given to friends. However, the high status of this line engraving meant that it was a print that was sought after by collectors. Having explored the market for single sheet portrait prints with Reynolds' mezzotint after von Breda, Boulton and his son considered the distribution of Sharp's print much more carefully. It suggested taste and quality in Boulton and in his manufactory and its products by association; it enhanced the whole Soho enterprise, not just Boulton's own standing.

[67]   BA&H, MS 3782/13/28/42 Sir George Chetwynd to MRB, 19 September 1838.

[68]   Birmingham Museums and Art Gallery files, letter from LBCL Muirhead, 20 December 1907. National Portrait Gallery, London (RP1532, NPG Archive) LBCL Muirhead to Cust, 10 February 1909.

Chapter 16

The Death of Matthew Boulton 1809: Ceremony, Controversy and Commemoration

Malcolm Dick

The ritual and commemoration of death are subjects that have attracted the interest of historians. Their attention has focused on changing attitudes, monuments, mourning and the deaths of the famous, especially in the Victorian period when the funeral industry in all its elaborate forms became a component of British culture and economics.[1] Much still needs to be done, particularly in exploring the business of death during the eighteenth and nineteenth centuries and the ways in which death was commemorated by different social classes, religious groups and ethnic communities. Matthew Boulton's life has been explored extensively, but the aftermath of his death has received little attention, though there are brief outlines in biographies by Samuel Smiles (1865) and H.W. Dickinson (1936), and Shena Mason has described the immediate reaction to his demise.[2] This chapter considers the ceremony, controversy and commemoration after his death through an examination of his funeral ritual and ways in which he was remembered through obituaries, texts and material culture. They reveal a great deal about how Boulton's family and associates, particularly his son and heir, Matthew Robinson Boulton

---

[1] Phillipe Ariès, *Western Attitudes toward Death from the Middle Ages to the Present* (London and Baltimore, 1974); Thomas Laqueur, 'Bodies, death and pauper funerals', *Representations*, 1/1 (1983), pp. 109–131; Alex Bruce, *Monuments, Memorials and the Local Historian* (Shaftesbury, 1997); Peter C. Jupp and Clare Gittings (eds), *Death in England: an Illustrated History* (Manchester, 1999); James Stevens Curl, *The Victorian Celebration of Death* (Stroud, 2000); John Wolffe, *Great Deaths: Grieving, Religion and Nationhood in Victorian and Edwardian Britain* (Oxford, 2000); Joseph Clarke, *Commemorating the Dead in Revolutionary France: Revolution and Remembrance 1789–1799* (Cambridge, 2007).

[2] Samuel Smiles, *Lives of Boulton and Watt* (London, 1865), pp. 477–478, 485–486; H.W. Dickinson, *Matthew Boulton* (Leamington Spa, 1999, first published 1936), pp. 192–194, Shena Mason, *The Hardware Man's Daughter: Matthew Boulton and his 'Dear Girl'* (Chichester, 2005), pp. 139–141.

(1772–1842),[3] wished him to be remembered. They also throw light on broader issues, such as attitudes to death and the deceased, social status and the cult of celebrity. This chapter is substantially based on letters, accounts, lists, drafts of obituaries and epitaphs in the papers of Robinson Boulton.[4] The Boulton family papers were, as Smiles noted, 'selected and arranged' by Robinson Boulton during his lifetime.[5] These papers provide a detailed but necessarily partial picture of the aftermath of the funeral. There are also published sources relating to the legal dispute between Robinson Boulton and the undertaker, George Lander, following the funeral. Boulton memorabilia are also examined, including the medals which were struck to honour him and the funeral monument which survives in St Mary's Church, Handsworth, Birmingham.

Death can be explored as a social and cultural phenomenon by addressing a number of questions. Firstly, what processes and rituals take place when the deceased is transmitted from their place of death to a final resting place? Secondly, how do family, friends, employees and other associates of the deceased react? Thirdly, how is the individual commemorated or remembered: via a funeral, obituaries, memoirs or physical objects, such as monuments? Fourthly, what do these activities reveal about the individual, their family or the wider society in which they lived? Julie Rugg has illuminated the culture of death and remembrance in late eighteenth and early nineteenth-century England.[6] In so doing, she has examined attitudes, which help in understanding the ways in which Boulton was commemorated after his death. She argues that during the Enlightenment, attitudes to death changed compared to previous periods. The process of death became a phenomenon that could be rationally understood: it could be observed, explained by medical science and eased by opiates.[7] Superstition and ignorance were progressively removed from the experience of death and there was a reduction of emphasis on hell and damnation in approaches to the afterlife.[8] Boulton's funeral service and the location of his memorial in St Mary's Church provide a religious frame for the ceremony, controversy and commemoration after his death. However, the dynamics of remembrance reveal Robinson Boulton's secular concern with his personal status and reputation as well as the long-term standing of his enterprising, industrialist father.

Matthew Boulton died on 17 August 1809 shortly before his 81st birthday, following a long illness which the post-mortem attributed to stones and pustules in

---

[3]   There is no satisfactory study of Robinson Boulton, but see Eric Robinson, 'Training captains of industry: The education of Matthew Robinson Boulton (1770–1842) and the younger James Watt (1769–1848)', *Annals of Science*, 10/4, 1954, pp. 301–313.

[4]   Archives of Soho collections in Birmingham Archives and Heritage (BA&H).

[5]   Smiles, p. vi.

[6]   Julie Rugg, 'From Reason to Regulation: 1760–1850' in Jupp and Gittings, *Death in England*, pp. 202–229.

[7]   Ibid., p. 203.

[8]   Ibid., p. 204.

his kidneys.[9] His funeral was held on Thursday 24 August 1809 and he was buried in a vault beneath the St Mary's[10] These are the facts of his death and burial, but the aftermath of his death is particularly interesting because of the attempts to preserve, celebrate and enhance his memory. Robinson Boulton, according to Smiles 'entertained the highest regard for his father's memory' and played an especially significant role in managing his father's reputation in the early nineteenth century.[11] This discussion is, in part, an exploration of how the son cemented the reputation of the father. An analysis of a number of incidents and activities provides the framework for considering the ceremony, controversy and commemoration which followed Boulton's death. Firstly, there was the funeral procession and service.[12] The planning of the funeral, its rituals and the subsequent public argument between Robinson Boulton and the undertaker, provide ways of exploring the importance of the event and its aftermath. Secondly, there are the contrasting ways in which two of Boulton's associates in the Lunar Society, James Watt and James Keir remembered him. Whereas Watt provided a dutiful and affectionate portrait, Keir was more critical. Thirdly, Boulton's death was remembered by physical objects, particularly, the monument which was commissioned for St Mary's Church. Although this memorial was located in a religious building, it transmitted secular messages, extending the ways in which other forms of remembrance represented the man. Finally, the aftermath of the commemoration of Boulton, including the events following the 100th anniversary of his death in 1909, provide evidence of whether Robinson Boulton was successful or not in creating a long-term legacy for his father.

The Funeral

Funeral rituals were shaped by the intellectual environment of the Enlightenment, but they also reflected social and economic changes. Funerals provided opportunities for display and consumption. Rugg notes that this was a phenomenon associated with the rise of 'a self-conscious middle class, eager to demonstrate its importance' and the commercialisation of funerals through the growth of the undertaking trade

---

[9]   BA&H, MS 3219/4/33/33, James Watt Jr (JWJ), to James Watt (JW), 17 August 1809; MS 3219/4/33/34, JWJ to JW, 18 August 1809.

[10]   Funeral Procession of the late Matthew Boulton Esq., in *Documents relative to an Investigation of the Manner in which the Funeral of the Late Mathew Boulton, Esq. was furnished* (Birmingham, 1811). BA&H, MS 3782/13/37/107, Matthew Boulton, Biographical Memoir, Decease, Funeral, Prints, Medals etc.

[11]   Smiles consulted the Boulton archives at Tew Park through the kindness of Robinson Boulton's son, Matthew Piers Watt Boulton, who probably provided Smiles with the quotation: see Smiles, pp. v–vi.

[12]   BA&H, MS 3219/4/49/88, John Furnell Tuffen (JFT) to JW, 29 August 1809.

'which both met and fed demand for funeral display'.[13] Boulton was an important member of the rising middle classes and a dominant figure in industrial society and Birmingham's civic life; his funeral enabled Robinson Boulton to demonstrate the family's affluence and celebrate his father as a national and local figure. The growth of a funeral industry was also important, which involved the production of mourning dress, funeral silk, coffins and coffin furniture. Birmingham, a metal manufacturing town, became a centre for the manufacture of decorative coffin plates and handles in the late eighteenth century.[14] Funerals could demonstrate family wealth and respectability: by 'affirming gentility through the correct adherence to mourning etiquette; the funeral could reinforce an individual and family's place in society; and expenditure on funerary monuments comprised conspicuous consumption indicating financial worth'.[15] Again, this was a dimension that was central to Boulton's funeral, as revealed in the legal dispute between Robinson Boulton and the undertaker over the quality of funeral silk which dressed the mourners and the interior of St Mary's.[16] Ritual, moreover, provided a medium through which individuals and families could demonstrate their place in society. Death, according to Rugg, could 'elicit a turnout from neighbours and colleagues in forming part of the procession and lining the streets of the procession route'. The numbers of participants and size of the audience provided, for 'middle-class civic worthies ... an indicator of civic worth'.[17] Boulton's funeral was a shared event which included family, friends and associates, workers employed at the Soho manufactories and observers who watched the conveying of the cortège.[18] It projected Boulton as an important figure in the local community, a respected employer and a man who was loved by his family and friends.

In the early nineteenth century, a precedent had been established for creating elaborate public funerals to honour celebrities. The best-known example was Vice-Admiral Lord Nelson's huge funeral in London in January 1806.[19] Boulton was part of the commemoration process through his production of several thousand medals for distribution to the marines and seamen who participated in the Battle

---

[13]   Ibid., p. 221.

[14]   Samuel Timmins, *Birmingham and the Midland Hardware District* (London, 1967; reprint of 1866 edition), pp. 292–293 and W.C. Aitken, 'Coffin Furniture', in Timmins, *Birmingham*, pp. 704–708.

[15]   Rugg, p. 221.

[16]   BA&H, MS3782/13/100, Boulton v.Lander Correspondence: Funeral of Matthew Boulton Esq. (hereafter BvLC).

[17]   Rugg, p. 222; See also Leonore Davidoff and Catherine Hall, *Family Fortunes: Men and Women of the English Middle Class 1780–1850*, 2nd edition (London and New York, 2002), p. 103.

[18]   BA&H, MS 3219/4/49/88, JFT to JW, 29 August 1809.

[19]   http://www.portcities.org.uk/london/server/show/ConNarrative.36/chapterId/306/The-state-funeral-of-Lord-Nelson-59-January-1806.html (accessed 29 December 2010).

of Trafalgar where Nelson died.[20] Robinson Boulton, it seems, was keen to provide a funeral for his father, a captain of industry, which would emulate, in a local context, the ceremony for an admiral of the fleet. Boulton wished to be buried in Handsworth but had not 'left any directions respecting his funeral'.[21] On 18 August, Robinson Boulton entrusted the arrangements to an undertaker, George Lander of Birmingham, indicating that the funeral should be 'on an extensive scale' and 'furnished in the handsomest manner avoiding ostentation'.[22] It was arranged for Thursday 24 August.[23] Making the arrangements within a week, tested Robinson Boulton's resources and caused him considerable anxiety. His papers contain a substantial correspondence with his father's friends and associates, inviting them to the funeral, several of whom, including Watt who was in Scotland, could not come.[24] There are also successive drafts of papers for the planned funeral procession as invitees signalled their ability or inability to attend.[25] A memorial medal was also rapidly produced for distribution to Soho workers who were present.[26] Robinson Boulton ordered a black mourning coat from London, which had not arrived by the date requested.[27] Moreover, the bier (moveable platform) for the coffin was built overnight on 23 and 24 August and only delivered to Soho House at 11.00am on the day of the funeral.[28] Robinson Boulton was not only mourning for his father, planning the funeral and supporting his invalid sister Anne,[29] but trying to create an event which celebrated and constructed his father's memory as an industrialist, benevolent employer and leading figure in the community. The stress he was under was considerable and it is this, perhaps, which explains his subsequent dispute with the undertaker.

---

[20]   Nicholas Goodison, *Matthew Boulton's Trafalgar Medal* (Birmingham, 2007), p. 13.

[21]   BA&H, MS 3219/4/33/33–4, JWJ to JW, 17 and 18 August 1809.

[22]   Declaration of Mr Matthew Robinson Boulton, in *Documents relative to the Late Matthew Boulton,* p. 9.

[23]   BA&H, MS 3219/4/33/35, JWJ to JW, 22 August 1809.

[24]   BA&H, MS 3782/13/37/48–100, Matthew Boulton, Biographical Memoir.

[25]   BA&H, MS 3782/13/37/101–103, Matthew Boulton, Biographical Memoir.

[26]   BA&H, MS 3219/4/49/88, JFT to JW, 29 August 1809; MS 3782/13/100/2–3, BvLC, Tokens; Dickinson, p. 193. This medal had a simple design, obverse: 'Matthew Boulton died August 17th 1809 aged 81 years'; reverse: 'In memory of his obsequies Augst 24th 1809' within a closed palm wreath. Birmingham Museums and Art Gallery (BM&AG) hold examples of the medal. Boulton's medals are featured in Sue Tungate, 'Matthew Boulton and the Soho Mint: Copper to Customer' (University of Birmingham, unpublished PhD thesis, 2011).

[27]   BA&H, MS 3782/6/138/122, William Cheshire to John Glynn, 21 August 1809. Cheshire was Robinson Boulton's agent.

[28]   B&AH, MS 3782/13/100/1, BvLC, MS 3782/13/100/1. Memorandum about the Bier.

[29]   BA&H, MS 3219/4/33/33–35, JWJ to JW, 17, 18 and 22 August 1809.

The funeral's importance is revealed by the size and scale of the procession and the service that followed. For James Watt Jr, the former was 'well-conducted, and the ceremony awful and impressive'.[30] The procession left Soho House at 12.00 noon to cover the journey of one mile to St Mary's.[31] A diagram survives which indicates the size of the convoy which carried Boulton to his last resting place. It was preceded by four mutes (professional mourners) who were followed by eight carriages carrying clergy and friends of the deceased, 24 under-bearers and the body on a bier carried by 'ten of the oldest men who had been from 30–50 years in his service',[32] and supported by eight gentlemen pallbearers. Robinson Boulton followed as chief mourner and then 600 employees of the Soho Manufactory and Foundry, including section heads and 500 male and 60 female workers. Servants, a family carriage and 14 private carriages made up the rest of the procession and two mutes walked behind at the rear. Twelve truncheon-men, six on each side, provided a ceremonial boundary between the procession and those who watched the funeral on the streets of Handsworth.[33] For mourners, the Anglican service was a multi-sensual experience: an oral and aural encounter which married words and music in a solemn visual setting where the church interior and the bier were bedecked with mourning silk. The service included a psalm, a lesson and Handel's *Funeral Anthem,* performed by the Choral Society of Birmingham, an indication of Birmingham's musical activity, which Boulton had helped to foster.[34] Immediately before the corpse was carried from the church, a verse from a Handel oratorio was 'sung by a Birmingham lad' who was 'considered to possess one of the finest voices in England'.[35] This was a particularly telling moment as the boy sang:

> Angels ever bright and fair,
> Take, oh take him to your care!
> Speed him in his heav'nly flight
> Take him to the realms of light

Boulton was not only buried according to the obsequies of the Church of England, he was transmitted from an Enlightenment earth to an Enlightenment heaven.

This ritual of remembrance was a shared experience which extended beyond the mourners to the Soho workforce and local inhabitants. James Furnell Tuffen claimed that the road to Handsworth Church 'was lined with spectators on foot, on horseback and in carriages, to the number, it is said of at least 10,000 persons. ... There never was perhaps a public funeral attended by so many real and respectful

---

30   BA&H, MS 3219/4/33/36, JWJ to JW, 25 August 1809.

31   BA&H, MS 3219/4/49/88, JFT to JW, 29 August 1809.

32   BA&H, MS 3219/4/33/36, JWJ to JW, 25 August 1809.

33   Funeral Procession, in *Documents relative to the Funeral.*

34   Mason, p. 70.

35   BA&H, MS 3782/13/37/101, Matthew Boulton, Biographical Memoir, Order of the Funeral; MS 3219/4/49/88, JFT to JW, 29 August 1809.

mourners'.[36] The conduct of the Soho workforce was also singled out for comment. Watt Jr. described their behaviour as 'proper and respectful' throughout and after receiving their memorial medal and drinking to 'the memory of their departed benefactor standing & in silence, they all repaired to their respective homes, and not a Soho man was to be seen upon the road for the remainder of the day'.[37] A local paper announced: 'Never have we witnessed a more affecting ceremony than the last sad tribute of respect paid with equal solemnity and sorrow to the remains of this excellent man'.[38] The funeral cemented respect for Boulton's memory in the minds of local inhabitants and workers and recording the event by the memorial medal and in newspaper reports contributed to transmitting his reputation. Robinson Boulton provided an appropriate conclusion to his father's life.

However, he was unhappy, despite the positive comments of Watt Jr., Tuffen and the local press. His discontent added controversy to the commemoration and probably influenced his desire to remember his father in additional ways. Through his solicitor, George Barker, he accused the undertaker, Lander, of making 'many distressing mistakes' and creating 'much inconvenience' during the ceremony.[39] During 1810 and 1811 a public conflict between the two men was played out in a war of letters, depositions, counter depositions, letters between Barker and Lander's attorney, W. Whateley, published pamphlets and a legal case, Lander v. Boulton.[40] Robinson Boulton's agent, William Cheshire provided critical comments about the quality of funeral silk which was used for the hatbands of mourners.[41] Robinson Boulton refused to pay Lander's bill of £544.17s.2d. in full, by holding back just over £54.[42] Others added their voice, including Rev. T.L. Freer, the Rector of Handsworth, who had conducted the service; he complained about the 'very inferior quality' of silk used to dress the pulpit.[43] Robinson Boulton

[36] BA&H, MS 3219/4/49/88, JFT to JW, 29 August 1809.
[37] BA&H, MS 3219/4/33/36, JWJ to JW, 25 August 1809.
[38] BA&H, MS 3219/4/49/88, JFT to JW, 29 August 1809.
[39] Declaration of Mr George Barker in *Documents relative to the Funeral*, p. 10.
[40] B&AH, MS 3782/13/100/1–144, BvLC. There were four publications: *Documents relative to an investigation of the manner in which the funeral of the late Matthew Boulton, Esq. was furnished* (Birmingham, 1811); George Lander, *An Answer to the Pamphlet of Matthew Robinson Boulton, Esq. of Soho, Relative to his father's funeral* (Birmingham, 1811); W. Whateley, Attorney, *A Reply to a Letter of Mr George Barker, Attorney, contained in Documents relative to an Investigation of the Manner in which the Funeral of the Late Matthew Boulton, Esq. was furnished* (Birmingham, 1811); George Barker, *An Answer to the letter of Mr W. Whateley* (Birmingham, 1811).
[41] Declaration of Mr William Cheshire, in *Documents relative to ... the funeral*, pp. 11–14.
[42] Lander, *Answer to the Pamphlet of Matthew Robinson Boulton*, p. 4.
[43] Declaration of the Rev. T.L. Freer, in *Documents relative to ... the funeral*, p. 53.

claimed that he had been overcharged. However, Lander disputed all of Robinson Boulton's claims[44] and eventually, he was paid in full.[45]

How can we interpret what appears to have been an unseemly and unnecessary public dispute which did not appear to enhance Robinson Boulton's reputation? As suggested before, he was under considerable stress before the funeral. But there was more: his reputation and that of his father were at stake. He believed he had to protect his father's memory and respect the mourners: 'in adopting any other course, he should be guilty of a dereliction of duty, and betray a culpable indifference to the opinion of those who had attended the remains of his father to the grave'.[46] Respect had also to be demonstrated to 'the memory of the deceased'.[47] In other words, it was about the integrity of Robinson Boulton's character and the repute in which he was held as the keeper of his father's memory. Despite the huge expenditure on the funeral,[48] in his own eyes, and, so it appears from the depositions of others, he had not provided an adequate farewell. His perceived humiliation over the quality of funeral silk may have propelled him to commemorate his father in other ways: through a published memoir, funeral monument and a new issue of medals.

The Memoir

Another form of remembrance is written text in the form of obituaries and memoirs. Little attention has been paid to these forms of evidence by historians, though collections of the obituaries of notable individuals have been published.[49] Obituaries provide an account and interpretation of the deceased's life and significance and by the time of Boulton's demise, newspapers recorded the death of individuals who made an impact on local or national life. There are manuscript drafts of obituaries which indicate the care which Robinson Boulton took to choose the right words to commemorate and celebrate his father's life.[50] Several obituaries were published in the local and national press.[51]

---

44   Lander, *Answer to Matthew Robinson Boulton.*

45   Dickinson, p. 193.

46   *Documents relative to ... the funeral*, p. 23.

47   Ibid., p. 23.

48   Expenditure also included coffin furniture, church and burial fees, the tolling of mourning bells and funeral biscuits and cakes for the mourners, BA&H, MS 3782/13/37/99–100, 107–110, Matthew Boulton, Biographical Memoir.

49   For example, Andrew Sanders and Ian Brunskill, *Great Victorian Lives: An Era in Obituaries* (London, 2007).

50   BA&H, MS 3782/13/37/98, Matthew Boulton, Biographical Memoir, Matthew Robinson Boulton. (MRB) to Ambrose Weston, 18 August 1809.

51   BA&H, MS 3219/4/33/37, JWJ to JW, 28 August 1809; BA&H MS 3782/13/37/150.

But Robinson Boulton wanted more than obituaries to commemorate his father's life. After Boulton's death, he asked James Watt and James Keir to write accounts of his father's life. Keir and Watt were slightly younger associates of Boulton since his early days as a manufacturer.[52] Watt had been unable to attend the funeral,[53] but within a month he had written a detailed biography of Boulton's life and importance.[54] Keir had attended the funeral and served as a pallbearer.[55] Subsequently, Robinson Boulton contacted Keir seeking an account of his father, which Keir wrote on 3 December 1809.[56] In a separate letter, Keir noted that Robinson Boulton was planning on producing a published account of Boulton's life.[57] He probably had in mind a book which was similar to the celebratory publication that Keir produced in 1791 to commemorate the death in 1789 of a fellow Lunar Society associate, Thomas Day.[58] Watt's and Keir's accounts were never published in Robinson Boulton's lifetime; Dickinson reproduced part of Keir's and the whole of Watt's in his biography, and fine copies of both were published by the City of Birmingham School of Printing in 1943 and 1947.[59] The question arises, why were these accounts not published by Robinson Boulton?

The texts themselves and the context in which they were written provide a probable answer. These two memoirs have been briefly discussed by David Miller as part of his assessment of how Boulton's reputation has changed since his death.[60] Watt's memoir is a description of Boulton's achievements and an evaluation of their relationship after 1768. He traced Boulton's career as a manufacturer of steam engines and coins and drew attention to his ingenuity and ability to translate ideas into practice:

---

[52]   Barbara M.D. Smith, 'Keir, James (1735–1820)', *Oxford Dictionary of National Biography*, online edition (accessed 29 December 2010); Jennifer Tann, 'Watt, James (1736–1819)', *Oxford Dictionary of National Biography*, online edition (accessed 29 December 2010).

[53]   BA&H, MS 3782/13/37/94, JW to JWJ, 22 August 1809. MS 3782/13/37/95: JWJ to MRB, 23 August 1809.

[54]   BA&H, MS3782/13/37/111, Biographical Sketch from James Watt, 17 September 1809.

[55]   BA&H, MS3782/13/37/68, James Keir (JK) to George Lander, 22 August 1809.

[56]   BA&H, MS3782/13/37/112, Memorandums for the Memoir of Matthew Boulton, 3 December 1809.

[57]   BA&H, MS3782/13/37/112, JK, Explanation of his connection with M. Boulton, 3 December 1809.

[58]   James Keir, *An Account of the Life and Writings of Thomas Day, Esq.* (London, 1791).

[59]   Dickinson, pp. 70–72, 203–208; *Memoir of Matthew Boulton by James Watt* (Birmingham, 1943); *Memoir of Matthew Boulton by James Keir* (Birmingham, 1947).

[60]   David Miller, 'Scales of Justice: Assaying Matthew Boulton's Reputation and the Partnership of Boulton and Watt', *Midland History*, 34/1, Spring 2009, pp. 61–62.

Mr Boulton was not only an ingenious mechanick, well skilled in all the
practices of the Birmingham manufacturers but possessed in a high degree the
faculty of rendering any new invention of his own or others useful to the publick
by organising & arranging the processes by which it could be carried on.[61]

His account also noted the personal encouragement that Boulton provided: 'Mr
B's. active & sanguine disposition, served to counterbalance the despondency &
diffidence which were natural to me' and the quality of his relationships with all of
those with whom he was connected:

> To his family he was a most affectionate parent. He was steady in his friendships,
> hospitable & benevolent to his acquaintances & indeed I may say to all who
> came within his reach who were worthy of his attention & to sum up humane &
> charitable to the distressed.[62]

Watt provided an extension to the positive accounts in the obituary notices,
but Keir's account was not so celebratory. Miller notes that Keir 'believed that
Boulton's willingness to entertain visitors drew him away from the business of
making money',[63] but Miller does not refer to Keir's most critical comments. Keir
drew attention to Boulton's capacity to flatter the rich and powerful: 'If we agree
with Horace, *Principibus Placuisse virisnon ultima laus est*, few men have been
more entitled to this commendation'. In translation, this Latin phrase means: 'to
please great men is not the last degree of praise'. Moreover, he did not see Boulton
as an original thinker, but a man who absorbed the ideas of others: 'It cannot be
doubted that he was indebted for much of his knowledge to the best preceptor, the
conversation of eminent men'.[64] Keir named and briefly described the importance
of 10 of these individuals: Erasmus Darwin, William Small, James Ferguson, John
Whitehurst, William Withering, John Smeaton, Joseph Black, Joseph Priestley,
Thomas Day and Benjamin Franklin, but interestingly, not Watt. In fact, Watt was
not mentioned at all in Keir's document. Keir's account is polite, but at best it
damns Boulton with faint praise and in terms which were very different from the
vocabulary Keir had used to describe Day in 1791. However, there is more to
Keir's account of Boulton than this brief biography. In an accompanying letter
to Robinson Boulton, he took strong exception to the way in which Watt had
portrayed him. Watt's memorandum described how Keir: 'being disengaged from
other Business, and Mr B. being obliged to be frequently absent, Mr K. gave his
assistance in the general superintendence of all the business at Soho during which

---

    [61]   BA&H, MS 3782/13/37/111, Biographical Sketch from James Watt, 17 September
1809.
    [62]   Ibid.
    [63]   Miller, p. 61.
    [64]   BA&H, MS 3782/13/37/112, JK, Memorandums for the Memoir of Boulton.

he made many valuable arrangements, & gave other assistance'.[65] Keir objected to these remarks and launched into a four-page criticism of Watt's account with a detailed commentary of the nature of his business relationship with Boulton. He claimed that he never had a general superintendence of the Soho business, but only acted in this role for a few weeks to supervise the 'execution of the second order for silk reels'.[66] Secondly, he noted that his role was small, but it required constant attention. Thirdly, Boulton had asked him to be in partnership with him, so in anticipation, he gave up his partnership in a Stourbridge glass works and moved to Winson Green, which was closer to Soho. In addition, Keir became aware that the business was facing financial problems and so he did not complete the partnership with Boulton. As a result, Keir claimed, he suffered financially, though he received a share of the silk reels profits at a later date of about £200.[67] Finally, Keir objected to Watt's implication that he was 'an agent or servant under Mr B'.[68] His anger with Watt is clear: he could not agree to a publication which referred to his connection with Soho in the way that Watt suggested. The letter revealed tension between Keir and Boulton and Keir and Watt. Keir did not want his reputation to be associated with a financially weak business or to be presented as a subordinate of the two men. With two of Boulton's surviving colleagues in disagreement about how he should be remembered, his son never proceeded with the literary celebration of his father, as Watt Jr. did after his father died in 1820.[69] Robinson Boulton concentrated on producing physical and therefore lasting remembrances of Boulton, including a range of medals and, most importantly, the Boulton Memorial.

The Monument

In 1813, Matthew Boulton's funeral monument was installed as a wall tablet in St Mary's Church in Handsworth.[70] As a form of 'text' as Alex Bruce points out, funeral monuments have various meanings.[71] For the living, those who create monuments for their own remembrance after their own death are, according to Lewis Mumford, part of 'a quest for vicarious immortality'.[72] For their descendants and those who look at monuments, they shape 'memories, attitudes, perceptions,

---

[65]   BA&H, MS 3782/13/37/112, JK, Explanation of his connection with Boulton.

[66]   Ibid.

[67]   Ibid.

[68]   Ibid.

[69]   Miller, pp. 66–68.

[70]   George. T. Noszlopy, edited by Jeremy Beach, *Public Sculpture of Birmingham including Sutton Coldfield* (Liverpool, 1998), pp. 67–68.

[71]   Bruce, p. 5.

[72]   Ibid., p. 5.

Figure 16.1   John Flaxman's monument to Matthew Boulton in St Mary's
              Church, Handsworth, c. 1813

Source: Parochial Church Council of St Mary's Handsworth, Birmingham. Photograph by
Jerome Turner.

even future actions'.[73] They therefore raise issues about how to interpret them. The meanings conveyed by any monument depend upon the interaction of several messages including the intentions of the commissioner and the artist, its setting and surroundings, the design and the materials which compose it and any inscription it contains.[74] Meaning is also constructed by those who look at the monument in the years after it was installed. Robinson Boulton commissioned the monument from John Flaxman in 1810,[75] after he had received Keir's less than celebratory account of his father. Flaxman was a fashionable sculptor of monuments to commemorate the great and the good, and he had been commissioned to produce the huge tomb of Lord Nelson in the crypt of St Paul's.[76] Moreover, Flaxman was known to Robinson Boulton as he had worked for his father for several years.[77] He quoted 120 guineas for a bust of Boulton, but, in total, the cost of the monument was 300 guineas which Robinson Boulton paid in 1813.[78] This was substantially more than the sum paid by the Wedgwood family (£86.12s.2d.) in 1803, for a simple bust portrait of Boulton's fellow West Midlands industrialist, Josiah Wedgwood.[79] It is arguable that Robinson Boulton was engaged in a game of monumental one-upmanship when he commissioned the sculpture to commemorate his father. Rugg suggests that Flaxman's funeral sculptures were imbued 'with a gentle Christian piety' and provided 'the most telling image of death' in this period.[80] His monuments were often carved using foliage, columns, ministering angels and weeping mourners in flowing garments.[81] This neo-classical iconography departed from the portrayal of skulls, skeletons and shrouds in sixteenth and seventeenth-century shrines of the dead towards an Enlightenment respite from the terrors of death and physical decay. In his monument to Boulton, Flaxman also moved away from the gentle piety of many of his sculptures by creating an overtly secular object. Broadly, his design contained four elements: a bust of the deceased, neo-classical motifs, an image of the Soho manufactory and an epitaph. The marble monument measures 128 cm from the top to the bottom, whilst the image of Boulton within this frame is 57 cm high.[82] George Noszlopy's description summarises its features:

---

73   Ibid., p. 5.

74   Ibid., p. 5.

75   BA&H, MS 3782/13/10/60: Matthew Robinson Boulton Correspondence, Folder F, 1787–1818, Letters 1809–1811, John Flaxman (hereafter JF) to MRB, 29 December 1810.

76   Sarah Symmons, 'Flaxman, John (1755–1826)', *Oxford Dictionary of National Biography*, online edition (accessed 29 December 2010).

77   Noszlopy, p. 68.

78   Ibid., p. 68; Nicholas Penny, *Church Monuments in Romantic England* (New Haven and London, 1977), p. 177.

79   Ibid., pp. 177, 223 note 13.

80   Rugg, p. 204.

81   Ibid., pp. 205–207.

82   Noszlopy, p. 67.

The classically inspired monument shows Flaxman's love of simplicity; the fine bust of Boulton in contemporary dress is set within a tondo below which are two winged baby angels. One of them holds a lighted torch, an emblem of immortality, and the other holds a topographical low relief of Boulton's famous Handsworth factory, inscribed SOHO, together with a branch of olive, an emblem of the peaceful arts.[83]

The location of the Boulton monument conveys meanings to contemporary observers which may not have been apparent in the early nineteenth century. Today the monument is set in the sanctuary of a church which was substantially reconstructed in the nineteenth century. Since the erection of the Boulton monument, the sanctuary was rebuilt and other monuments were placed close to it, including one to William Murdock and, in an adjacent chapel, a huge statue of James Watt.[84] A visitor to Handsworth Church is from one standpoint, viewing a mortuary chapel dedicated to distinguished residents in late eighteenth- and early nineteenth-century South Staffordshire. An early print of the church in Smiles' biography showed a larger space with fewer monuments.[85] Given the relative absence of memorial clutter in the early nineteenth century, the observer would have been more conscious then of the impact of Boulton's shrine than in the twenty-first century. It does, however, remain a striking image, despite the much larger statue to of Watt. There is one other point about the location. It is in a church, close to the altar and forms a visual component of the messages that Anglican ceremonies would convey, including the role of the church in sanctifying rites of passage, such as the funeral service and in supporting the established order, in this case celebrating Boulton's status as a leading local figure. From the observer's standpoint, the memorial's location next to the altar suggests an intimate connection between the ritual and culture of the church and Boulton's reputation at the time of his death and thereafter.

However, the design and inscriptions on the memorial do not convey a religious message: the iconographical language and the words of the epitaph say something else. The angels are one of only two religious representations on the monument, but they resemble chubby classical cherubs rather than spiritual beings. They are located at the base of a classical column topped by a bust of Boulton, who looks more like a proud and austere Roman general than a mortal man about to meet his maker. The angels also hold an image of Boulton's most important monument in his lifetime, the Soho Manufactory. They are clearly in a supporting role, literally and metaphorically, to the secular representations of Boulton and his factory. The text of the epitaph also conveys an essentially non-religious message. Though

---

[83]   Ibid., p. 68.

[84]   Noszlopy, pp. 68–69; Penny, pp. 182, 198–201. In 1824, James Watt Jr paid Sir Francis Chantrey £2,000 to produce the statue of his father, ibid., p. 198.

[85]   Smiles, p. 478.

quoted by Smiles and Dickinson in their biographies, it has not been interpreted.[86] When read, the words provide a rhythmical incantation, asserting Boulton's intellect, taste, enterprise, benevolence and honourable and esteemed reputation in Britain and throughout the world:

Sacred to the Memory of
Matthew Boulton F.R.S.
By the skilful exertion of a mind turned to philosophy and mechanics,
the application of a taste correct & refined,
and an ardent spirit of enterprize,
he improved, embellished, and extended
the arts & manufactures of his country;
leaving his establishment of Soho a noble monument of
his genius, industry and success.
The character his talents had raised, his virtues adorned and exalted.
Active to discover merit, and prompt to relieve distress.
His encouragement was liberal, his benevolence unwearied.
Honoured and admired at home and abroad.
He closed a life eminently useful the 17th August 1809 aged 81.
Esteemed, loved and lamented.

The epitaph also contains the second religious representation on the memorial, the word 'sacred', but it is the only one with a religious meaning in 110 words of text. Boulton is described as a Fellow of the Royal Society, interested in science and technology, a successful businessman, improver of arts and manufacturing and a virtuous man who recognised talents in others, but there is no mention of any religious qualities. Despite its location close to the altar of the church, the memorial is an overtly secular monument which celebrated Boulton as a *savant* and *fabricant,* a thinker and a doer and a benevolent and cultured Enlightenment figure.[87] This was the message which Robinson Boulton promoted and Flaxman executed. The patron and the artist selected images and text to create an enduring stone obituary to Boulton which continues to celebrate his memory in the present.

The Aftermath

How effectively did Robinson Boulton sustain the cultural memory of Matthew Boulton? In 1819, he produced a set of medals to commemorate the tenth

---

[86]   Ibid., p. 477; Dickinson, p. 193.

[87]   Peter Jones, *Industrial Enlightenment: Science, Technology and Culture in Birmingham and the West Midlands 1760–1820* (Manchester, 2008) discusses Boulton's significance as a regional Enlightenment figure.

anniversary of his father's death.[88] Unlike the rapidly produced item which was minted in 1809, this example was larger, had a more elaborate design and included Latin wording.[89] In 1820, it was distributed to Boulton's close associates and admirers in the arts, engineering, politics and science, including, Charles Babbage, Sir William Beechey, John Flaxman, Lord Liverpool and John Rennie.[90] It was, in effect, a medal for the middle and upper classes to remind them of their enhanced reputations because of their association with the great man. Robinson Boulton must have been gratified by the recipients' positive responses. Flaxman's reply was particularly effusive:

> I shall preserve it with particular respect and regard, both on account of the giver and the person represented, the portrait is alive! and the arrangement noble, the composition so simple and truly elegant that they demonstrate Art and Mechanics directed by the creation of Filial piety will produce a memorial not equalled by the wealth of nations or the power of kings ...[91]

By 1820, Robinson Boulton had probably felt he had wiped out the perceived stain on his reputation of the poor quality mourning silk and his failure to publish a substantial account of his father based on the words of Watt and Keir. However, before long, Boulton's reputation began to suffer in comparison with that of his partner, Watt.[92] The 1830s and 1840s marked the beginning of 'the cult of Watt as a heroic inventor'.[93] Though mid and late nineteenth-century texts continued to record Boulton's importance as an industrialist, he emerged as a 'facilitator' to Watt.[94] Miller attributes this to the 'filial project' of Watt Jr. (1769–1848), which he claimed, was in marked contrast to Robinson Boulton, whose enthusiasm for their joint management of the firm of Boulton and Watt faded and led to the end

---

[88]   BA&H, MS 3782/13/37/117, Matthew Boulton, Biographical Memoir, List of persons to whom it is intended to present medals of the late Matthew Boulton Esq.; BM&AG hold copies.

[89]   Obverse: 'Matthaeus Boulton' with a bust of Boulton; Reverse: 'Inventas avt qvi vitam excolvere per artis' (who embellished life through the skills they discovered) in an open laurel leaf with single-branched stem. 'Patris amicis M.R.B.' (to the friends of his father, M.R.B.), was engraved on the edge. BM&AG hold copies.

[90]   BA&H, MS 3782/13/37/117–45, Matthew Boulton, Biographical Memoir, List of persons ... and various letters from recipients.

[91]   BA&H, MS 3782/13/37/123, Matthew Boulton, Biographical Memoir, JF to MRB, 15 November 1820.

[92]   Miller, pp. 58–76.

[93]   Ibid., p. 60.

[94]   Ibid., pp. 66–68.

of their partnership in 1840.[95] Perhaps this is a little unfair given what Robinson Boulton had achieved by 1820, but Watt Jr. worked particularly hard to sustain and shape his father's memory by engaging James Patrick Muirhead as the 'family historian', which led to three substantial publications about his father, culminating in *The Life of Watt* of in 1858.[96] Boulton was not so well served and according to Muirhead's daughter, Beatrix Sturt, his descendants were happy to leave matters as they were:

> at the time when my father wrote his *Life of Watt* the Boulton family did not wish that any publicity should be given to the life and work of Watt's great partner. They considered that the few notes and references in my father's book to be all that need to be told on that subject.[97]

In 1909, at the time of the 100th anniversary of his death, an attempt was made to revive Boulton's reputation. A scrapbook of letters and printed material has survived, which charts the process of local commemoration, compiled by William Henry Harris, Vice Chairman of Handsworth Urban District Council.[98] On the centenary of Boulton's death on 17 August 1909, a local newspaper published a celebratory article: 'Today we offer our homage to this man – truly the "Father of Birmingham", whose rare powers deserve national honour, and here in his native city especial reverence'.[99] Was the writer suggesting that Boulton had not received the attention he believed he deserved in Birmingham and Britain? When the commemoration took place, it had a Handsworth rather than a Birmingham focus. The centenary memorial service, held a month later at St. Mary's Church, was attended by representatives of the different denominations 'in the district', representatives of Handsworth Urban District Council and 'several great grandchildren of Boulton'.[100] Prebendary Burn, in his sermon also noted, using words indicating that the commemoration was purely parochial, that the congregation had 'come from opposite ends of Handsworth' to pay their respects

[95]   Ibid., pp. 66–67. On Watt Jr, see Eric H. Robinson, 'Watt, James (1769–1848)', *Oxford Dictionary of National Biography*, Oxford, 2004: http://www.oxforddnb.com/view/article/28881 (accessed 4 January 2011).

[96]   Miller, p. 66.

[97]   BA&H, LF78.1, Matthew Boulton Centenary: Correspondence, portraits, newspaper cuttings etc. 1788–1911, collected by William Joseph Harris, p. 15, Beatrix Sturt (hereafter BS) to William Harris (hereafter WH), 15 March 1910.

[98]   BA&H, LF78.1, Matthew Boulton Centenary: Correspondence, etc.

[99]   Arthur W. Clarke, 'Matthew Boulton: The Centenary of his Death', newspaper cutting, in ibid., p. 9.

[100]   BA&H, LF78.1, Matthew Boulton Centenary Memorial Service, newspaper cutting, in ibid., p. 9.

'to our great fellow citizen'.[101] Harris also tried to produce a publication about Matthew Boulton, which would be distributed to Handsworth school children.[102] It is clear that he faced difficulty securing source material. Boulton's great grandson, M.E. Boulton, at Tew Park, agreed to send Harris a memoir of Boulton, presumably Watt's 1809 account,[103] but he did not encourage Harris to use the Boulton papers. Perhaps the family felt that Smiles book was sufficient. In a letter to Harris, Beatrix Sturt wrote that Smiles' *Life of Boulton*, 'has told all that there is to tell; including the few family stories that survive'.[104] Harris compiled a few notes and the scrapbook included some photographs, but he abandoned the project and focused instead on raising £1,200 to £1,300 to erect a bronze statue to Boulton in Handsworth.[105] A planned road widening scheme near Soho House provided a possible location.[106] However, in 1911, the Birmingham Extension Act was passed which led to Handsworth's incorporation within a much enlarged Birmingham, and Harris was advised not to proceed with his scheme without the authorisation of Birmingham Corporation.[107] A short article in *The Times* noted that two other additional commemorative suggestions were proposed, the establishment of a mechanical engineering scholarship at the University of Birmingham for a student from Handsworth Technical School and the acquisition of land to form Boulton Square, a permanent open space in Handsworth.[108] These schemes were not implemented either. Harris's scrapbook is a testament to a limited and unsuccessful local attempt to celebrate Boulton in 1909. Substantial work on Boulton was not published until later, facilitated by the public availability of the Boulton papers and an interest in the history of entrepreneurship.[109] Birmingham's first statue of Boulton, apart from the monument, was not unveiled until 1956.[110]

---

[101]   *Matthew Boulton Centenary, Sermon preached at the Memorial Service in the Parish Church of St. Mary, Handsworth by the Prebendary Burn, D.D., 15th September, 1909* (Birmingham, 1910).

[102]   BA&H, LF78.1, Matthew Boulton Centenary: Correspondence, etc., Note, p. 10.

[103]   Ibid., p. 13, M.E. Boulton to WH, 21 September 1909.

[104]   Ibid., p. 11, BS to WH, 21 September 1909.

[105]   Ibid., p. 31, Letter from WH, 31 March 2010.

[106]   Ibid., pp. 35–36, Two Maps of the proposed Junction of Soho Hill, Soho Road and St Michael's Road, n.d.

[107]   Ibid., p. 34, Thomas Silver and Ernest Way, Secretaries of Handsworth Urban District Council to WH, 8 November 1909.

[108]   *The Times*, 25 August 1909. I am grateful to Jack Kirby of Thinktank, Birmingham Science Museum, for this reference.

[109]   Two early examples are Eric Roll, *An Early Experiment in Industrial Organisation: Being a History of the Firm of Boulton and Watt 1775–1805* (London, 1930) and Dickinson, which was published in 1936.

[110]   Noszlopy, p. 16.

Conclusion

The events after Matthew Boulton's death illuminate the processes of ceremony, controversy and commemoration. Robinson Boulton wanted a public funeral on an 'extensive scale' to remember his celebrity father and construct a positive lasting reputation. It was a large public event which cost a lot of money, but which he believed did not honour his father adequately. Clearly, he found the planning and process of the funeral and its aftermath extremely stressful. The procession, service and interment did not lay Boulton to rest and he failed to publish a substantial written memorial of his father because neither James Watt, nor James Keir agreed on Boulton's importance and reputation. He was more successful in remembering his father through material objects. The production of high-quality medals was a secure long-established activity of the firm of Boulton and Watt, and he could delegate the production of a memorial to John Flaxman, a successful English sculptor with a national reputation. By 1820, Robinson Boulton had achieved his filial project, but by the late nineteenth century, Watt's reputation overshadowed Boulton's. Despite the attempts in 1909 to celebrate his memory, little was achieved beyond the production of ideas, newspaper articles and an anniversary service, which, in reality, did not extend beyond the boundaries of Handsworth and Boulton's remaining family. The demise of Handsworth Urban District Council in 1911 put paid to any civic commemorations. Even Boulton's descendants were lukewarm in their support: they seem to have been happy to leave Boulton's memory within the past.

The history of remembering Boulton after 1809 reveals that reputations can be constructed, decay and be reconstructed again. It also reveals a great deal about how celebrities were commemorated through ritual, writing, medals and memorials. In Boulton's case, his reputation was framed within an Enlightenment culture which celebrated individual genius, artistic and mechanical improvement and public benevolence. During his lifetime, Boulton was active in promoting his own products and personality, and for 10 years, Robinson Boulton upheld the cult of his father. As a successful businessman he had the authority, wealth and status to underwrite and lead this cult, resources that were not available to those families who might have wanted to commemorate less affluent individuals. Moreover, if Boulton had not had a son, it is unlikely that he would have been publicly celebrated. After a decent funeral, it is inconceivable that Anne Boulton would have engaged in a project to promote her father's public memory. Not only was she a woman and expected to fulfil her life within a domestic setting, but she was not physically robust.[111] The commemoration of Boulton's death and the memorialisation of his reputation sustained a public recollection of his significance during the late nineteenth and early twentieth centuries, but attempts to revive the memory had limited success. A resurrection of interest did not take place until the 1930s and culminated 200 years after his death in the commemorative events of

---

[111]   Mason, p. xiv.

2009, which included a major exhibition in Birmingham Museum and Art Gallery, several publications, an international conference and recognition of Boulton by the state through the issuing of a first-class stamp and, in 2011, a 50 pound note.[112] His reputation has extended beyond Handsworth. Boulton is remembered, but in ways not anticipated by his family and associates in the early nineteenth century.

---

[112]   http://www.matthewboulton2009.org/ (accessed 5 January 2011).

Appendix

'Where Genius and the Arts Preside' Matthew Boulton and the Soho Manufactory 1809–2009

Friday 3 – Sunday 5 July 2009

All lectures and panels were held at The University of Birmingham apart from Session Four, which was held at Birmingham City University.

Conference Programme

Friday 3 July

Keynote lecture

> *Matthew Boulton: Enlightenment Man*, **Peter Jones**, University of Birmingham

Session One

Panel A: Assessments

> *The Boulton-Watt Relationship*, **Margaret Jacob**, University of California, Los Angeles, USA
> *'Do you want the dust?' Matthew Boulton and the Long Shadow of James Watt*, **Ben Russell**, Science Museum, London
> *The Death of Matthew Boulton 1809: Ceremony, Controversy and Commemoration*, **Malcolm Dick**, University of Birmingham

Panel B: Coining for Nations

> *Competition, Monopoly and Great Britain's Big Problem of Small Change*, **George Selgin**, University of Georgia, USA
> *Matthew Boulton and the Gold Coinage of 1773–1776*, **David Symons**, Birmingham Museums & Art Gallery
> *Soho goes Global: New Mints, New Coins and the Limits of Success*, **Richard Doty**, Smithsonian Institution, Washington DC, USA

Session Two

Panel C: Employer and Entrepreneur

> *Matthew Boulton: Enlightened Employer?* **John Griffiths**,
> Linnaean Society
> *Dark Satanic Millwrights: Foremen at the Soho Foundry*,
> **Joseph Melling**, University of Exeter
> *Matthew Boulton and James Watt: Empowering the World*,
> **Roger Williams**, independent scholar

Panel D: Sites of Industry

> *The Layout and Archaeology of the Soho Manufactory and Soho Mint*,
> **George Demidowicz**, Coventry City Council
> *Portraying an Industrialist*, **Valerie Loggie**, Arts and Humanities
> Research Council PhD student, University of Birmingham
> *Stars at Soho: Matthew Boulton's Lost Observatories*, **Andrew Lound**,
> The Society for the History of Astronomy

Saturday 4 July

Session Three

Panel E: The Medium is the Message

> *Matthew Boulton and the Art of Making Money*, **Richard Clay**,
> University of Birmingham
> *Messages for the Masses: Trade Tokens and Medals as Mass Media for
> Radical Ideas*, **Carolyn Downs**, University of Salford
> *Collecting Boulton: Sarah Sophia Banks and her Coins and Medals*,
> **Catherine Eagleton**, British Museum

Panel F: Visitors and Spies

> *Russian Visitors to Soho and the Transfer of Steam Technologies at
> the End of the Eighteenth Century*, **Dmitri Gouzévitch**, Centre d'études
> des mondes russe, caucasien et est-européen, EHESS, Paris, France
> *Swedish Views on Urban and Industrial Development in Eighteenth-
> Century Birmingham*, **Göran Rydén**, Uppsala University, Sweden
> *Matthew Boulton and Augustin de Betancourt: Enlightened
> Entrepreneur Versus 'Philosophical Pirate', 1788–1809*, **Irina
> Gouzévitch**, Centre Maurice Halbwachs, EHESS, Paris, France

Keynote Lecture

Matthew Boulton: Creative Pragmatist, **Jennifer Tann**, University of Birmingham

Session Four

Panel G: Patron of the Arts

> *Matthew Boulton and Neo-Classicism*, **Nicholas Goodison**,
> independent scholar
> *In Matthew Boulton's Orbit: the Wolverhampton Artist Joseph Barney*,
> **Olga Baird**, Wolverhampton Art Gallery
> *The 'Mechanical Paintings' of Matthew Boulton and Francis Eginton*,
> **Barbara Fogarty,** MPhil student, University of Birmingham

Panel H: Metals

> *Matthew Boulton and the Meaning of Steel in the Eighteenth Century*,
> **Chris Evans**, University of Glamorgan
> *Matthew Boulton's Copper*, **Peter Northover and Nick Wilcox**,
> University of Oxford
> *The Soho Mint: Copper to Customer*, **Sue Tungate**, Arts and
> Humanities Research Council PhD student, University of Birmingham

Sunday 5 July

Keynote Lecture

> *Was Matthew Boulton a Scientist? Operating between the Abstract and
> the Entrepreneurial*, **David Miller**, University of New South Wales,
> Australia

Panel I: Steam, Science and Technology

> *Boulton, Watt, Wilkinson: the Birth of the Improved Steam Engine*,
> **Jim Andrew**, Thinktank, Birmingham Science Museum
> *Ideas Embodied in the Whitbread Engine: a Tribute to Matthew
> Boulton*, **Debbie Rudder**, Powerhouse Museum, Sydney, Australia

Panel J: Silver and Plated Wares

> *Early Boulton and Fothergill Silver, 1763–1773*, **Kenneth Quickenden**,
> Birmingham City University

Matthew Boulton and the Production of Sheffield Plate, **Gordon Crosskey**, Royal Northern College of Music
Real Knowledge and Occult Misteries: Matthew Boulton and the Establishment of the Birmingham Assay Office, **Sally Baggott**, The Birmingham Assay Office

Panel K: Trade in an Age of Revolution

Matthew Boulton's Jewish Partners: Oppenheimer, Baumgartner and Hyman, **Liliane Hilaire-Pérez**, Conservatoire national des arts et métiers, Paris, France
Boulton & Watt and their first Engine Sale to France, 1778–1781, **Paul Naegel**, Centre Francois Viète, Nantes, France
Big Business in a Time of War: Matthew Boulton, the Kingdom of Denmark And the British Bombardment of Copenhagen, 1807, **Dan Christensen**, Roskilde University, Denmark

Panel L: Networks of Innovation

Thomas Jefferson: Looking at Industrial England, **Maurizio Valsania**, University of Turin, Italy
William Small: a Spark of Revolution, **Martin Clagett**, College of William and Mary, USA
Gilded Greatness: Boulton, Boyle and Birmingham, **Sally Hoban**, PhD student, University of Birmingham

Sunday afternoon visits

Soho House, Handsworth, Birmingham[1]
Thinktank, Birmingham Science Museum[2]
Gas Street Basin, Birmingham in the hydrogen fuel cell long boat 'Ross Barlow'[3]

---

[1]   http://www.bmag.org.uk/soho-house

[2]   http://www.thinktank.ac/

[3]   http://www.midlandsenergyconsortium.org/Case%20Studies/Hydrogetn%20Fuel%20 Cell%20Powered%20Barge.pdf

Exhibitions

Matthew Boulton: Selling What All the World Desires, Birmingham Museums &
Art Gallery
The Art of Making Money, Barber Institute of Fine Arts, University of Birmingham

Concert

Hark! I hear Musick! An Evening with Matthew Boulton and Friends, City of
Birmingham Symphony Orchestra Baroque Ensemble, St Philip's Cathedral,
Birmingham

Select Bibliography

This bibliography provides a list of printed works. It is composed of published primary sources, collections of primary material or secondary articles, books or theses, which provide significant information and knowledge about Boulton or an insight into his context. There is also a list of relevant websites. Sources cited by authors in this volume to frame or illuminate their approaches, but which do not contain significant references to Boulton, are excluded from this bibliography. Manuscript primary materials, paintings, prints and objects are also excluded. A list of these would fill a book in itself. Readers should consult the Introduction to this volume to identify the locations of significant archives and collections.

Andrew, J.H., 'The Costs of Eighteenth Century Steam Engines', *Transactions of the Newcomen Society*, 66 (1994–1995), 77–91.

Ballard, Phillada, Loggie, Valerie and Mason, Shena, *A Lost Landscape: Matthew Boulton's Gardens at Soho* (Chichester: Phillimore, 2009).

Berg, Maxine, *The Age of Manufacturers 1700–1820: Industry, Innovation and Work in Britain (1985)* (2nd edition, London and New York: Routledge, 1994).

Berg, Maxine, 'From Imitation to Invention: Creating commodities in Eighteenth Century Britain', *The Economic History Review*, New Series, 55/1 (2002), 1–30.

Berg, Maxine, *Luxury and Pleasure in Eighteenth-Century Britain* (Oxford: Oxford University Press, 2005).

Birmingham Gold and Silver, 1773–1973: An Exhibition Celebrating the Bicentenary of the Assay Office, exhibition catalogue (Birmingham: City Museum and Art Gallery, 1973).

Bradbury, Frederick, *History of Old Sheffield Plate* (Sheffield: J.W. Northend Ltd., 1912).

Bushrod, Emily, 'The History of Unitarianism in Birmingham from the Middle of the Eighteenth Century to 1893' (unpublished MA dissertation, University of Birmingham, 1954).

Chambers, Neil (ed.), *The Letters of Joseph Banks: A Selection 1768–1820* (New Jersey: River Edge, 2000).

Chapman, R.W. (ed.), *James Boswell: Life of Johnson (1791)* (Oxford: Oxford University Press, 1980).

Clarke, Desmond, *The Ingenious Mr. Edgeworth* (London: Oldbourne, 1965).

Clay, Richard and Tungate, Sue (eds), *Matthew Boulton and the Art of Making Money* (Studley: Brewin, 2009).

Court, W.H.B., *The Rise of the Midland Industries, 1600–1838* (Oxford: Oxford University Press, 1938).

Cule, J.E., 'The Financial History of Matthew Boulton, 1759–1800' (unpublished M Comm. thesis, University of Birmingham, 1935).

Cule, J.E., 'Finance and Industry in the Eighteenth Century: The Firm of Boulton and Watt', *Economic History Supplement to Economic Journal*, 4 (1940), 319–325.

Cule, J.E., 'The Bill Account: Industrial Banking for 'Matthew Boulton'', *The Three Banks*, 63 (1964), 42–51.

Craven, Maxwell, *John Whitehurst of Derby: Clockmaker and Scientist 1713–88* (Ashbourne: Mayfield Books, 1996).

Crosskey, Gordon, *Old Sheffield Plate:A History of the Eighteenth-century Plated Trade* (Sheffield: Treffry Publishing, 2011).

Darley, Gillian, *Factory* (London: Reaktion Books, 2003).

Darwin, Erasmus, 'The Economy of Vegetation, Canto I, lines 259–288', in Darwin, E., *Cosmologia (1791)*, edited by Stuart Harris (2nd edition, Sheffield: privately printed, 2004).

Dawson, Warren R., *The Banks Letters* (London: University of London, 1958).

Delieb, Eric and Roberts, Michael, *The Great Silver Manufactory: Matthew Boulton & the Silversmiths of Birmingham, 1760–90* (London: Studio Vista, 1971).

Demidowicz, George, *The Soho Foundry Smethwick West Midlands: A Documentary and Archaeological Study* (Sandwell: Sandwell Metropolitan District Council, 2002).

Dick, Malcolm (ed.), *Joseph Priestley and Birmingham* (Studley: Brewin, 2005).

Dick, Malcolm, 'Discourses for the New Industrial World: Industrialisation and the Education of the Public in Late Eighteenth-century Britain', *History of Education*, 37, 4 (2008), 567–584.

Dick, Malcolm (ed.), *Matthew Boulton: A Revolutionary Player* (Studley: Brewin, 2009).

Dickinson, H.W., *James Watt and the Steam Engine (1927)* (Reprinted London: Encore editions, 1981).

Dickinson, H.W., *Matthew Boulton (1936)* (Leamington Spa: TEE Publishing, 1999).

Dolan, Brian, *Josiah Wedgwood: Entrepreneur to the Enlightenment* (London: HarperCollins, 2004).

Doty, Richard, 'Early United States Copper Coinage: The English Connection', *British Numismatic Journal*, 57 (1987), 54–76.

Doty, Richard, *The Soho Mint and the Industrialisation of Money* (London: Spink, 1998).

Edgeworth, Richard Lovell, *Memoirs of Richard Lovell Edgeworth Esq., Begun by Himself and Concluded by his Daughter Maria Edgeworth* (London: Richard Bentley, 1844).

Fogarty, Barbara, 'Matthew Boulton and Francis Eginton's Mechanical Paintings: Production and Consumption 1777 to 1781' (Unpublished MPhil thesis, University of Birmingham, 2011).

Fox, Celina, *The Arts of Industry in the Age of the Enlightenment* (New Haven and London: Yale University Press, 2009).

Gale, W.K.V., *Boulton, Watt and the Soho Undertakings* (Birmingham: Birmingham Museums and Art Gallery, 1952 and subsequent editions)

Glanville, Philippa, *Silver in England* (London: HarperCollins Publishers Ltd., 1987).

Goede, Christian August Gottlieb, *The Stranger in Great Britain*, 3 vols (London: Mathews and Leigh, 1807).

Goodison, Nicholas, *Ormolu: the Work of Matthew Boulton* (London: Phaidon Press, 1974).

Goodison, Nicholas, *Matthew Boulton: Ormolu* (London: Christie's, 2002).

Goodison, Nicholas, *Matthew Boulton's Trafalgar Medal* (Birmingham: Birmingham Museums and Art Gallery, 2007).

Hankin, Christiana C. (ed.), *Life of Mary Anne Schimmelpenninck, Vol. 1. Autobiography* (London: Longman, Brown, Green, Longmans and Roberts, 1858).

Harris, J.R., *Industrial Espionage and Technology Transfer: Britain and France in the Eighteenth Century* (Aldershot: Ashgate, 1998).

Heal, Felicity, *Hospitality in Early Modern Europe* (Oxford: Oxford University Press, 1990).

Hilaire-Pérez, Liliane, *L'invention Technique au Siècle des Lumières* (Paris: Albin Michel, 2000).

Hills, Richard, *James Watt*, 3 vols (Ashbourne: Landmark Publishing Ltd., 2002–6).

Hopkins, Eric, *The Rise of the Manufacturing Town: Birmingham and the Industrial Revolution* (Stroud: Sutton, 1998).

Hutton, William, *An History of Birmingham (1782)* (Wakefield: EP Publishing, 1976).

Ingamells, John, *National Portrait Gallery. Mid-Georgian Portraits 1760–1790* (London: National Portrait Gallery, 2004).

Jackson, Charles James, *An Illustrated History of English Plate*, 2 vols (London: Country Life and Batsford, 1911).

Jones, Peter M., 'Living the Enlightenment and the French Revolution: James Watt, Matthew Boulton, and their sons', *The Historical Journal*, 42, 1 (1999), 159–160.

Jones, Peter M., 'Matthew Boulton's "Enchanted Castle": Visions of the Enlightenment in the English Midlands c. 1765–1800', in Mortier, Rowland (ed.), *Visualisation* (Berlin: Verlag, Arno Spitz GmbH, 1999).

Jones, Peter M., 'Industrial Enlightenment in Practice. Visitors to the Soho Manufactory, 1765–1820', *Midland History*, 33, 1 (2008), 68–96.

Jones, Peter M., *Industrial Enlightenment: Science, Technology and Culture in Birmingham and the West Midlands* (Manchester: Manchester University Press, 2008).

Kanefsky, J. and Robey, J., 'Steam Engines in Eighteenth Century Britain: A Quantitative Assessment', *Technology and Culture* (1980), 161–186.

Keir, James, *Memoir of Matthew Boulton by James Keir F.R.S., December 1809* (Birmingham: City of Birmingham School of Printing, 1947).

King-Hele, Desmond, *Erasmus Darwin: A Life of Unequalled Achievement* (London: Giles de la Mare, 1999).

King-Hele, Desmond, *The Collected Letters of Erasmus Darwin* (Cambridge: Cambridge University Press, 2007).

Labaree, Leonard W. and others (eds), *The Papers of Benjamin Franklin*, 39 vols (New Haven: Yale University Press, 1959–2008).

Loggie, Valerie, 'Soho Depicted: Prints, Drawings and Watercolours of Matthew Boulton, his Manufactory and Estate, 1760–1809' (unpublished PhD thesis, University of Birmingham, 2011).

Lord, J., *Capital and Steam Power 1750–1800* (London: Frank Cass & Co., 1923).

Lound, Andrew P.B., *Lunatick Astronomy: The Astronomical Activities of the Lunar Society Odyssey* (Birmingham: Odyssey DL, 2008).

MacLeod, Christine, *Heroes of Invention. Technology, Liberalism and British Identity, 1750–1914* (Cambridge: Cambridge University Press, 2007).

Mantoux, Paul, *The Industrial Revolution in the Eighteenth Century (1928)* (London: Methuen, 1964).

Mare, Margaret, and Quarrell, W.H. (trans), *Lichtenberg's Visits to England as described in his Letters and Diaries* (Oxford: Clarendon Press, 1938).

Margolis, Richard, 'Matthew Boulton's French Ventures of 1791 and 1792: Tokens for the Monnerons Freres of Paris', *The British Numismatic Journal*, 58 (1988), 102–109.

Mason, Shena, *Jewellery Making in Birmingham, 1750–1995* (Chichester: Phillimore, 1998).

Mason, Shena, *The Hardware Man's Daughter: Matthew Boulton and his 'Dear Girl'* (Chichester: Phillimore, 2005).

Mason, Shena (ed.), *Matthew Boulton: Selling What All the World Desires* (New Haven and London: Yale University Press, 2009).

McKendrick, Neil (ed.), *Historical Perspectives: Studies in English Thought and Society in Honour of J.H. Plumb* (London: Europa Publications, 1974).

McKendrick, Neil, Brewer, John and Plumb, J.H. *The Birth of a Consumer Society: the Commercialization of Eighteenth-century England* (London: Europa Publications, 1983).

Miller, David Philip, *James Watt, Chemist: Understanding the Origins of the Steam Age* (London: Pickering and Chatto, 2009).

Miller, David Philip, 'Scales of Justice: Assaying Matthew Boulton's Reputation and the Partnership of Boulton and Watt', *Midland History*, 34, 1 (2009), 58–76.

Moilliet, Andrew (ed.), *Elizabeth Anne Galton, 1808–1906: A Well-Connected Gentlewoman* (Hartford: Léonie Press, 2003).

Moilliet, J.L. and Smith, Barbara M.D., *A Mighty Chemist: James Keir of the Lunar Society* (Worcester and Birmingham: Privately Printed, 1982).

Mokyr, Joel, *The Gifts of Athena: Historical Origins of the Knowledge Economy* (Princeton, New Jersey: Princeton University Press, 2002).

Mokyr, Joel, 'The Intellectual Origins of Modern Economic Growth', *Journal of Economic History*, 65, 2 (2005), 285–351.

Mokyr, Joel, *The Enlightened Economy: an Economic History of Britain 1700–1850* (New Haven and London: Yale University Press, 2010).

Mokyr, Joel, 'The Great Synergy: The European Enlightenment as a factor in Modern Economic Growth', http://www.crei.cat/activities/sc_conferences/23/papers/mokyr.pdf (n.d.), 8–9.

Money, John, *Experience and Identity: Birmingham and the West Midlands 1760–1800* (Manchester: Manchester University Press, 1977).

Morgan, Kenneth (ed.), *An American Quaker in the British Isles. The Travel Journals of Jabez Maud Fisher, 1775–1779* (Oxford: Oxford University Press, 1992).

Moritz, Carl, Philipp, *Travels of Carl Philipp Moritz in England. A Reprint of the English Translation of 1795* (London: Humphrey Milford, 1924).

Muirhead, J.P., *The Origin and Progress of The Mechanical Inventions of James Watt*, vol. 2 (London: J. Murray, 1854).

Musson, A.E. and Robinson, Eric, *Science and Technology in the Industrial Revolution* (Manchester: Manchester University Press, 1969).

Noszlopy, George T., ed. by Beach, Jeremy, J., *Public Sculpture of Birmingham including Sutton Coldfield* (Liverpool: Liverpool University Press, 1998).

Peck, T. Whitmore and Wilkinson, Douglas, *William Withering of Birmingham M. D.; R. R. S.; F. L. S.* (Baltimore: The Williams and Wilkins Co., 1950).

Pollard, J.G., 'Matthew Boulton and J.P. Droz', *The Numismatic Chronicle*, 8 (1968), 241–265.

Pollard, Sidney, *The Genesis of Modern Management: A Study of the Industrial Revolution in Great Britain* (Cambridge, MA: Harvard University Press, 1965).

Porter, Roy, *Enlightenment: Britain and the Creation of the Modern World* (London: Penguin, 2001).

Porter, Roy, 'Matrix of Modernity? The Colin Matthew Memorial Lecture', *Transactions of the Royal Historical Society*, 12 (2002), 245–259.

Pratt, Samuel Jackson, *Harvest-Home: Consisting of Supplementary Gleanings, Original Drama and Poems, Contributions of Literary Friends and Select Republications*, 3 vols (London: Richard Phillips, 1805).

Quickenden, Kenneth, 'Boulton and Fothergill Silver: Business Plans and Miscalculations', *Art History*, 3, 3 (1980), 274–294.

Quickenden, Kenneth, 'Boulton and Fothergill Silver' (unpublished PhD thesis, University of London, Westfield College, 1989).

Quickenden, Kenneth, *Boulton Silver and Sheffield Plate: Seven Essays by Kenneth Quickenden* (London: The Silver Society in association with Birmingham City University, 2009).

Quickenden, Kenneth and Kover, Arthur J., 'Did Boulton Sell Silver Plate to the Middle Class? A Quantitative Study of Luxury Marketing in late Eighteenth-century Britain', *Journal of Macromarketing*, 27, 1 (March 2007), 51–64.

Ransome-Wallis, Rosemary, 'Matthew Boulton and the Toymakers, Silver from The Birmingham Assay Office', exhibition catalogue (London: The Worshipful Company of Goldsmiths, 1982).

Reilly, Robin, *Josiah Wedgwood 1730–1795* (London: Macmillan, 1992).

Ridgway, Maurice H., *Chester Silver 1727–1837* (Chichester: Phillimore & Co. Ltd., 1985).

Robinson, Eric, 'Training captains of industry: The education of Matthew Robinson Boulton (1770–1842) and the younger James Watt (1769–1848)', *Annals of Science*, 10, 4 (1954), 301–313.

Robinson, Eric, 'The Lunar Society and the Improvement of Scientific Instruments', *Annals of Science*, 12 and 13 (1956, 1957), 296–304, 1–8.

Robinson, Eric, 'Matthew Boulton's Birthplace and his Home at Snow Hill: A Problem in Detection', *Transactions of Birmingham & Warwickshire Archaeological Society*, 75 (1957), 85–89.

Robinson, Eric, 'Boulton and Fothergill, 1762–1782, and the Birmingham Export of Hardware', *University of Birmingham Historical Journal*, 7, 1 (1959), 60–79.

Robinson, Eric, 'The Lunar Society, its Membership and Organisation', *Transactions of the Newcomen Society*, 35 (1963), 153–157.

Robinson, Eric, 'Eighteenth Century Commerce and Fashion: Matthew Boulton's Marketing Techniques', *Economic History Review*, 16, 1 (1963), 39–60.

Robinson, Eric, 'Matthew Boulton and the Art of Parliamentary Lobbying', *Historical Journal*, 7, 2 (1964), 209–229.

Robinson, Eric, 'The Origins and Life-span of the Lunar Society', *University of Birmingham Historical Journal*, 11, 1 (1967), 5–16.

Robinson, Eric and Thompson, Keith R. 'Matthew Boulton's Mechanical Paintings', *Burlington Magazine*, 112, 809 (1970), 497–507.

Roll, Eric, *An Early Experiment in Industrial Organization* (London: Longmans, Green & Co. Ltd., 1930).

Rosen, William, *The Most Powerful Idea in the World: A Story of Steam, Industry and Invention* (Chicago: The University of Chicago Press, 2010).

Rowland, Peter, *The Life and Times of Thomas Day, 1748–1789: English Philanthropist and Author* (Lampeter: The Edwin Mellen Press, 1996).

Scarfe, Norman, *Innocent Espionage: The La Rochefoucauld Brothers' Tour of England in 1786* (Woodbridge: The Boydell Press, 1995).

Schofield, Robert E., *The Lunar Society of Birmingham: A Social History of Provincial Science and Industry in Eighteenth-century England* (Oxford: Clarendon Press, 1963).

Schofield, Robert, E., *The Enlightened Joseph Priestley: A Study of his Life and Work from 1773 to 1804* (University Park, PA: Pennsylvania State University Press, 2004).

Selgin, George, *Good Money, Birmingham Button Makers, the Royal Mint and the Beginnings of Modern Coinage 1775–1821* (Michigan: University of Michigan Press, 2008).

Shaw, Stebbing, *The Antiquities and History of Staffordshire*, 2 vols (London: J. Nichols and Son, 1798–1801).

Smiles, Samuel, *The Lives of the Great Engineers: Boulton and Watt* (London: John Murray, 1865).

Smith, C.U.M. and Arnott, Robert (eds), *The Genius of Erasmus Darwin* (Aldershot: Ashgate, 2004).

Tann, Jennifer, *The Development of the Factory* (London: Cornmarket Press, 1970).

Tann, Jennifer, 'Boulton and Watt's Organisation of Steam Engine Production before the opening of Soho Foundry', *Transactions of the Newcomen Society*, 49 (1977–78), 41–53.

Tann, Jennifer, 'Marketing Methods in the International Steam Engine market: The Case of Boulton and Watt', *Journal of Economic History*, 38, 2 (1978), 363–391.

Tann, Jennifer (ed.), *The Selected Papers of Boulton & Watt. Vol. 1, the Engine Partnership, 1775–1825* (London: Diploma, 1981).

Tann, Jennifer, *Birmingham Assay Office 1773–1993* (Birmingham: The Birmingham Assay Office, 1993).

Tann, Jennifer, 'Steam and Sugar: the Diffusion of the Stationary Steam Engine to the Caribbean Sugar Industry', *History of Technology*, 19 (1997), 63–84.

Tann, Jennifer and Breckin, M.J., 'The International Diffusion of the Watt Engine, 1775–1825', *Economic History Review*, 31 (1978), 541–564.

Thomason, Edward, *Sir Edward Thomason's Memoirs during Half a Century*, vol. 1 (London: Longman, Brown, Green and Longmans, 1845).

Tomory, Leslie, *Progressive Enlightenment: The Origin of the Gaslight Industry 1780–1820* (Cambridge, Massachusetts and London: The MIT Press, 2012).

Torre di Rezzonico, Carlo Castone della, *Viaggio in Inghilterra di Carlo Castone della Torre de Renzzionico Comasco* (Venice: Tipografia di Alvisopoli, 1824).

Uglow, Jenny, *The Lunar Men: The Friends Who Made the Future* (London: Faber and Faber, 2002).

Watt, James, *Memoir of Matthew Boulton by James Watt, September 17 1809* (Birmingham: City of Birmingham School of Printing, 1943).

Westwood, Arthur, *The Assay Office at Birmingham Part 1: Its Foundation* (Birmingham: The Birmingham Assay Office, 1936).

Wilcox, Nicholas, 'Copper in the Industrial Revolution' (unpublished MEng thesis, Department of Materials, University of Oxford, 2009).

Wilson, Philip K., Dolan, Elizabeth A. and Dick, Malcolm (eds), *Anna Seward's Life of Erasmus Darwin* (Studley: Brewin Books, 2010).

Wolfe, J.J., *Brandy, Balloons and Lamps: Ami Argand, 1750–1803* (Carbondale: Southern Illinois University Press, 1999).

Websites

http://www.bankofengland.co.uk
http://www.bmag.org.uk
http://www.bmagic.org.uk
http://www.britishmuseum.org
http://www.digitalhandsworth.org.uk
http://www.matthewboulton2009.org
http://www.oxforddnb.com
http://www.powerhousemuseum.com
http://www.revolutionaryplayers.org.uk
http://www.royalmail.com
http://www.sciencemuseum.org.uk
http://www.speedmuseum.org
http://www.theassayoffice.co.uk
http://www.thinktank.ac
http://www.vam.ac.uk
http://www.visitheartofengland.com

Index

References such as "138–9" indicate (not necessarily continuous) discussion of a topic across a range of pages. Wherever possible in the case of topics with many references, these have either been divided into sub-topics or only the most significant discussions of the topic are listed. Because the entire volume is about "Matthew Boulton", the use of this name as an entry point has been restricted. Information will be found under the corresponding detailed topics.

Matthew Boulton

For Product Safety Concerns and Information please contact our
EU representative GPSR@taylorandfrancis.com Taylor & Francis
Verlag GmbH, Kaufingerstraße 24, 80331 München, Germany